ROUTLEDGE LIBRARY EDITIONS:
POLLUTION, CLIMATE AND CHANGE

Volume 16

ACID RAIN

ACID RAIN

Edited by
J. ROSE

LONDON AND NEW YORK

First published in 1994 by Gordon and Breach Science Publishers

This edition first published in 2020
by Routledge
2 Park Square, Milton Park, Abingdon, Oxon OX14 4RN

and by Routledge
52 Vanderbilt Avenue, New York, NY 10017

Routledge is an imprint of the Taylor & Francis Group, an informa business

British Library Cataloguing in Publication Data
A catalogue record for this book is available from the British Library

ISBN: 978-0-367-34494-8 (Set)
ISBN: 978-0-429-34741-2 (Set) (ebk)
ISBN: 978-0-367-36481-6 (Volume 16) (hbk)
ISBN: 978-0-367-36487-8 (Volume 16) (pbk)
ISBN: 978-0-429-34637-8 (Volume 16) (ebk)

Publisher's Note
The publisher has gone to great lengths to ensure the quality of this reprint but points out that some imperfections in the original copies may be apparent.

Disclaimer
The publisher has made every effort to trace copyright holders and would welcome correspondence from those they have been unable to trace.

Acid Rain

Current Situation and Remedies

Edited by

J. Rose

GORDON AND BREACH SCIENCE PUBLISHERS
Switzerland • Australia • Belgium • France • Germany • Great Britain
India • Japan • Malaysia • Netherlands • Russia • Singapore • USA

Gordon and Breach Science Publishers

Y-Parc
Chemin de la Sallaz
CH-1400 Yverdon, Switzerland

Post Office Box 90
Reading, Berkshire RG1 8JL
Great Britain

Private Bag 8
Camberwell, Victoria 3124
Australia

3-14-9, Okubo
Shinjuku-ku, Tokyo 169
Japan

12, Cour Saint-Eloi
75012 Paris
France

Emmaplein 5
1075 AW Amsterdam
Netherlands

Christburger Strasse II
10405 Berlin
Germany

820 Town Center Drive
Langhorne, Pennsylvania 19047
United States of America

Library of Congress Cataloging-in-Publication Data

Acid Rain : current situation and remedies / edited by J. Rose.
p. cm. -- (Environmental topics, ISSN 1046-5294 : v. 4)
Includes bibliographical references and index.
ISBN 2-88124-850-0
1. Acid rain--Environmental aspects. I. Rose, John, 1917–
II. Series.
TD195.42.A242 1994
363.73'86--dc20 94–1936
CIP

To my wife Fay, for her help and attention.
With love and gratitude.

Contents

Preface

The acidity of rain and surface water exerts a damaging effect on the environment by two main processes. Firstly, the acid-causing gaseous constituents of the rain are harmful to human and animal health as well as to vegetation and increase the rate of metal corrosion. Secondly, acid deposition affects soils, fresh waters and forests. Over the last three decades, considerable research efforts in many countries have been devoted to a thorough study of the acid rain problem and its environmental effects. The recognition of the dangers posed by air pollution has led to various legislative measures in a number of countries in order to control emission of pollutants from industry, vehicles and other sources. This volume is a contribution to the current discussion of this important problem.

This volume comprises ten chapters contributed by authors from five different countries, who examine various aspects of the topic with special reference to salient points of theoretical and practical interest, such as forest decline, transport, legal considerations, regional cooperation, the role of science and technology, trends in rainfall and possible remedies. While there is bound to be some overlap in certain areas with respect to authors' views, the book as a whole endeavours to present a balanced and rational approach to this serious and complex problem, which is in dire need of a rapid and effective solution.

The first chapter on long-distance transport concentrates on long-range aspects of acid rain rather than on local pollution problems. Professor Scorer deals with two main areas, namely concentration of obnoxious materials at a distance and long-term and global effects of air pollution. Some theoretical points are considered, such as the dispersion and mechanics of dilution with distance, ambient turbulence and mechanisms that may produce large local concentrations of polluted air and deposition of toxic materials in the soil. The author then examines the climatic effects of acid rain, arguing that this problem is very complex and little understood at present. In effect, the hypothesis is advanced that acid rain may actually oppose the greenhouse effect. Above all, caution in dealing with global warming is necessary in view of the uncertainties involved.

The potential for reversal of the acid rain phenomenon and mitigation of its effects is examined in great detail by Dr Howells in the second chapter. The author studies the perspectives for possible remedies, the time-scale involved and the economics of such an enterprise. This chapter deals with the acid rain phenomenon, its biological effects, some relevant case studies, modelling and the results of such activities and the results of various studies. The author concludes that after three decades of scientific endeavours, one can now judge the real effects and economic implications of acid rain.

A different aspect of the topic is analysed in the third chapter, which discusses the role of science and technology by comparing the attitudes adopted by Germany and the UK towards the reduction of pollution arising from energy-producing processes. Dr Boehmer-Christiansen first discusses the regulation of acid emissions in Europe, domestic policies that affect international positions and the single market implications. Policy developments in Germany and the UK are then reviewed. Other aspects considered include abatement technology and fuel competition, investment and technological progress, policy strategies and the limitation of a science-led policy. The author isolates three factors underlying the different policies in the two countries and argues that more open and competitive policy-making processes would prove fruitful in realizing the potential of environmental strategies for dealing with pollution.

The important effect of acid rain on forest decline in Europe is dealt with by Professor Nilsson in Chapter 4. In particular, he concentrates on modelling this decline and examines the results of such extensive efforts with respect to the future wood supply in Europe, in the absence and presence of air pollution. The author employs a special dose-response model to estimate the forest decline at different rates of air pollutant deposition. The political implications of this study are then briefly described: these are not altogether optimistic.

Chapter 5 deals with the changes in nutritional status of forests which have been affected by acid rain. The author first examines the new type of forest damage and nutrition-related damage types from the point of view of past history, diagnostic tools and deficiency symptoms. Dr Huettl then considers various nutritional disturbances in declining forests, namely deficiencies in magnesium, potassium, phosphorus and manganese. To explain this situation, the author analyses direct and indirect damage pathways. In conclusion, it is suggested that forest decline could be mitigated by changes in nutrient supply, provided the damage is due to nutritional disturbances.

Another aspect of the acid rain problem is surveyed by Dr Jakobowicz in his chapter on regional cooperation in Europe, with special reference to the Long-Range Transboundary Air Pollution Convention of the United Nations Economic Commission for Europe. The chapter is divided into two parts: the first part stresses that the acid rain issue can only be addressed at a national level; the second part advances the idea that an international instrument of control, such as the above-mentioned convention, should evolve both in time and spirit, otherwise such an international instrument would become obsolete in a relatively short time. Finally, the author elaborates on the above thesis and examines the evolution of the various protocols to the convention. Some potential developments for establishing abatement strategies for acid rain are outlined, with particular reference to the critical load approach.

Chapter 7 turns to another subject: trends in precipitation composition in North America in the 1980s. The main aim of this contribution is to identify the locations of maximum deposition of pollutants in North America in a period of great interest.

The authors survey the trend analysis methodology and establish criteria for the selection and analysis of a single concentration and wet deposition database for the investigation of spatial and temporal phenomena relating to the precipitation of major inorganic ions. The authors conclude that the areas of maximum deposition of individual inorganic chemical species are located in the northeastern USA and southwestern Canada, and that these locations have not changed significantly in the period analysed. However, actual deposition and other relevant values have changed annually as well as seasonally.

In Chapter 8, Dr Holman examines in detail acid emissions caused by road transport. The principal pollutants, such as nitrogen oxides, volatile organic compounds, carbon monoxide, sulphur dioxide, ozone, carbon dioxide and particulates, are described initially. Control strategies for pollution caused by petrol and diesel vehicles are surveyed. The author examines projected traffic growth for the next decades and argues that technical solutions cannot achieve environmental improvement in the absence of other measures.

Chapter 9 addresses some legal questions with particular attention to the US Clean Air Act of 1990. The authors first discuss the US perception of the acid rain problem and the history of past failures to deal with this environmental hazard. The 1990 Act endeavours to achieve success by using market-oriented regulations. Ms Archer and Professor Brooks, of Vermont Law School, attempt to evaluate the progress in pollution control based on the new regulations. They maintain that success in the field of acid rain may permit an application of similar regulations to pesticides, recycling, water contamination and vehicle emission. A new economics and law literature is emerging in this field of "buying and selling the sky".

The final chapter devotes much attention to legislation developments in various countries with respect to vehicular traffic. The author maintains that it is essential to tackle the huge increase in motor vehicles by various measures, including road pricing, better public transport and other legislation. Ms McHarry stresses the problem of acid rain in developing countries and advocates technology transfer from developed countries to assist developing countries.

The aim of this volume is to present objectively the most important aspects of the acid rain problem. Though not all of the many facets of this topic are discussed here, the authors have endeavoured to deal with some important aspects and indicate rational solutions to this difficult problem. In dealing with such emotive issues, one has to preserve a sense of balance and cool appraisal: the contributors have done this with success.

I am grateful to the authors for their goodwill and patience.

List of Contributors

ARCHER, M.B.	Vermont Law School, South Royalton, Vermont, 05068, USA.
BOEHMER-CHRISTIANSEN, S.A.	Science Policy Research Unit, University of Sussex, Brighton, East Sussex BN1 9RF, UK.
BOWERSOX, V.C.	Illinois State Water Survey, 2204 Griffith Drive, Champaign, Illinois 61820, USA.
BROOKS, R.O.	Vermont Law School, South Royalton, Vermont 05068, USA.
HOLMAN, C.	Environmental Consultant, 12 St. Oswald's Road, Redland, Bristol BS6 7HT, UK.
HOWELLS, G.	Department of Zoology, University of Cambridge, Downing Street, Cambridge CB2 3EJ, UK.
HUETTL, R.F.	Technische Universität Cottbus, Fakultat für Umweltwissenschaften, Lehrstuhl für Bodenschutz und Rekultivierung, PO Box 101344, 03013 Cottbus, Germany.
JAKOBOWICZ, J.M.	United Nations Economic Commission for Europe, Palais des Nations, CH-1211, Geneva, Switzerland.
McHARRY, J.	Environmental Research and Information, PO Box 893, London E5 9RU, UK.
NILSSON, S.	International Institute for Applied Systems Analysis, A-2361, Laxenburg, Austria.

OLSEN, A.R.
US Environmental Protection Agency, Corvallis Environmental Research Laboratory, 200 SW 35th Street, Corvallis, Oregon 97333, USA.

SCORER, R.S.
Department of Mathematics, Imperial College, 180 Queen's Gate, London SW7 2BZ.

SISTERSON, D.L.
Environmental Research Division, Argonne National Laboratory, 9700 South Cass Avenue, Argonne, Illinois 60439, USA.

1. LONG-DISTANCE TRANSPORT

R. S. SCORER

It was long thought that when gases and smokes were released into the atmosphere they became diluted in an irreversible way, and the building of tall chimneys was considered to be a solution to pollution problems because it reduced ground-level concentrations to acceptable magnitudes.

However, there are mechanisms which may still produce large local concentrations: fumigation, scavenging by trees on mountains, deposition by showers (exceptionally illustrated by the deposition of identifiable debris from Chernobyl), and accumulation in winter snow pack. But that is not all, for in some cases acidity is stored in the ground and one particular result is the release of aluminum into water in the absence of calcium

The ultimate disaster is when, in spite of being dispersed throughout the whole atmosphere, there is enough polluting gas to produce a change of climate through the mechanism known as greenhouse warming. Unlike local pollution problems which depend on special vagaries of the weather and can be understood and even predicted or avoided, the global mechanics are not sufficiently subject to numerical modelling, and the result is that prediction of global warming and climatic change are so uncertain that the effects are seen as possible risks rather than real problems.

It could even be that what we know as pollution (acid rain) has an opposite global effect and may so far have cancelled the warming so that no climatic change has yet been observed for certain, which turns the cleaning up of acid rain into a risk.

1. CONCENTRATION AT A DISTANCE

1.1. *Definition*

Being the combination of two ordinary simple words the term "acid rain" captured the public imagination, and has attained a much wider meaning than was at first intended. Originally it was thought that there had been a serious increase in the acidity of rain because of the great increase in the output of industrial waste to the atmosphere. Rain is the most important mechanism which cleans the atmosphere; indeed, anything that descends from the atmosphere to the ground may be said to have been "rained" out, this being the general term for ultimate deposition by gravity, whether liquid rain drops were involved or not. As a result, and because no other catch-all phrase has been invented to describe the process, acid rain is now thought of as referring to all the mechanisms whereby acidity is deposited on objects on the ground.

It was not enough to divide it into wet- and dry-deposition, because this did not satisfactorily divide the mechanisms into mutually exclusive groups, there being many more different mechanisms than two. In particular we should note that the deposition of atmospheric acidity is not significantly achieved by rain which was not previously acid, picking up the acids as it falls. Rain has always been acid to some extent, and much ground water is more acid than the rain which fell to supply it. The acidity of rain is not well correlated with the acidity of the air at the ground where it falls and could have originated from a quite different source and travelled in quite different directions at quite different altitudes.

As happens in every air pollution episode, the weather on that particular occasion and in the preceding hours or days since the pollution was emitted is all important in causing the events. Understanding the episodes is dependent on understanding the weather, and it must be emphasized that a purely statistical description or analysis which introduces randomness into the argument adds nothing to understanding the mechanism. All the statistical arguments can do is suggest that what we have experienced in the past may be the best available guide to the future, and cannot be any use in predicting serious episodes. In this chapter we are concerned to understand the occurrence of the unexpected and what has to be done to avoid its recurrence. Know thine enemy! (For a full discussion of the title see *Acid Rain and Acid Waters*, Gwyneth Howells.[1])

1.2. *Classical Ideas of Dispersion*

It has been natural to assume that the world is very large and that pollution put into the air would be "carried away" and diluted to the extent that it became more or less undetectable. This is the procedure with every breath we take. It has become the procedure every time we drive a car or any other combustion-driven engine, but there the differences begin: first, the high temperatures and pressures of internal combustion engines create oxides of nitrogen which are not present in animal breath; and second, many fuels are volatile and either escape inadvertently or remain unburnt in the combustion and escape with the exhaust of an engine.

While the commonest fuels provide most of their heat by oxidizing carbon and hydrogen, many, and in particular wood, coal and oil (and some mineral gases) contain sulphur and also produce incompletely burnt compounds which create smoke and tarry substances. Combustion also emits fine ash when fuel is burnt in large quantities requiring forced draught and producing large amounts of heat.

The great increase in human population which is now 6 times what it was a mere 200 years ago, has been facilitated by an enormous consumption of fossil fuels with a great increase in the emission of these unnatural products, which we call pollution. The emissions from volcanos are equally polluting but are carried by thermal convection to much greater heights in the atmosphere from where they are ultimately rained out on a worldwide basis. In some cases the sulphurous fumes have a damaging and long lasting local effect: the global effects will be discussed briefly later in this chapter.

The combustion of hydrogen makes no difference to the atmosphere which already contains of the order of 0.1% of water vapour, an amount which is determined by the presence of the ocean occupying about six sevenths of the earth's surface. But the combustion of carbon has more than doubled the emission of carbon dioxide (CO_2) since the beginning of the industrial era. Although this has produced no detectable adverse effect, and neither has the accompanying emission of carbon monoxide (CO), it appears that, together with the greater emissions of other gases which absorb the earth's emissions of thermal radiation, these carbon compounds may be sufficient to cause a rise in the surface temperature by a very few degrees with a consequent change in climate. This is commonly called the "greenhouse effect."

We are concerned in this chapter with the deposition of the acid components of these increased emissions at large distances from their source. Most pollution from man-made sources causes its damage close to the source. This is usually reduced by the use of tall chimneys which undoubtedly greatly reduce the maximum ground level concentrations of the pollutant in the air. Indeed the reduction of concentration with distance from the source is such a commonplace experience that we need a new understanding when this "law" seems to be flouted.

There are two possibilities. First, there may be cases when serious levels of pollution still in the air occur at a large distance from the source. As the distance increases it becomes more uncertain which source was the emitter of the observed nuisance and so distant sources tend to be exonerated from blame. However the disastrous accident at Chernobyl in April 1986 released pollution which could be identified without question, for there was no other source of radioactive particles to which it could be attributed. Also the higher levels of pollution were patchy, and did not indicate a persistent and gradual dilution, so that we have a good example of the flouting of the presumed dilution laws.

The second possibility is that secondary products resulting from chemical changes in the original (primary) pollution may accumulate. Thus road traffic exhaust may cause the build-up of serious concentrations of ozone (O_3) and PAN (peroxyacetyl nitrate), which are secondary products from photochemical changes caused by prolonged sunshine. There is also the possible contribution to climatic change on the continental scale due to this secondary pollution.

1.3. *Mechanics of Dilution with Distance*

The emissions from a chimney are seen to be "carried away" by the wind. As a consequence almost all formulae for the concentration contain the factor 1/U, where U is the wind speed. This dilution by the wind simply stretches the plume so that there is less pollution being carried sideways, upwards or downwards, in unit length of plume.

The plume becomes spread over an increasing cross section area by what is usually called "turbulence," although almost no theories are explicit about the nature of the mechanism even when they are quite explicit about the effect of it. The usual approach is to invent a convenient formula for the spread of the pollution in the widening plume, leaving one or more coefficients to be determined by comparison of the formula with observations. Such a procedure is easily faulted by noting that the spreading varies greatly from one occasion to another. Nature is so varied that it is not easily tied down by imposed simplicity, and weather forecasters are very familiar with this problem and tackle it on a case by case basis. Yet people appear so impatient to have a formula for dispersion that they cannot wait for the weather forecast which would tell them about the differences from case to case.

To circumnavigate this difficulty it is usual to argue that any formula will only be regarded as a statistical average of probable events. But this creates additional difficulties in finding observations with which to compare the formula to give values to the coefficients in it. Furthermore it is usually recognised that the agreement of a formula with the observed values of pollution concentration on any particular occasion is acceptable if it is within a multiplicative factor of two or three. Indeed it is obviously advisable to accept the best available, rather than wait for the impossible.

Nevertheless the literature is full of offers of supposedly improved formulas, but these inevitably contain additional disposable coefficients which are presumed to be adjusted to suit the weather of the day. This makes the formulae unhelpful for planning and design purposes because they can only be used after the coefficients have been chosen to conform to the weather forecast; but they can be helpful in calculating trajectories after an event, and may be used in attributing observed pollution to a particular source.

In order to have some sort of generality the occasions chosen for making the observations, from which values can be given to the undetermined constants in a formula, have to be particularly simple. For example the wind has to be fairly constant for several hours with neutral stratification and a fairly uniform terrain over a large area. Even so, formulas have the disadvantage that they always specify a dilution process.

Dilution is regarded as inevitable, as a sort of unavoidable example of the arrow of time, as sacred as the second law of thermodynamics. Yet there are many examples in nature in which components of a mixture are sorted and separated. This, of course, requires an outside source of energy. A particular wind speed may select out a particular size of sand particle for transport along the ground, smaller ones becoming airborne and larger ones remaining scarcely moved. Migrating barchan dunes are a good example of this. Almost any beach provides cases where the movement of water has a similar effect. Crystallization of solutes often follows cooling, and salt lakes illustrate how dissolved material may be concentrated by evaporation of the water. So it should not, in principle, be surprising, to find that air pollution may be "carried away" and subsequently become concentrated at a distant site.

1.4. *Ambient Turbulence*

Any formula describing the spread of pollution from the source recognises the factor $1/U$ for dilution by the wind. Formulas used to calculate trajectories of particles carried with the wind have to recognise that the wind velocity varies with height and that the pollution particles may move vertically relative to the ground, but we shall not discuss these problems in this chapter. However it is usual to recognise the lateral spread of a plume, and this is done by assuming the pollution to be diluted by the eddy motion of the air.

Eddies are necessarily caused by the passage of the air around obstacles such as buildings, trees, and even hills with steep sides. The wakes of these constitute what we call mechanically induced turbulence, or simply "mechanical turbulence." Inevitably these eddies are less intense at greater heights above the ground, even though at the same time their linear dimension may increase with height.

In recognition of this "ambient turbulence" many formulas are designed to give the plume a gaussian distribution of pollution concentration which decreases like $\exp(-ky^2)$, where k is related to the turbulent intensity, and y is the distance from the centre line of the plume. This is done for mathematical convenience, although it cannot be strictly correct because in the formulas there is no distance beyond which the concentration is strictly zero. The real plume has an edge at a finite distance, but the plumes of the theories have no edge and fade to zero at infinite distance. The turbulent intensity defines a characteristic eddy velocity as being a factor multiplying the wind speed U.

Reynolds's mathematical representation of momentum transport by turbulent eddies was such a great breakthrough in thinking about turbulence that almost every treatment of the transfer of pollution since has treated the effects of turbulence as being mathematically identical with molecular transfer mechanisms. The mathematics is the same as that for heat conduction or fluid friction (viscosity), and is irreversible as a consequence.

This chapter, by contrast, deals with cases in which there is an absence of dilution, or even a mechanism of subsequent concentration of airborne pollution, so that higher concentrations than expected are found at large distances.

The essential difficulty of making progress by using statistical theories of turbulence for the atmosphere is that there is no mean value to which we may regard the turbulent fluctuations as being added. Reynolds's theory requires this mean value to be clearly defined. As we increase the time for which the mean is calculated the mean does not home in on a steady value. Anyway if a long time is taken, as in climatological studies, it completely misrepresents the realities of cases when the wind is in a direction quite different from the average. The disaster at Chernobyl occurred on such an occasion, and the radioactive pollution travelled against the direction of the so-called prevailing (long term mean) wind.

Formulae for the concentration of air pollution are such crudely simplified representations of reality that it is no cause for surprise that we are satisfied with a calculation which agrees with observation only within a factor of two or three; and that, even after the event when the observed wind is known with as much accuracy as it ever will be, and is included in the calculation!

Many plumes are warmer than the surrounding air and rise accordingly. They can be seen to widen in a similar way to an isolated mass of warm air or a cumulus cloud as they rise. (Studies of this process are described in detail by Scorer, 1978, esp. Chapter 8.[2]) Some plumes are seen to be carried up and down by the motion of the air in which they are embedded, and this is usually due to the convection in the air caused by the warming of the ground by sunshine.

Other plumes are less buoyant and are seen to be widened by the "turbulence" already in the air, conveniently called the "ambient turbulence." This is readily represented by a factor which is the ratio of the mean velocity fluctuation to the mean wind speed, and this is usually around 0.1. But the magnitude of this ratio varies enormously, and does not distinguish between mechanical turbulence and eddies due to buoyant convection. Generally, the buoyancy-driven fluctuations increase in intensity with height, but the rate of increase (or decrease) depends very much on the thermal stratification of the air.

The difference between occasions is illustrated by a day on which thermal convection is vigorous and is accompanied by showers of rain, hail, or snow, when the factor may well be in the region of 1.0, and could be higher if the mean wind were light and the strongest gusts were due to the spreading out of downdraughts produced by the showers. At the opposite end of the scale the factor may be close to zero on an evening when the air near the ground is being stabilised by cooling of the ground but the emission from tall chimneys is rising under its buoyancy up to the level of a strong inversion, which is a layer of air through which it cannot rise further, and which is typical of the weather in many an anticyclone.

1.5. *Fumigation*

In this section we describe an atmospheric process, and are not discussing experimental work in what are called fumigation chambers designed to study the effect of certain gases on vegetation placed in them.

Fumigation in the present context is the stirring up of the lowest layers of air by convection in the morning sunshine with the result that pollution which previously only existed in the higher levels is brought down to the ground by the mixing. The ground is then said to be fumigated.

The example just mentioned at the end of the previous section is typical: the buoyant emissions from tall chimneys accumulate under an inversion as soon as the ground begins to cool in the evening. This cooled layer does not reach above the tops of tall chimneys for several hours after cooling begins as the sun goes down; and even when it has deepened to enclose the stacks the effluent is buoyant enough to rise up to the top of what had been the mixed layer of the previous day.

The warming of the layer in the morning after sunrise is a much more rapid process because the warming is from the bottom. At first thermals penetrate only a small distance up from the warmed ground; but they fairly quickly rise higher as the lowest layers get hotter until the layer containing the accumulated over-night pollution is reached. Then some of it is displaced downwards as the thermals rise through it. Fumigation begins when these downward moving masses begin to reach the ground, compensating the upward displacement of thermals. On a warm summer morning fumigation may begin two or three hours after sunrise, but in winter in high latitudes it may not have become warm enough by the time of sunset for fumigation to occur by this mechanism. But that is the classical winter smog situation in which radiation from the top of the stable layer cools it by downward convection, and this produces the same effect as continuous fumigation: it is most effective when there is a cloud layer at the inversion, for cloud emits long wave radiation to space much more strongly than smoke or haze.

Fumigation is the most serious when the pollution accumulates in the neighbourhood of the source, which is when there is only a light wind or none at all, for that is when the accumulated pollution is most dense.

If the source is in a long narrow valley the accumulation can spread only along it if the inversion is below the tops of the valley walls and the wind can only blow along it. It was in such a case that the phenomenon of fumigation was first recognised and given this name by Wendell Hewson.[3] The source was a smelter which operated throughout the night at Trail in the valley of the Columbia River in British Columbia, and a layer of pollution was carried down the valley under the inversion for a distance of over 50 km. In the morning several places along the valley experienced fumigation at the same time, and at each it was thought to have come from a local source which had just started up. The plume had crossed the border into the state of Washington USA. Hewson's investigation showed that this was not an isolated incident and that the concentration of the fumigation at a distance of 50 km might even be higher than much nearer to the source. In this classic case there had been no dilution while the plume was being transported under the inversion, and no lateral spread because of the valley walls.

Figure 1 A view of smog trapped beneath an inversion. The pool of cold air shut in by mountains loses heat radiated from the cloud top to space, while the cloud reflects much of the sunshine. The smog is trapped until a change of wind moves it away. This picture shows Northern Bohemia in winter in the early 1960s, as seen from the Milesovka Observatory. Photo by Milan Koldovsky.

1.6. *Concentration by Showers*

A rain shower is a mechanism through which a much greater volume of air passes than is contained in it at any one moment. Moist air enters at the base and water is condensed in the up-current and falls out as rain or sometimes hail or snow. Usually the storm is moving with the wind and so the fallout is left behind as a long trail on the ground.

Shower clouds which remain stationary often cause local floods. In one such classic case a hailstorm at Tonbridge (England) one summer sunday afternoon the main street was filled with hailstones to a depth of three feet.

The explosion which wrecked the nuclear power station at Chernobyl shortly before midnight on 25 April 1986 caused the emission of radioactive particles and gases into an atmosphere which had a very stable layer under an inversion which covered a large part of Europe. The emissions were warm enough to be carried up to that inversion, although some were mixed with much greater dilution into all of the layer up to it. This more dilute part was easily monitored by air samplers at the ground as it travelled in a very tortuous path across the continent. It was first detected outside the USSR in Stockholm and shortly after-

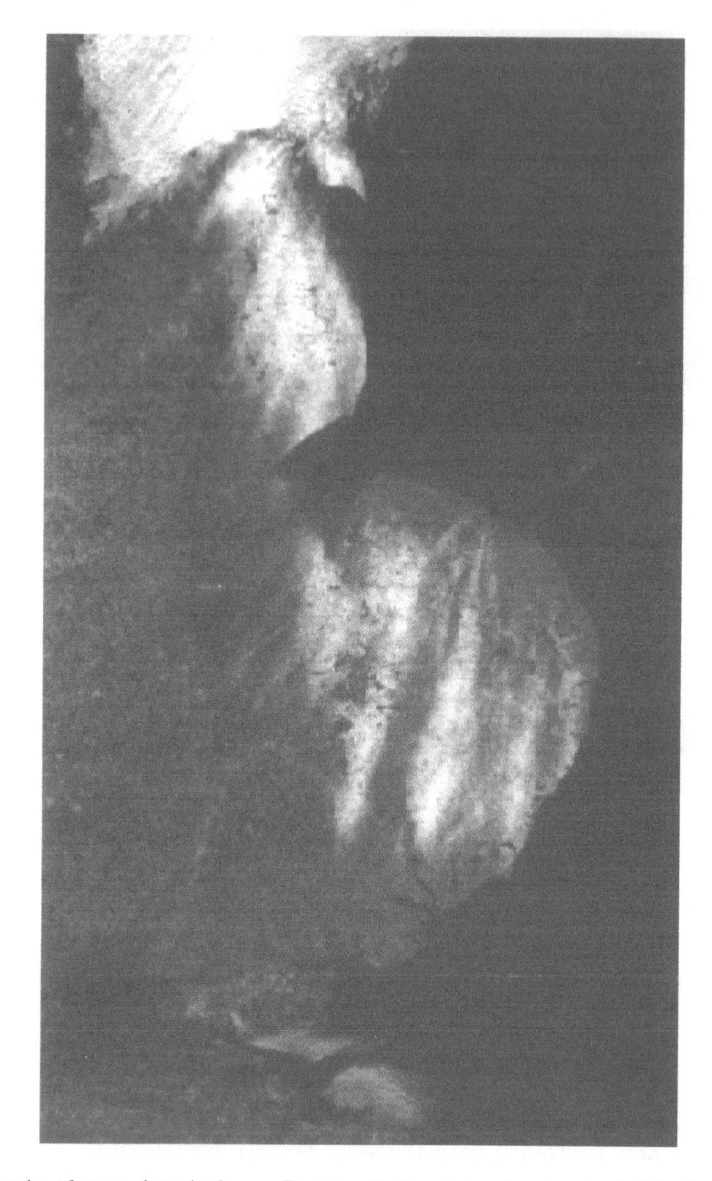

Figure 2 Streaks of snow deposited over East Anglia by wintry showers carried by the wind from the North Sea. Slower moving showers leave denser streaks which include any pollution contained in the fallout. A shower which remains over the same place, with no wind, leaves a very concentrated deposit. Picture by Coastal Zone Colour Scanner, 1105, 8.12.80, Chan 5 (red). University of Dundee.

wards serious depositions occurred in rain in northern Finland and other parts of Scandinavian Lapland. Of course the worst deposits were in Ukraine and Byelorussia but we are concerned here with deposits at large distance, and because of the radioactive nature of the pollution it could be traced and identified with certainty for a very long distance.

Figure 3 The pollution from the nuclear explosion at Chernobyl (marked by **O** in N. Ukraine) was carried by a very devious route and ultimately deposited by showers in mountainous areas of the British Isles on the day after this picture. Here we see shower clouds over Bavaria 8 days after the explosion. The cyclone seen here south of Ireland came from the **WSW** and moved away later into the Norwegian Sea. Composite of pictures by NOAA-9 at 1233 and 1413, 3.5.86, Channel 2 (deep red/IR). University of Dundee.

A change of wind carried some of it south-westwards from the Baltic to southern Germany, where the concentrating power of showers was demonstrated. Although it had been carried over 2000 km, much of it remained with very little dilution in the layer just beneath the inversion. The convection drew it horizontally into the showers and it was scavenged out of the upcurrent by rain formation and deposited on the ground in a few places with no pollution of any potential danger at places in between where no rain happened to fall.

Actually, some of the pollution-laden air was drawn into the cyclonic circulation of a frontal depression which advanced from the Atlantic across the British Isles. Although many showers occurred over a wide area and did not deposit any of the pollution, there was serious fallout of radioactive material in the mountainous parts of Wales, Scotland and Northern England, with some even more serious cases in Ireland. These events took place between six and ten days after the emissions from Chernobyl, and at the end of a tortuous journey of the order of 3000 km. The source of the pollution could not possibly have been identified if it had not had the unique characteristics of radioactivity. Indeed, on the basis of climatological average winds or a serious attempt to forecast for the week following the explosion, depositions in France and the British Isles would have been considered extremely unlikely.

Like showers, cyclonic storms are travelling mechanisms which process air through the cloud systems of which they are composed, and very little of the air in it at any one time travels with it for more than two days. The fronts in the system are easily identified on the current weather charts and on satellite pictures of the clouds of the region, but the substance of the clouds seen in the pictures is continuously changing as air flows through them. The air flowing out has been thoroughly cleansed of its air pollution in the process of generating rain. The inflow takes place at and near the bottom, and the outflow at and near the top. Almost all the higher level air has got there by ascent in storms of one kind or another, and is therefore much cleaner and drier than air at the ground. Whether the rain is polluted or not depends on the contents of the air in which it is formed as it is drawn into the storm.

It must be remembered that air enters a storm from many different directions, and it is typical of the cyclonic storms of temperate latitudes that the air over any one point may be moving from many different directions at different heights at the same time. It is therefore very difficult to track any air that may be producing rain with a greater than usual acidity back to the source of the acidity.

1.7. *Deposition by Scavenging*

Let us suppose that air pollution rises from its source up to an inversion where there is a layer of cloud. It is most usual for there to be an inversion at the top of a cloud layer because the cloud radiates like a black body into space, and is cooled like the ground at night. This cooling continues by day for the cloud is not warmed directly by sunshine because most of the energy of sunshine is either transmitted through to the ground or sea below or is refracted and scattered back into space; very little indeed of the direct solar energy being absorbed by the cloud itself. This radiative cooling causes clouds very commonly to have flat tops and to form horizontal layers by spreading out under the warm air above.

The continuous cooling of the cloud top causes downward convection from the top of the cloud to the surface (land or sea), which means that the layer becomes well mixed with continual exchange of polluted air through the cloud base. Pollution emitted at the earth's surface is soon mixed into the cloud, which then becomes smog.

Most pollution provides very good nuclei on which water droplets condense, and in the course of time the amalgamation of droplets forms drops of drizzle and rain. In this way the cloud collects the pollution in the droplets. Anyone who has cycled through a smog, (natural fog made dirty by smoke, or other air pollution) knows that dirty wet particles are collected on hair or other objects of small dimension carried along through the fog. This is the mechanism of scavenging. It is a process of sifting the air and capturing solid or liquid particles from it.

The objects which collect polluted cloud most effectively are conifer trees on mountain slopes which place them in the clouds. These trees also make the ground beneath them acid even when there is no pollution involved. When it rains the runoff is usually more acid than the rain falling from cloud which may be polluted and which is at a much higher altitude than the trees. This effect is greatly enhanced when the trees collect polluted cloud directly by scavenging. Conifers collect more effectively than broadleaf trees because of the small dimension of their needle-like foliage.

Broad leaf trees such as beech do not naturally make the ground beneath them acid to the extent that conifers do. Conifers growing at a lower altitude, which are therefore never in cloud, acidify the ground much less than those that are frequently in cloud.

There are many places in Europe and North America where scavenging of smog-like cloud deposits more acidity (and more water) on the trees than all other mechanisms of deposition together. Thus there are places where trees close to industries appear very healthy in the valley bottoms where industry is situated while trees up on the nearby mountains are severely damaged, even though they are much more distant from the pollution source. This may be seen as the result of the increase in size of the pollution particles which act as condensation nuclei when the cloud is formed as the air rises up the mountain slopes. Below cloud, pollution is much less effectively captured by the trees because the unwetted pollution particles are too small. Some forests are continually in cloud for long periods, and if the mountain is the first to be encountered by the polluted air after leaving the pollution source a considerable amount of pollution may be collected while almost none is deposited on the territory in between. The climatological frequency of the stratus cloud on the mountain in the appropriate wind direction determines the seriousness of the effect, rather than the distance from source to point of deposition.

There are several other factors which are very important in determining the extent of the damage to the forest. The mountain tops are the sites of the worst frost damage. This does not produce the most obvious effect in selecting trees on mountain tops which are most resistant to frost. Survival is as much determined by growth during the warmer months, and the most serious damage is to the buds in the spring. The trees which bud early have a longer growing period; those which bud late suffer less damage by frost. The fittest, in the Darwinian sense[4] have made a compromise between reducing frost damage and increasing growing time by early budding. The dates of the latest severe frosts vary by several weeks over the years and the result is that even the fittest suffer frost damage in about one year in four.

It is argued by botanists that there is some synergy in the various stresses to which the trees are subjected. The trees are well protected naturally against destructive insects (particularly bark bettle) but suffer severe infestation if they are subjected to attack when under severe stress. The beetle may achieve plague proportions if their attack occurs at an unfortunate time for the trees, but might avoid any serious damage in more favourable climatic circumstances, or if the beetle is not present when the stress is at its worst.

A forest may be reported as having been destroyed by "acid rain", while a neighbouring forest survives merely because the bark beetle never took hold. Such was the fate of the Jizerske Hory (in Northern Bohemia) while the forests of the Krkonose (less than 100 km distant to the south-east) remained almost undamaged, only to suffer the same fate of almost complete destruction by beetle a few years later. The "acid rain" and frost experience of the two regions differed only in the presence of a plague of bark beetle. Young trees appear to grow up very healthily beside the dying forest in the Jizerske Hory case, and may have had a more fortunate freedom from frost than their predecessors during their early growing years. The pollution in the cold episodes appears to have come from the then East Germany, and it is noteworthy that similar spruce in the valley around Liberec (in the same area) did not suffer similarly.

A particular example of scavenging is the case of so-called acid rain in south-eastern Norway. Initially the industrial areas of the Netherlands, northern Germany, and rather more particularly England were blamed on the basis of a sim-

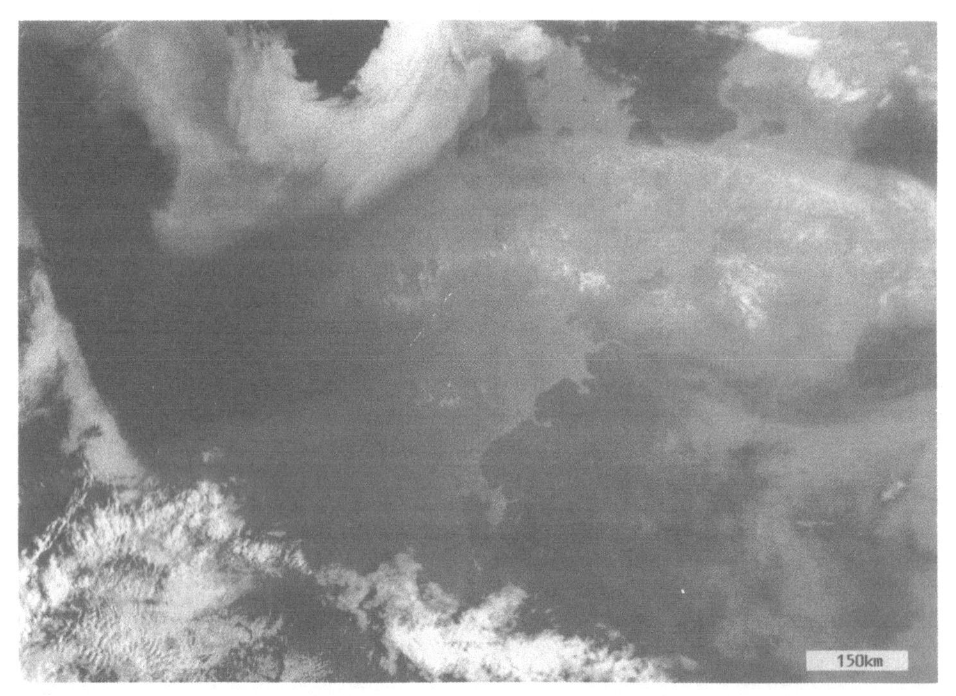

Figure 4 A wind from the south-east carries pollution from central Europe towards Scandinavia under an inversion. Low cloud is formed over the slopes of the mountains of Norway after the air has been made more moist over the sea. As the pollution moves into the cloud its particles will be scavenged out of the air by trees on the slopes. The coastline of Holland is clearly seen through the haze near the centre of the picture. Picture by NOAA-8 at 0846, 20.8.83, Channel 1 (vis-red). University of Dundee.

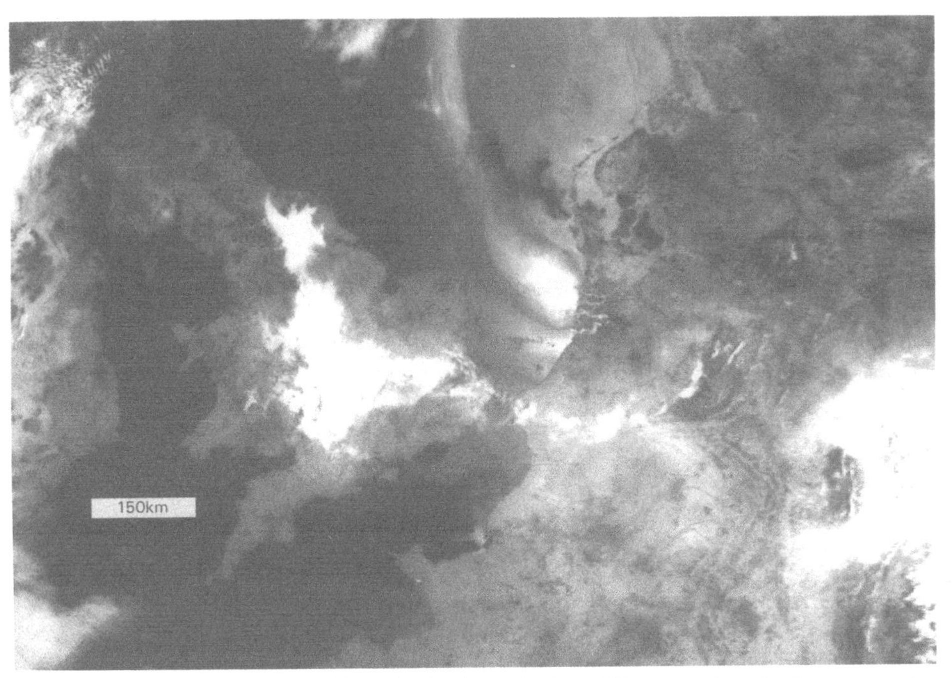

Figure 5 An exceptional case in which pollution from the lower Rhine area is made damp on passing on to the North Sea. The plume of polluted cloud is effectively scavenged from low cloud on the mountain slopes of Norway. Picture by NOAA-10 at 0852, 27.4.87, Chan 2 (deep red/IR). University of Dundee.

plistic concept of the "prevailing" wind. However satellite pictures showed many examples of winds from the south-east carrying pollution from Poland, East Germany, Czechoslovakia and Hungary, which was readily deposited on the tree-covered slopes of the Norwegian mountains from low clouds which had either travelled all the way from Eastern Europe or had been formed over the Baltic and narrow seas of the Skagerrak and Kattegat.[5]

Episodes of black snow in Eastern Scotland have usually been attributed to sources in Britain, especially since the midlands of England and the lowlands of Scotland gained reputations for dirty air in the nineteenth century. But since many of these traditionally black areas have been cleaned up, microscopic analysis of the black particles goes some way to revealing their source. We now know that pollution may be carried for thousands of km before it is scavenged out of low cloud on mountain slopes or by showers, after a long passage over sea and cool low-lying land. One event of this kind in 1984 is discussed in a paper by Scorer.[6]

The most significant feature of the transport of pollution from Eastern Europe is that when it is transported westward the weather is very likely to be anticyclonic with a strong inversion preventing the pollution from being carried upwards and dispersed in many different directions. Of course it still requires a mechanism like scavenging of clouds on mountainsides or deposition in showers to extract it from the air in concentrated form while intermediate places over which the air passes may suffer very little, if at all.

2. LONG-TERM AND GLOBAL

2.1. *Rainout and Snow Accumulation*

Reasoning complementary to that of the last paragraph of the last section may be applied to pollution from Western Europe carried eastwards. This takes place in usually much more unstable air so that it is spread into a greater depth of air and carried in many different directions at different altitudes. Although much of it is subsequently deposited in frontal rainfall there is no mechanism which concentrates it in a few places. Nevertheless the seasonal accumulation of snow on mountains may store the pollution deposited in frontal rain over a few months even though it occurs only on a few rainy days.

The acidity may then be concentrated in the spring melt. In a strong thaw, such as may be produced by the passage of a warm front, the melt water dissolves a large proportion of the acidity present in the whole deposit, leaving the later melts to produce less acid water. This initial surge of acidity is often enough to cause large kills of fish, and these have been reported in Norwegian rivers in the very early years of the twentieth century as well as more recently. It would be more correct to describe this phenomenon as due to "acid snow" because the acidity is much greater than in any single rain episode, such as might be produced by the slow passage of a single frontal depression.

The mountains on which the snow is collected in the winter are a cause of greater rainfall than on the lower areas, both upwind and downwind. Furthermore falling rain or snow is a mechanism for scavenging cloud droplets from the lower cloud layers; but in a typical region of frontal rain of a cyclone the different layers of cloud have different geographical origins. Thus, when the pollution from a source is dispersed upwards the air does not move as a solid block, and rain generated in the upper layers does not fall through layers polluted by the same source. Thus rain falling in Sweden might be generated in air containing pollution from Scotland but falls through lower layers which received their pollution from Germany and Holland. The pollution from both sources would be diluted by being spread horizontally before deposition, and only causes serious damage because it is stored over many weeks or months in snow. The collection of the acidified water in several tributaries of a river at the same time probably causes a larger stretch of river to become affected and more likely to cause a fish kill.

While there was plenty of evidence that acidity in stream and river water was associated with fish kills, in which most of the fish in some rivers were simultaneously killed over a considerable section of the river, there was much less evidence that air pollution was directly responsible for the acidification, or that it was the acidification as such that killed the fish. It was known that the cultivation of conifers for timber made the soil water more acid, particularly in streams which drained an area from which the trees had recently been harvested, but also at other times. In many places the trees did not appear to suffer any damage either through the roots or directly from air pollution, and fish did not always suffer from increased acidity. In some lakes the fish survived even though the acidity was greater than in other lakes where the fish had been killed. The kills occurred typically in the spring and had been recorded in the very early years of the century. Not all kinds of fish were equally affected, and the economic damage was confined to fish associated with sport rather than to fish caught to be eaten.

On the chemical side it had been thought that sulphur dioxide was converted into sulphate in water droplets, which was presumed to make it soluble enough in water to supply all the observed acidity in rain water. Even so in winds from the south-west the rain in south-west Scotland contained as much as 80% of that observed in Norway, which implied that much of the sulphur arrived from the Atlantic. Furthermore, acidic rain was from air that had been over the Atlantic for from two to four days previous to arrival in Scandinavia and Scotland, and this meant that the sulphate came from the sea, from America, or from air from Europe via a long passage over the ocean. Thus the older ideas that attributed sulphur deposition to the last major source area that might have emitted it, have been called in question.

Many attempts to define the sources and destinations of sulphur deposited in Europe, and emitted in Europe, lacked adequate models and were not claimed to have more than 50% accuracy; but many thought even that claim to be very misleading if taken as a guide. One reason for uncertainty was that SO_2 is not readily enough dissolved in cloud water whereas if it is oxidised it will be easily rained out as a sulphate, and even if it is oxidised by ozone or hydrogen peroxide in the air not enough was known about the extent or the presence of these oxidants.

2.2. *Problems become Global*

Awareness that there were problems with fish and trees which had their origins in air pollution spread throughout Europe and north-eastern USA during the period 1960–1980. Political and xenophobic prejudices rendered public discussion according to scientific criteria very difficult. It was said that the scientists ought to arrive at a consensus about what was happening in order that the politicians could enact and apply (or enforce?) appropriate remedies. The onset of the industrial revolution was seen as the origin of all the harm and the public mood assumed that a "cleaning up" process would be applied, without having to halt any of the desirable modern trends such as everyone owning a car and electricity being in unlimited supply.

The picture emerging more recently after serious research particularly in northern and western Europe and North America will now be briefly described. This means omitting many of the multitudinous arguments and suggested explanations (which were jokingly said to be as numerous as the individual scientists involved in the research) of the forest decline and fish kills.

It is worth remarking, in advance, that the early horrors of the industrial revolution which made poets and artists remind us of the rapidly disappearing beauty of the countryside, were in obvious fact due to the raining down of acids which killed vegetation and trees. This led to the establishment of the Alkali Inspectorate about a century ago to control the manufacture of alkalis such as sodium bicarbonate from which hydrochloric acid was a main waste product. That kind of pollution has been got under strict control and was anyway confined to the immediate neighbourhood of the industries responsible. The Clean Air Acts of the 1950s and since have tended to be aimed at the same problem of cleaning up the industrialised areas. This fitted in with the dispersion philosophy and the tall stack policy which have actually been blamed as if they had merely transposed the same problem to places more distant. One of the main arguments in favour of tall stacks was that the evidence clearly showed that their main effect was to increase the background level of pollution only very slightly and in many cases to an extent that it was not even detectable against the already existing background. That argument still remains valid, but what it ignores is the enormous increase of the output in all industrialised countries during the last two centuries. This is a very serious omission in view of the much-sought prospect that the so-called under-developed and developing countries will add as much as the developed countries already emit when they raise their living standards to the same level.

In the case of fish kills it has been established that the fatality is due to aluminium salts being dissolved from the rocks, or soil grains, even when the acidity reached levels commonly achieved in rivers during the spring melt; but that this only happened when the calcium content of the soil available to be leeched out was too small to neutralise the acid in the water. Thus the deposition of limestone dust or small granules, on the runoff areas or in the streams in which the trout (the most affected species) habitually spawn their young, has been demonstrated to make the streams and lakes safe for them. The obvious suggestion that aluminium poisoning might be a cause of forest decline has been shown not to be the case.

This all reminds us that human "progress" has already reached the magnitude at which we must regard the whole world as the human city with even the remotest areas as its immediate suburbia. Thus it could be said that if the world's population had remained at the number it had achieved in the year 1800, namely about one thousand million, we would still have only the old "urban" pollution problem to face, and we do know how to solve that. By the year 2000 the population will be about six thousand million, and this means that even if it ceased to grow any further we would have only one seventh of the time before fossil fuel becomes too scarce to support it in which to replan our global lifestyle than we would have had in the year 1800 if we had realised then what was becoming of us. Even Malthus, writing in the early 1800s, only thought of one country at most, and the pollution aspect had not troubled him, for food supply was seen as a far more pressing problem.

Sulphur compounds are emitted into the atmosphere by life (phytoplankton) in the ocean, and that includes the North Sea where the processes may have been intensified by the industrial effluents in the main rivers (Rhine, Elbe, Thames, etc.). Forestry practices, particularly the mechanical planting and harvesting of conifers, cause serious acidification of rivers. The great increase in the use of cars and combustion of fuel in industry has caused many important photochemical reactions in sunshine which follow the production of oxides of nitrogen (NO_x) and emission of hydrocarbons (HCs) and produce oxidants, ozone (O_3) and hydrogen peroxide (H_2O_2) in particular, and other compounds which cause haze of which the most important is PAN (peroxyacetyl nitrate, which was discovered to be the main secondary pollutant in the Los Angeles smog).

Thus the slowness with which SO_2 becomes oxidised meant that the origin of any particular deposition could be thousands of km away and not at the first upwind opportunity. The deposition of snow might accumulate small depositions over a season and release most of it at the spring melt. Insects could, if present when the opportunity occurs, infest a forest, particularly where frost and air pollution are present together. While these can be long distance effects, the effects on fish seem to be no worse than seasonal and might therefore be treated by the spread of limestone dust on the land or even deposition of limestone rock in rivers and lakes.

2.3. *Ground Storage*

Calcium compounds are not particularly soluble in water, by comparison with sodium salts, for example. But the plentiful supply of calcium in many surface rocks is quickly dissolved in acid waters, to the extent that trout have been observed swimming in the water immediately downstream of limestone rocks which had been thrown into a river to counteract effects of acidity.

Over a large part of Western Europe the rock at the surface is secondary (e.g. Jurassic) or later, while in Norway and most of Sweden it is pre-cambrian, primary or intrusive, of which granite is typical. In the former, e.g. Jurassic, case the natural acidity of rain, due largely to dissolved carbon dioxide, is sufficient to dissolve calcium compounds so that water that has percolated through the ground is not harmful to fish; it also means that the soil-making processes have proceeded further. In the older rocks the calcium component is very much less in proportion and has been quickly dissolved at the surface so that almost none is quickly available and aluminium is therefore dissolved. Sulphur compounds of as yet unknown composition have accumulated, starting apparently from the level of the bed rock probably at the time when forest management began. In some areas this accumulation reached up to the surface during the last two decades.

In particular the high acidity of the spring melt, or of water draining from areas of harvested forest where the trees are no longer taking up nitrate, would not now be neutralized.[7]

A similar situation is found in the soil of the north-east of the United States, New Brunswick and Newfoundland, but not in the Canadian provinces of Ontario and Quebec where the soil is not similarly sensitive. Furthermore the wind most commonly blows from Canada towards the states of north-east USA, rather than in the opposite direction. In order to recover the original healthy state both in America and in Northern Europe it is necessary to reduce the input of man-made acidity for at least a few decades. Of course it must be remembered that the input of sulphates from the sea cannot be halted. Acidity might also be reduced by growing other kinds of tree than conifers.

Thus, the problem of acid emissions in Britain and Canada was for a long time thought of as a local problem of high aerial concentration near to the sources. It was therefore dealt with, correctly, by having very tall chimneys so that the effluent would be much diluted before reaching the ground. There had been no comparable complaints from Denmark where agriculture is much more highly developed, and it is nearer to the other major sources of acidity in Germany. But Denmark has no high mountains on which to accumulate snow, and it has soil similar to that in northern Germany, where the tall stack policy was also being pursued.

The message of this section, therefore, is that a small rate of deposition at a site distant from the source may have properties which cause a slow, but continual, additive accumulation of pollution which, after many years of being hidden under ground, may begin to cause significant damage at the surface. Deposits on the beds of lakes are known to force evolutionary changes to the life forms in them. It seems that human presence and interference forced a rate of change of an evolutionary nature so rapid that it has become noticeable within a human lifetime.

2.4. *Global Warming by Greenhouse Gases*

The acid components of air pollution have not tended to be implicated in damage on the global scale up to now because they are readily removed by the natural mechanisms which clean the air. Carbon dioxide (CO_2) is, of course an acid gas, although no places are normally threatened with concentrations which might be a risk to health in the out-of-doors context. (A special case might be found in the plume of a large source of CO_2 where men might be working on a tall pylon and a slight effect on their breathing might have dangerous consequences for their safety.)

The contemporary worry about CO_2 on the worldwide scale is global warming. This is mostly concerned with gases whose molecules consist of three or more atoms, and which consequently absorb and emit selected narrow bands of infra-red radiation. The earth emits radiation over a very wide band which includes these narrow bands, and this absorption reduces the escape of heat from the earth out to space. Since they keep the earth warmer than it would be without them they are called "greenhouse gases." Then, if the temperature remained the same at all levels but the amount of these absorbing gases was increased, a viewer from outer space would receive less radiation because it would be more predominantly coming from the outer layers which are always cooler and which would now be absorbing more of the radiation from the warmer lower layers nearer the earth's surface.

According to the simple aspects of the theory, the earth is assumed at present to be losing as much heat as it is gaining from the sun because its temperature has remained fairly constant over thousands of years. The incoming sunshine is mostly in the visible and the very near infra-red wavelengths: these are of much shorter wavelength than the earth emissions and are not absorbed by the greenhouse gases but pass through without loss. Sunshine received at the ground is therefore not reduced if they are increased. Consequently the earth's temperature is reasonably expected to rise slowly, and this will increase its heat loss by emission to make it equal to the input from the sun.

The result would be that when the earth and the layers of air close to it got warmer it would emit more energy in those wavelengths which are not absorbed by the greenhouse gases, and so the outer layers would actually get cooler to make up the same total emission as before. Thus one of the phenomena looked for as evidence of global warming is a cooling of the lower part of the stratosphere, and this may seem paradoxical to those not versed in meteorological complexities. The stratosphere is far removed from local influences which might change its temperature, and so stratospheric temperatures might reveal climatic change more readily.

The greenhouse gases are many. The most important are water vapour and carbon dioxide. The amount of water vapour is determined mainly by the temperature of the sea and we cannot do anything to control that. The amount of carbon dioxide has been increased by about 30 per cent as a result of the burning of fossil fuels and forests since the beginning of the industrial revolution about 250 years ago. The most important other greenhouse gases are methane (NH_4), nitrous oxide (N_2O), and the chlorofluorocarbons (which are, in effect, methane with the hydrogen atoms replaced by chlorine or fluorine). There are other minor greenhouse gases present; some are soluble in water, others are very reactive chemically, so that they do not remain in the air in sufficient quantity to have an important effect. Sulphur dioxide (SO_2) and ammonia (NH_3) are the two most important of these.

Global warming may, at first sight seem like a good idea, especially as all vegetable foods would grow more prolifically if the concentration of CO_2 and the temperature were increased. The two main worries would be first a rise of sea level due mainly to the expansion of the water of the sea, and secondly a change in the distribution of climate as a result of which some areas which now have good rainfall might become drier, and areas now having a high production of food might become more like deserts. There might also be repercussions in the animal world when climates become unfamiliar. These effects might force large movements of population, causing great political and administrative difficulties needing advance planning to avoid great hardships.

2.5. *Feedback*

The physical principle is very simple. There are many possible effects of a temperature rise which might reduce, or even enhance, it. These are called feedback mechanisms, positive feedback if they increase the effect; negative feedback if they reduce it. It is possible that the negative feedback ones may be so powerful that the temperature rise might be too small to be detected in among the seasonal and natural variations such as have been recorded in the past and must be presumed to continue.

Some feedback mechanisms are well understood. For example, snow on the ground has positive feedback because the more lying snow there is the more sunshine will be reflected and not absorbed and hence the colder it will get and snow will tend to become more extensive. Water vapour, being a "greenhouse gas" because its molecules have three atoms each, is also positive in its feedback because if the air and the sea become colder the amount of water vapour in the air will decrease and so its "greenhouse warming" will decrease. Likewise, if the air and ocean become warmer, the amount of water vapour will increase by evaporation from the sea and the greenhouse effect will be enhanced. These mechanisms cannot cause a run-away cooling (or warming) because when too much of the ocean became frozen the amount of sunshine absorbed by the unfrozen part would be enough to stop further cooling. Likewise the opposite effect of too much warming cannot stop the loss of heat by radiation in the wavebands not absorbed soon rising to be equal to the sun's input. The atmosphere therefore has a very strong stability mechanism preventing very large changes.

Ever since life became established on earth, and probably for many millions of years before that, the temperature over a large part of the earth has been between the freezing point of the sea (around −20 degrees centigrade) and the maximum temperature which life can withstand which cannot be much above 50 degrees centigrade. During all this time there have been places where creative evolution of life can continue. So from that point of view we are not thinking about the premature extinction of life on earth.

2.6. *Has the Warming begun yet?*

In their examination of climates in the more distant past climatologists have often looked into the effects of volcanic dust, believing it probably to be a cause of climatic cooling. But it must be remembered that smoke and dust, as well as sulphur gases, have been produced in excess by the activities of industrial man, although most of it has remained in the lower layers of the air. Volcanic dust has only had an effect when it has persisted for a year or more in the stratosphere, because it is quickly washed out from the troposphere. Industrial dust and smoke from forest fires is also quickly washed out, but industry creates a new pall of haze and smoke every day, and therefore may have a similar long term effect of similar magnitude to the greenhouse effect.

The size of haze particles and of any dust which takes a year or more to fall out of the stratosphere, or be carried down by descending air, is of the order of 0.1 μm. This means that it scatters (reflects) a significant part of the incoming sunshine back into space. Such particles do not have much effect on the infra-red radiation from the earth or the greenhouse gases because they are much smaller than the wavelength.

In the 1920s the cooling effect of volcanic dust was well appreciated, but unfortunately the effect of the greenhouse gases was greatly underestimated. It was argued, by the two leading authorities Humphries[8] and Brunt[9] that in the case of the two most important greenhouse gases, water vapour and carbon dioxide, there was enough already in the atmosphere to absorb completely all the radiation emitted by the ground in the absorbed wavelengths; therefore having more of those gases in the air would make no difference because there was no more energy in the rays to be absorbed. The fallacy lay in the fact that the greenhouse gases also emit radiation in those particular wavelengths and half of that radiation is downwards: the atmosphere blocks the leakage of energy to space and only the uppermost layers emit to space. These are the colder layers and therefore emit less than the ground would if there were no greenhouse gases, and so the lower layers and the earth's surface get warmer. This makes the ground and ocean emit more in the unabsorbed wavelengths so that the total emission to space still equals the sun's input.

We can proceed no further without discussing the effect of clouds because they intercept all the wavelengths emitted by the earth's surface. A layer of low cloud traps it all and emits nearly as much back to the surface because its temperature is not much less than that of the surface. At the same time it also emits from its upper surface only a little less than the surface would emit with a clear sky. But the tops of the clouds reflect a large amount of the incoming sunshine back up into space, and they look very bright, shining upwards in the sunshine, when viewed from above from an aeroplane. Thus, low clouds reduce the incoming sunshine but only slightly reduce the outgoing radiation from the earth. Low clouds therefore make the earth cooler than it would have been without them.

If warming by greenhouse gases caused an increase in the amount of low cloud it would produce a negative feedback, making the warming less. Warming would certainly increase the humidity of the air, and this could well increase the amount of low cloud. But it has to be admitted that we have no satisfactory model of the clouds which clearly tells us, via the computer, how much the increase of the low cloud would be.

At this point "acid rain" enters the mechanisms. It is well established that air pollution generally increases very significantly the number of condensation nuclei in the air. It is possible that, with an increase of humidity, these nuclei might increase the amount of low cloud, and would certainly increase the size of haze particles to make them behave more like low cloud. We also know that if the number of droplets in a cloud is increased it will have smaller droplets at its top and more of them too. This makes the cloud look brighter in the sunshine and it reflects more back into space. This effect is illustrated in a startling manner by the exhaust trails of ships which are much brighter than the surrounding cloud. Thus air pollution has a double effect in reducing sunshine, and opposes the greenhouse effect (see Figure 6 and Scorer[5,10]).

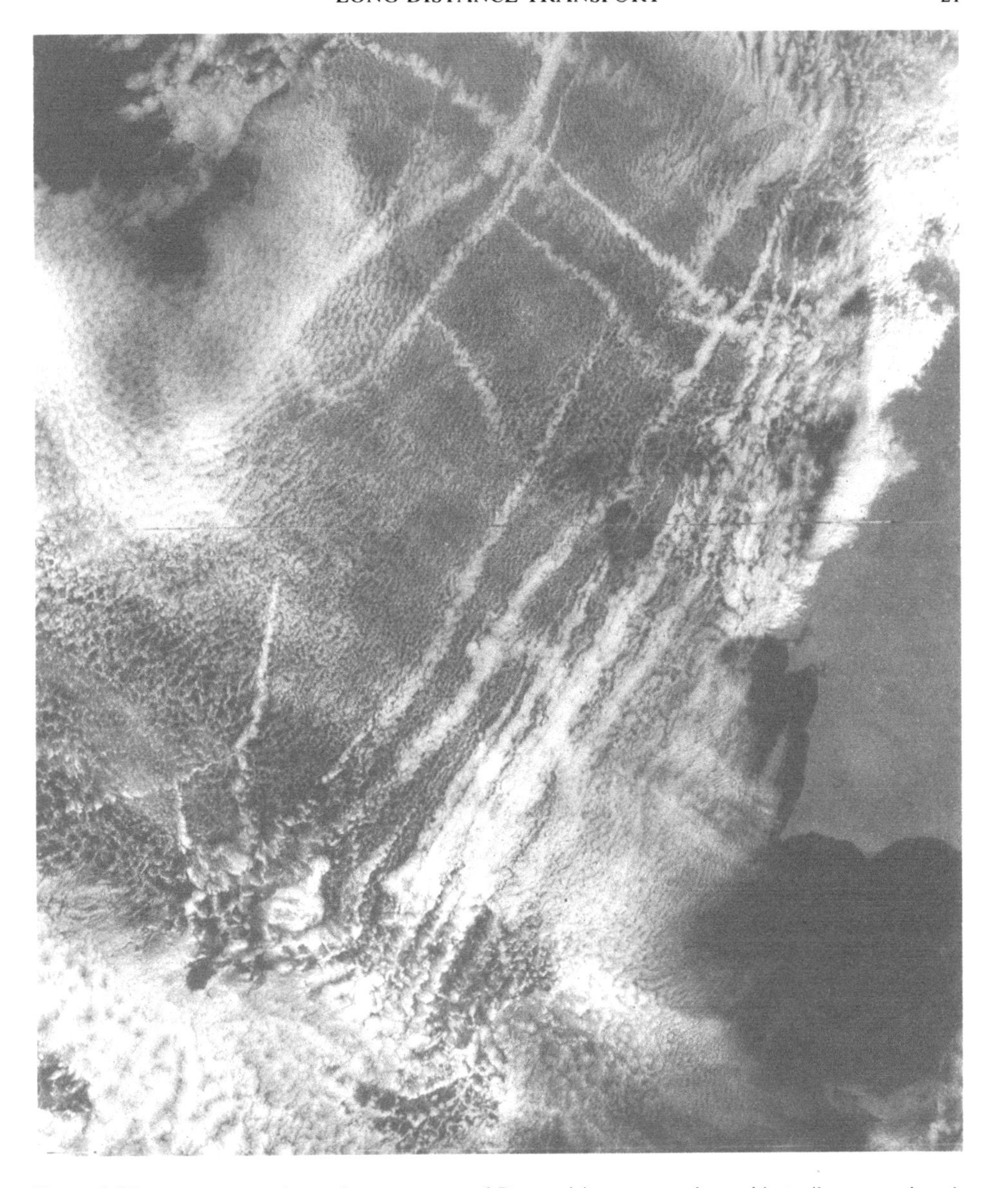

Figure 6 The ocean route down the west coast of Portugal is an area where ship trails are produced in clean air that has come from the west side of Greenland.

This picture shows trails formed one morning, with some new ones in the course of formation. They are formed only in a shallow layer of cloud under an inversion in which a small number of large drops are replaced by a large number of small droplets containing about the same amount of condensed water but having a much greater albedo, which is the whiteness, or the reflectivity to sunshine. They are a factor contributing to global cooling. Picture by CZCS (on Nimbus 7) at 1233, 8.2.84, Chan 5 (red). University of Dundee.

The greenhouse effect is not a new concept. It was given that name by Humphreys[8] and it could well have been used long before that. In the 1950s it was often a topic of discussion among meteorologists[11] and we often wondered why there was no evidence for it in the world temperature figures. Indeed between 1940 and 1970 there was much discussion of the possible approach of a new ice age because the world temperature seemed to be going down and the positive feedback mechanisms were emphasized. It was therefore thought that the production of industrial haze was the cause of the cooling which had cancelled the warming. However the 1980s seem to have wiped out all the cooling since 1940 so that now the worry is the warming!

Uncertainty reigns. The variations which have worried people since 1920 are of the same magnitude as climatic variations of previous centuries, and since we are rather unsure of the causes of each of those we cannot with any certainty say that there is evidence that the greenhouse effect is producing a warming.

Recent studies have shown that the computer models of climate suggest that warming might be detected by observing the cooling that would be occurring in the lower stratosphere. In an article entitled "Hints of Another Gremlin in the Greenhouse: Anthropogenic Sulphur," Slinn[12] indicates that while there are some increases of temperature, reversing the 1940–1970 downward trend, they seem to be greater in the southern hemisphere where there are fewer sources of greenhouse gases than in the north. This may be due to the greater emission of acid pollution in the north, although a greater reduction of stratospheric ozone may have something to do with the greater warming in the lowest layers in the south. He also points out that different model atmospheres using similar physics and numerical methods, but different spacing of the grid points at which the atmospheric parameters are calculated, give predictions which differ from each other much more than any of the effects attributable to physical effects expected to be producing climatic change. "It may be that the most troublesome gremlin in the greenhouse is the numerics."

It may be misleading to correlate recent temperature effects with sulphur emissions, because that may be an indirect way of saying it goes with something which goes with sulphur emissions, and that might well be smoke and dust. The particles may produce a greater radiation effect than the condensation nuclei.

In view of the unsuitability of the rest of the solar system as an abode for life, and the fact that we cannot learn from the behaviour of the atmospheres of the other planets because none has the same balance between the ocean and the atmosphere, it is important that we should not risk the destruction of the very suitable conditions on earth. We have available to us, for perhaps only the next hundred years or so, supplies of fossil fuel created by earlier life on earth. If we use it injudiciously we may do great harm to our own and other species by making the environment chemically inhospitable. Life certainly flourished during the carboniferous era, when there must have been very large amounts of carbon dioxide available. But never before has life itself been responsible for changes so rapid that evolutionary biological advance to react to the change was impossible. Now dangerous changes seem to be threatened as a direct result of our unprecedented success.

For if our numbers had remained the same as at the time when the industrial revolution began, when there were less than one thousand million (nowadays called one billion) people, there would be no talk of climatic change caused by the flue gases and exhaust fumes of our machines because they would be far fewer. Nor would the forests be threatened by our consumption of paper and the need to make room for more cultivation of food crops. As the forests disappear, and many of our fellow species become extinct—not through the slow process of gradual elimination of all but the fittest in which they would be replaced by their more advanced and fitter cousins; but through our predations, in our mad rush to increase our numbers regardless of their quality. We have multiplied our numbers by six in two centuries after three million years as very modest residents of this earthly paradise.

Between 1990 and the year 2000 we shall add yet another billion, making the total increase since we became drunk with mechanical power sevenfold in seven generations. Thus our inclination to save every child that is born and assert that it is a human right to have as many children as we might wish is a positive feedback to disaster in this decidedly finite world. This creates a moral imperative which is not yet widely recognised.

As scientists and engineers we must halt this meaningless multiplication of unedifying mediocrity which has not grown up to carry its responsibilities with due dignity but only with sensuous and sentimental greed. How better than to put fear into their imaginations, threaten them with climatic change and famine caused by their own acquisitiveness, with environmental toxins, and with the spread of deserts and a rise of sea level?

The indication of this section is that a warming of the air may increase the amount of low cloud, which would cause a negative feedback; while acid and dust emissions may already have caused a cooling which may have counteracted a warming that would otherwise have occurred.

2.7. *Have We a Safe Model of Climate?*

As it happens there is great uncertainty in these predictions of climatic change of our own making, and indeed the credibility with which they are met may only be a sign of our incorrigible conceit which sees all environmental change as due to our activity, as caused by our power. We call these "anthropogenic" consequences (when "man-made" blames us too obviously). And if our science seems, to those of us who are working at the frontiers of human knowledge, to be telling us that the predictions greatly exaggerate the probable severity and seriousness of the outcome, is it our duty to utter comforting words?

Should we even say, for example, that theoretically the world can comfortably accommodate another seven billion people; that the extinction of other species is as natural as the disappearance of Neanderthal Man and the Dodo, and is a natural evolutionary process; that it is our genius which is making evolution go faster and this is to be expected with clever people like us around?

Would that not be as sincere as those who felt contrarily, when their mathematical models gave predictions of a doom-laden future, that if they were to be honest they could not withhold this revelation? For it was their duty to reveal the results of the research which the people had paid for when they made the giant computers available. Anyway matters of such importance should not be kept secret. But how do we convey our uncertainties?

What we need perhaps is to avoid the predicament, rather than seek to solve the problem. And even if our computations are unreliable, should we not avoid what seems likely to happen if we do nothing to change course?

There have been some eras when the climate was warmer, and others when it was colder, than at present. Many theories have been put forward to explain the ice ages which are known to have occurred since the species to which we belong evolved in its present form around 3 million years ago. The last ice age ended about 10,000 years ago, and there is still uncertainty as to the cause of its beginning and end, and almost everything we know about the history of our species has occurred after the warming that ended it. Indeed it can be said that history did not begin until after that warming, because before it people do not appear to have been self-conscious enough to have recorded anything about themselves: they had no concept of distant origins, or of their (and our) future. They were like other animals in that respect.

How true are the predictions made by these giant computers using very sophisticated mathematical models programmed by the latest algorithms and computed by the most modern numerical analysis? After all, we know that by these machines we can perform calculations which would, in the days of Isaac Newton, have taken all the clerks in the world their whole lifetimes even if they had all been as clever as he was.

The computer, which has been substituted also for the forecaster's art which was based on (long) experience in the meteorological profession, carries out a succession of arithmetical operations on the starting point for a forecast which is deduced from the routine observations of the worldwide weather observing system. At present this does not include any input from reports of air pollution, although the forecaster may use such information in framing his weather forecasts.

In principle every country tells every other country what the weather is over its own territory at regular intervals. Information gathered by satellite is mostly broadcast in such a way that anyone wanting to receive it can do so. Indeed the World Meteorological Organisation can validly claim to be the world leader in international cooperation. Meteorologists become the most internationally minded group of people in the world and they are most friendly to each other without racial, xenophobic, or other prejudice. They speak a common professional language, and exchange forecasts of their own weather with each other. Collectively they perform an invaluable service to travellers, particularly by air and sea. They keep no trade secrets and freely share their discoveries. Their science is not yet spoiled by commercial considerations.

The spirit of international idealism is, unfortunately, threatened by politicians and their favourite institution—the market. As usual, the market takes most notice of the people with the most money and the most property, which in this case means the most aeroplanes, the most ships, the busiest airports and trade routes. These people, unlike the meteorologists, have enemies in their own business, and do not have the modesty of the meteorologists. They are not used to waiting for nature to serve up the next ration of weather, always on time even though not always what was desired or expected in the forecast. Wars and commercial jealousies seriously interfere with the free exchange of information which is both the conversation and livelihood in the world of the weather. Excuses for this interference in so-called peacetime (but which is understandable in war) are mostly based on the widespread contemporary philosophy that everything, including the free interchange of weather information, becomes more efficient if it has to be paid for. This extends to the crumbs falling from the rich man's table as soon as the rich man realises first that the crumbs exist and secondly that they are important to the poor man. The rich even have resort to a perversion of Darwinism which states that fitness for survival is the same as economic success. The poor, like animals, have no money so that their existence as sentient beings does not have to be recognised and they can be exploited like any other resource in this world (of "ours"). Their minerals can be appropriated if they have not learned to exploit themselves.

Thus there are moves by some rich governments to institute the selling of weather information to other countries, and even to their own citizens, who, they say, will not realise the true value of what they have paid for through taxation unless a price is put upon it. Other excuses abound, which do not always represent unavoidable obstacles to free exchange, but serve to hide real reasons.

For example, the geostationary satellite INSAT, which covers central Asia, Siberia, China, and Indian sub-continent, the Indian Ocean and surrounding countries, is one of the five internationally planned geostationary satellites whose overall surveillance covers the whole world except for the polar regions. Immediate access to the pictures obtained by INSAT is denied to Pakistan and Bangladesh and some other interested countries. I have heard it said that the transmission and reception facilities have to be used for commercial purposes at the insistence of the country which paid for the satellite to be built and set in orbit. This cannot be the real or full reason because it is obviously a difficulty easy to overcome technically when the satellite message is received in India. Clearly, too, this failure to share the information is contrary to the wishes of meteorologists and is imposed by government action somewhere that is not readily revealed.

In spite of such difficulties, global information is regularly received at a few centres which have big computers dedicated to operating world models of the weather. They convert it into values of the winds, temperatures, etc., at a set of grid points covering the world as the starting values from which the thermodynamic model atmosphere will calculate the future state from which the future weather is then deduced. The models are based on differential equations which state the rate of change of the quantities which were used to define the starting values. From this the values after a specified time interval are calculated, and then the whole process is repeated. Provided that the time interval specified is much greater than the time taken to perform one step forward in the calculation, we obtain a prediction of the state of atmosphere at a time in the future.

If the computer program is swift enough it may be made to print out the forecast for the future at six hour intervals for the next week, and do this every twelve hours, or every day, so that the prediction for a week ahead is always based on the latest incoming information. The differential equations used in the model are, in some cases very accurate versions of reality, although some are rather crude approximations. But even in the best of them the computer has to use what are called finite difference approximations to do the arithmetic to carry it forward to the next time step. Thus it uses the "rate of change" to calculate the change that occurs in the finite time interval, which is chosen in advance and could be anything from a minute to a day. An equally important aspect of the calculations is the space interval between the points of the grid used for defining the starting values for each time step.

The use of these approximations may have an effect similar to one of the physical mechanisms represented by one of the other equations. The equations for the motion are reversible, but those for the physical processes may be irreversible. For example, the equation for the transfer of heat must express irreversibility because heat must always travel from the hotter to the colder region, and never the other way. The same applies to the finite difference approximations in the arithmetic. They may be of irreversible form: for example, if the starting values at two neighbouring points are very different, which would indicate a situation close to a discontinuity, such as we might find at a "weather front," the numerics would make them somewhat less different in the next time step, but almost never more different. These numerical processes may thus have an effect which amounts to the same as one of the physical processes. The parameters which represent the physical process can therefore be adjusted so that the answer comes out with less error than if no adjustment were made. These adjustments are made as a result of comparing many forecasts with the actual weather that did follow the starting values. Thus there is always a lot of "bodging up" to be done to prevent the numerics representing a false physical effect. The magnitude of this false effect can be gauged by changing the space interval between the points in the grid or the length of the steps forward in time. Some quite startling results have emerged which represent much bigger changes than are produced by altering the physics in a way similar to the greenhouse effect.[12]

Consequently if the climate is changing the bodging up may need to be changed, and we do not have real cases of the pattern of weather in the changed climate because it hasn't happened yet. And if we study the effect of changed physical processes (which is what the study of climatic change is all about) we haven't got the strictly correct parameters representing them in the first place because we have used them to massage the numerics. How then do we know exactly the changes needed to represent the changed reality? If we put what may politely be call rubbish into the computer we will certainly get out what may accurately be called rubbish!

The correct values of the parameters representing some of the physical mechanisms in a changed climate cannot be obtained until we have recorded some experience of the new climate, and possibly done some trial runs in forecasting some of the recorded extremes which did not occur in the previous model climate.

A good computer model can be used to generate a climate by simply running it for a few seasons. That is to say it generates a series of weather experiences which indicate the sort of weather which would occur in an atmosphere like the one which the model represents. The models in use for weather forecasting have done this and some have provided quite good samples of the real climate as we have experienced it. Sometimes the averages are too low or too high but the main fault is that they do not produce as wide a range of weather as actually occurs. That is to say that they are very bad at forecasting the frequency of extremes of weather, and don't usually predict enough extremes. Usually the predictions are not extreme enough in actual cases.

When climatologists examine the real weather of the past and find what seem to be climatic trends, in which several years are warmer, or colder, than the nor which came before, they look for causes. Whether they find them or not they learn a lot about the variations and trends that have been going on in the most recent decades, and much research has been carried on to discover these variations in the decades, centuries, and eras of the past. These are regarded as the natural fluctuations of climate. If man-made effects are producing a climate change it ought to be distinguishable from the natural fluctuations. So far nothing has occurred which can be said certainly to lie outside the range of natural variations.

Some changes have features which seem to rule them out as being caused by the greenhouse effect. Thus, as already mentioned, recent changes have shown some effects which are greater in the southern hemisphere, whereas the greenhouse theory indicates that they would be greater in the north. If this is eventually explained it will probably be the result of some influence which has not yet been measured and understood. Indeed this particular anomaly has been correlated with the greater production of man-made sulphur compounds in the northern hemisphere. Sulphur oxides are, of course, the most important group of effluents we have in mind when we use the phrase "acid rain".

2.8. *High Clouds*

The high clouds of weather systems, which are the highest clouds in the troposphere and are usually called cirrus, may act sometimes like dust when the particles are very small, but more often like low clouds, but even then not quite like them. The differences depend on the pollution present in them.

Most commonly cirrus clouds are formed in the same way as other clouds, which is by the lifting of their air until it is cooled to its dew point and is saturated so that further cooling causes a cloud of water droplets to be created. But the air at the top of the troposphere is usually at a temperature between −35 and −55 degrees centigrade so that it is not very long before these droplets freeze, and become ice crystals. But the vapour pressure over ice is less than that over water, and so the air is supersaturated for ice, and the ice crystals grow rapidly until they have absorbed all the excess vapour. The crystal structure and its scattering of sunshine depend on the rate at which the crystals grow, and this depends on the type of cloud in which they grow.

The air usually contains a fairly plentiful supply of condensation nuclei on which the droplets condense, but no nuclei on which ice crystals form out of clear air. Consequently the cloud first forms as a collection of water droplets, and then these freeze if crystals are nucleated within the droplets. Ice nucleation is rapid if the temperature is below −40 degrees, but is progressively slower up to 0 degrees, and indeed cloud droplets sometimes remain unfrozen for hours at temperatures warmer than about −15 degrees. Those ice crystals that are nucleated will grow much more rapidly if some of the droplets around them have not frozen, because the growth of the ice reduces the pressure of the surrounding water vapour and the droplets then begin to evaporate; ultimately more water is condensed on to the cloud particles than if they had not frozen but remained liquid.

After freezing has begun there will be fewer ice crystals than there were droplets because some of the droplets (those which did not freeze) had disappeared: the crystals are therefore larger, with larger gaps between them; which means that visible light rays may pass through the cloud without finding any crystals. From above it is therefore sometimes possible to see details on the ground through the ice particle clouds, which is unusual with water droplet clouds. The mode of growth of the crystals, whether they are like needles or plates, has a big effect on these optical and radiational properties.

Likewise it is often possible to see clearly the disc of the sun or the moon through cirrus cloud. Even those sun's rays which do fall on a cloud crystal are mostly refracted forwards to the earth, so that the clouds look bright white when seen from below, but often look dark compared with the rest of the sky when seen from above. This is important because high clouds do not reflect back into space as high a proportion of the sunshine falling on them as low, water droplet clouds. Because of their greater size, ice cloud particles scatter (reflect) back to space much less of the sunshine than any small dust particles, which may be either in the stratosphere or close to the ground like industrial haze. These last results refer to needle-shaped crystals, but are less applicable to plate-shaped forms which grow dendrites and form snow flakes by aggregation, and reflect more sunshine upwards.

But let us suppose that a cloud of ice particles was very thick and did reflect a lot of sunshine; would it then act in the same way as low cloud which reduces the warming effect? The answer to this is "no." That is to say that clouds at different levels do not have the same effect. For the cloud is a good absorber of the heat (infra red) rays emitted from the earth's surface. At the same time they are good emitters of the same kind of rays from their tops, but since the temperature of cirrus clouds is around 60 degrees colder than the earth's surface their emissions are only about two fifths of the power of the surface emissions, and so they have a warming effect on the layers below them. If the amount of high cloud were increased by the greenhouse warming they might enhance the effect but at present we have no good reasons to suppose that the greenhouse effect either increases or decreases the cover of high cloud, so that the feedback of such cloud remains uncertain.

That is not all there is to be said in the present context because pollution does have an effect on high clouds. Just as ship trails in clean ocean low clouds give a useful indication of an effect of pollution near the surface, so aircraft trails, usually called "contrails" (an abbreviation of "condensation trails," first used by Ernest Gold early in World War II when they became a clear indicator of the presence of aircraft), provide useful information about the size of particles in high cloud.

The exhaust of aircraft contains water vapour and a copious supply of condensation nuclei. It is also much hotter than the surrounding air so that when they are mixed into it supersaturation occurs if the outside air is nearly saturated already. Much of the time the ambient air is too dry for a trail cloud to be formed, or it is soon evaporated again after further mixing. The most interesting results occur when the temperature is below -40 degrees because then most, if not all, of the cloud droplets formed in the trail will freeze. A persistent trail is produced if the ambient air is already saturated for ice, but has no cloud in it because no water droplet cloud has yet been formed naturally. Thus the size of the ice particles depends on the ambient temperature, and on the extent to which it is supersaturated for ice—which determines how large they will grow after being frozen. If they grow much larger than about 1 μm they will have a warming effect on the lower atmosphere as just described, but if they remained much smaller then they could operate like dust particles and contribute to cooling.

As it happens the meteorological satellites have had cameras operating at an intermediate wavelength, namely Channel 3 at around 3.75 μm, and when seen by the satellite some contrails look "white," the same as they do in visible light at wavelengths less than 1 μm (channels 1 and 2); while others equally often look dark and cold, giving weak emissions like high clouds and contrails seen at long infra red wavelengths (channels 4 and 5). It also happens that some natural high clouds show the same variations in Channel 3, and look like low clouds sometimes and like high clouds on other occasions.[5]

A few decades ago when contrails (at 10–14 km) caused by commercial jet aircraft became much more common, it was suggested that they might become an influence for climatic change. At that time it was common for the prospect of another ice age, which the climatic change might induce, to be a controversial topic of discussion. That was typical of much argument of this kind, for it exaggerated the influence of man-made effects on the dynamic aspects of the envi-

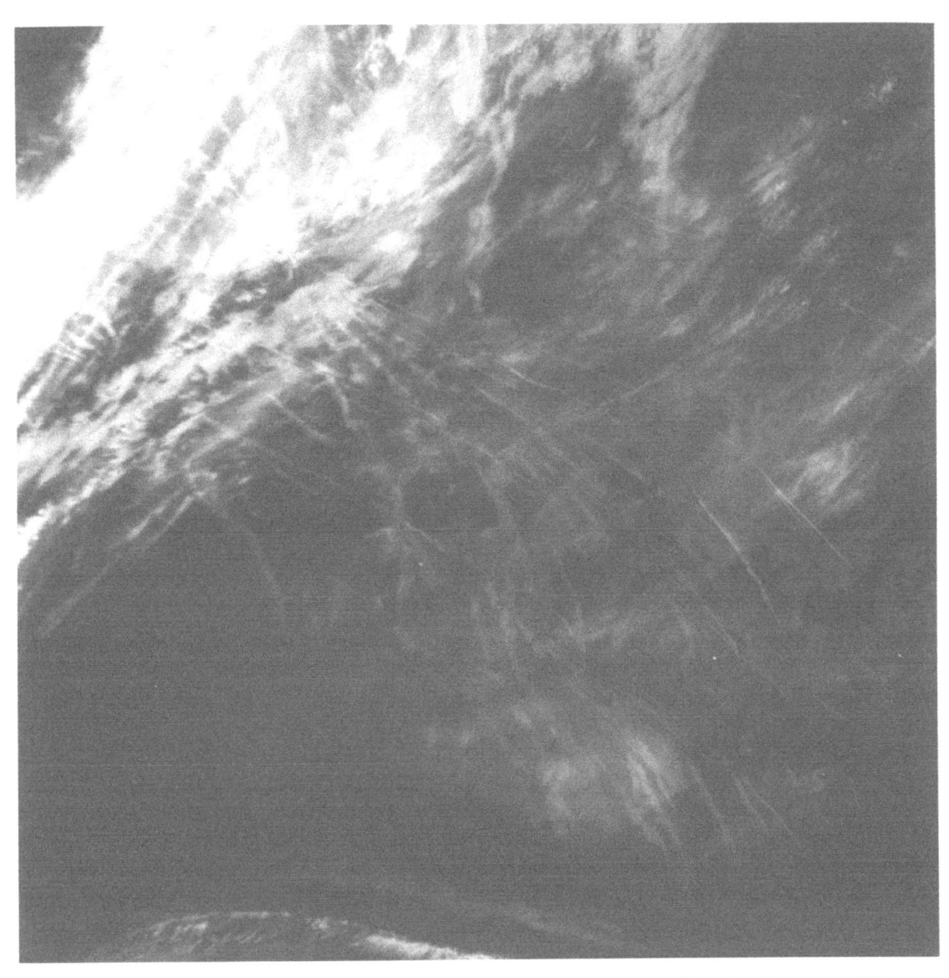

Figure 7 Almost every day areas of contrails are seen over the North Atlantic ocean. They are formed by the civil air liners following one another across navigation points where the wind on the west side is different from that on the east side of a front. On the approach to a front the humidity is greater than further away so that very frequently the air is saturated for ice but not for liquid water. At temperatures below −40 degrees conditions are ideal for the formation of persistent trails. Picture by Tiros-N at 1533, 7.6.79, Chan 4 (IR). University of Dundee.

ronment (weather and seasonal growth of vegetation), and awful effects were suggested as a likely consequence of the latest technological inventions. Thus at that time there was extensive propaganda against super-sonic air liners on the grounds that their exhaust would destroy the so-called ozone layer. Since then the only very important effect on the ozone has been the so-called hole in the layer around the South Pole, and that is now seen as due to the chlorofluorocarbons, and is not an effect of aircraft pollution. That hole, incidentally, has occurred in an unexpected manner, for the mechanism requires the presence of stratospheric clouds (at around 18–25 km), and these are extensive enough around Antarctica to produce a significant additional cooling of that bit of the lower stratosphere. It is remotely possible that they have become more extensive as a result of the cooling of the lower stratosphere due to the greenhouse effect to which their own cooling is therefore a positive feedback; but that idea cannot be tested because we have no evidence as to their extent before the recent decades in which the greenhouse gases have increased.

To summarize, we may say that natural high clouds certainly have, on the whole, a warming effect on the air at the earth's surface. But air pollution at their level may cause some of them to have a cooling effect. Because of the uncertainty of the radiative effect combined with the more important fact that we have no adequate models of high cloud formation mechanisms which will give an acceptable prediction of changes in high cloud amount or crystal structure due to the greenhouse effect, it is unlikely that high cloud amount will be considered as an important feedback mechanism to greenhouse warming.

It is certainly true that the possible effect of an increase in low cloud amount is thought to be an important negative feedback mechanism; but even in that case weather models on the global scale are very far from the precision required to predict the magnitude of any feedback.

2.9. *Complexity and Safety*

In order to assess the dangers which are widely discussed in gloomy terms it is important to see the role of danger and the ways in which life forms deal with them as part of our evolutionary history. The dangers due to industrial activity are not completely new, and it is agriculture, urbanisation, and human domestic activity which are responsible for most of the extinction of other animal species. The effects of industry are confined to close neighbouring territory, rivers and confined seas, and concentrated depositions from the air. Those whose vision is doom-oriented may take the thought that creative evolution has required the destruction of the majority of all offspring so that those having contemporary fitness could flourish. It is now clear that there is room on earth for only a minority of potential progeny. Evolution has never had an objective which it has relentlessly pursued; and there must have been many flukes of environmental change which favoured varieties which would have been written off on the basis of their early contemporary environmental prospects but turned out to be winners through newly discovered friends or phenotype extensions.[4] The industrialist view that "we can cope, and clean up if necessary" has plenty of substance but only if the population growth can be overlooked (or overcome).

The biological world is full of negative feedback mechanisms which enable animals to carry on after suffering bodily damage. Some curative mechanisms restore completely the previous healthy condition; others set about creating alternative ways of achieving the same purpose as the damaged or lost elements. None of the enormous geological calamities to the face of the earth which have occurred during the last 2,000 million years, and probably more, including ice ages, volcanic desolation, mountain building and continental drift, have hazarded the earth as a place for the progressive advance of life forms. Indeed mild hazards have probably speeded up the creative processes by exaggerating the differences in survival potential between the fittest and the least fit. In the present human mood of planning how to manage our problems, we have to remember that all evolutionary successes were flukes of mutation and were not planned. Even adaptations were at the expense of the unadapted but previously successful.

The stresses imposed by the physical environment are not necessarily more creative in the progressive evolution of the species than the competitive pressures of other species or other varieties of the same species. Nature is as red in earthquake, lava flow, flood and drought as it is in tooth and claw. We know that to cull the forest deer produces good results; good, that is, for the next generation. The mere reduction of population has always been a moment of opportunity and growth for the offspring of the survivors. Therefore in the geologically trivial disasters of advancing ice or ocean (of which there have been at least 17 known to geological students of North America since the last significant drift of the continents) we need not fear for our own species and the achievements of civilisation. We have to look beyond the probably extravagant but gracious thoughts of those who wish to preserve the white rhinoceros in surroundings which it would enjoy only with unnatural difficulty, the purpose being their possible appreciation by the posterity of their destroyers-become-captors. Our assumption that benefit is for humans only shows us to be slaves of our selfish genes. Such disinterested thoughts are not the stuff of today's politics, nor of those who do not see "environment-friendly" and "mankind-friendly" as rather limited concepts.

True to the precedent of evolution we have loyally tried out any new ideas. Now, as the world becomes full, our inventiveness has put upon us the burden of foreseeing any disasters we may create. Meteorologists try to be useful and forecast the weather and to use the same skill to provide such warnings. It is well known that weather forecasts cannot be relied upon after about three days. The climates represented by the output of computers using good weather forecasting models are not bad and can improve upon crude predictions by statistics.

But the computer has no random element of input like volcanoes, very dry or very wet expanses of land, unusual ocean currents, changes in the human input, or such things as the passage of the sun through a denser than usual cloud of inter-stellar gas. To excuse difficulties caused by things we cannot predict, some meteorologists have asserted that nature itself has an essential element of unpredictability because that is what the models in the computer have. The output of models is always affected by very tiny changes in the starting conditions. These are smaller than could be specified because the observations are not made with the required accuracy; nor could they be meaningfully made more accurate.

The picturesque example sometimes given is that of a butterfly somewhere in Brazil which, by flapping its wings, causes a major change in the behaviour of an Atlantic cyclone. Actually it is really meaningless to argue along such lines because there are millions of other possible events of comparable magnitude and we cannot possibly know whether they took place or not. Even if we had perfect information about such an event it would be too small to include in the calculation. However the model is unduly sensitive, and probably would be perceptibly affected by the butterfly's wing flap if it were calculated to that accuracy with everything else left exactly the same. Even then it would only show that the model contains hidden factors which make its forecasts invalid after a fairly short time: it tells us nothing about the real atmosphere in the long run.

It has been estimated that the atmospheric system has friction and other dissipative mechanisms which would bring it to a standstill within two or three weeks if the continuous uneven input of energy which keeps the motion going were to cease. A million butterflies could not make as much difference as a small bit of the friction systems, and their efforts must therefore be damped out of existence before they have had time to have any influence as much as a city block away from themselves. If the forecast by the model did have the alleged butterfly effect it could be a good model only up to the moment when the effect begins to appear. In fact it is not protected against the (much greater) interference of its own arithmetical approximations. After only one human generation of computer modelling it is not surprising that a model which is designed for forecasting for tomorrow and the next day has serious failings when used on problems of climate and changes of climate.

There are many factors which have been overlooked in global models, possibly because even if they were included there would still be many mechanisms of comparable magnitude which are still too unknown in detail to be modelled properly anyway. The strength of many living bodies lies in the many alternative ways of keeping going. This applies to the curative and healing processes in particular. But any real system which has been going for many millions of years has a stability inherent in the interactions of the many component mechanisms and processes which have, collectively but without coordination, accommodated the present regular sequences of hours, days, seasons and centuries.

The concept of essential unpredictability, which is a popular thesis in some quarters, is in contradiction with the presumption of causality which is basic to the scientific approach. It is also in conflict with all pragmatic human philosophies which ascribe real character to things and living beings. Character is an essential cause; and if character changes there is another cause of that, and we would like to know what it is and how it works. If we fail, that is never taken by us to mean that there is or was no cause: we have to be content to say "we do not know how it works!"

This is an area of philosophical dispute (e.g. Scorer[13]): it may be said that all scientific laws are proven only in the same sense that an exception to the law cannot be predicted. Exceptions are possible but are so unlikely statistically that they can be ignored. However, it is said that there exist random events which initiate unpredictable causes. The randomness is said to occur at a level of minuteness that makes recording of the events impossible. There is said to be an analogy between Heisenberg's principle of uncertainty at the atomic level and this randomness at the level of small events, just as there is between the molecular phenomena of viscosity and diffusion, and the phenomena called eddy viscosity and diffusion according to certain theories of turbulence. The analogy is theoretical but fails completely in practice. At the atomic level there are certainly uncontrollable events, such as mutations which have very important macroscopic consequences, and we can accept Heisenberg's limitation because all the material we can control is composed of structures made of atoms, which cannot be made to guide individual atoms. But the analogy fails when applied to butterfly wings because nothing larger is made out of butterfly wings like proteins are made out of thousands of atoms selected individually out of food, or blood. So randomness is simply an aspect of complexity we cannot, or do not intend to try to, penetrate. It certainly is not a limit to causality and determinism, but it is a bar to practical predictability. The introduction of randomness into theories does nothing to clarify causality or the understanding of mechanism.

In one sense the only satisfactory model is the events themselves which tell us what had been about to happen. Thus learning and parameter choosing is done by comparing model-based forecasts with recorded behaviour of the real thing to obtain a model which would have made a good forecast at least on those occasions. This means that we cannot hope for a good model of climatic change until we have some changes or absence of changes on record in, say, another 20 years from now. In so far as we have any advice from the records so far it includes the statement that acid and particulate aerosols act to cool the climate and reduce the greenhouse effect. Acid rain may have saved us from some catastrophic warming already!

References

1. G. Howells, *Acid Rain and Acid Waters* (Ellis Horwood, Chichester, 1990).
2. R. S. Scorer, *Environmental Aerodynamics* (Ellis Horwood, Chichester, 1978).
3. E. W. Hewson, "The meteorological control of atmospheric pollution by heavy industry" *Quart. J. R. Met. Soc.* **71** 266 (1945).
4. R. Dawkins, *The Extended Phenotype* (Oxford University Press, Oxford, 1982).
5. R. S. Scorer, *Satellite as Microscope* (Ellis Horwood, Chichester, 1990).
6. R. S. Scorer, "Deposition of concentrated pollution at large distance" *Atmospheric Environment* **26A**, No. 5, 793–805 (1992).
7. P. F. Chester, "Acid lakes in Scandinavia—the evolution of understanding" CEGB Report TRPD/L/PFC/010/R86 (1986).
8. W. J. Humphreys, *Physics of the Air; Part V Factors of Climatic Control*, Ch II (McGraw Hill, 1920, 2nd Edn., 1940, and Dover, 1964).
9. D. Brunt, *Physical and Dynamical Meteorology* (Cambridge University Press, Cambridge, 1939).
10. R. S. Scorer, "Ship Trails" *Atmospheric Environment* **21**, 1417 (1987).
11. J. S. Sawyer, "Man-made carbon dioxide and the "greenhouse" effect" Nature **239**, No. 5366, 33 (1972).
12. W. G. N. Slinn, "Hints of another gremlin in the greenhouse: Anthropogenic sulfur" *Atmospheric Environment* **25A**, No. 11, 2473–2489 (1991).
13. R. S. Scorer, "Lessons from observations of chaotic flow in the atmosphere" *Journal de Mécanique Théorétique et Appliquée*. Special issue, supplement No. 2 to Vol **7**, 147–165 (1988).

2. ACIDIFICATION: POTENTIAL FOR REVERSAL OR MITIGATION?

GWYNETH HOWELLS

"Acid Rain" has been a major issue in scientific and political circles for the last three decades. Recent decisions on the control of industrial emissions of acid generating gases have been taken in EEC and USA. What is the perspective for the reversal of acidification, its time-scale, and the extent of expected benefits and possible costs? Will acidification continue unless such action is taken? Are there effective alternatives? Recent scientific analyses of acidification are reviewed, including relevant cases and modelling exercises from which the feasibility of reaching the desired objective may be judged.

1. INTRODUCTION

"Acid Rain" has been a scientific, economic, political and emotional issue for the past three decades. After extensive research in many countries, followed by much debate and discussion, recent decisions have been taken on gaseous emission control in the EEC, and in USA and other nations, and will have repercussions for the decades ahead. It is worth considering what is the perspective for reversal of acidification by emission control, its time-scale, and possible alternative strategies.

1.1 *What is Acid Rain?*

The acidity of rain is measured by its pH (the negative logarithm of the hydrogen ion concentration) representing the balance of hydrogen (acidity) and basic ions (and alkalinity) in solution. The pH scale was conceived in 1909 by a Danish chemist, Sorenson, who first recognized the importance of the hydrogen ion in biochemical processes, and devised practical colorimetric and potentiometric methods for its measurement. The conventional (logarithmic) scale ranges from 0 to 14 (Figure 1) with the mid-point value (pH = 7) where $[H^+]$ and $[OH^-]$ are equal in water with no solutes. Where solutes contribute acid or alkaline ions, this equilibrium is displaced to a pH of less than 7 in acid solutions and to a pH of greater than 7 in alkaline ones. Most of the world's fresh waters lie within a range of pH from 5 to 10, reflecting not only the acidity of deposition, but also the weathering of soils and underlying geological materials, of vegetative interactions in the flow path, and the hydrological pathways and dynamics which govern the extent of these interactions (Figure 2). While a circumneutral pH is reasonably well maintained in solutions with some buffering capacity, very dilute waters (such as rain or upland lakes and streams draining a terrain of slow-weathering rocks and soils) have little buffering capacity, and the balance of H^+/OH^- (i.e. pH) is unstable; such natural waters are better represented by a range rather than a single pH value. In contrast, the pH of sea water is remarkably

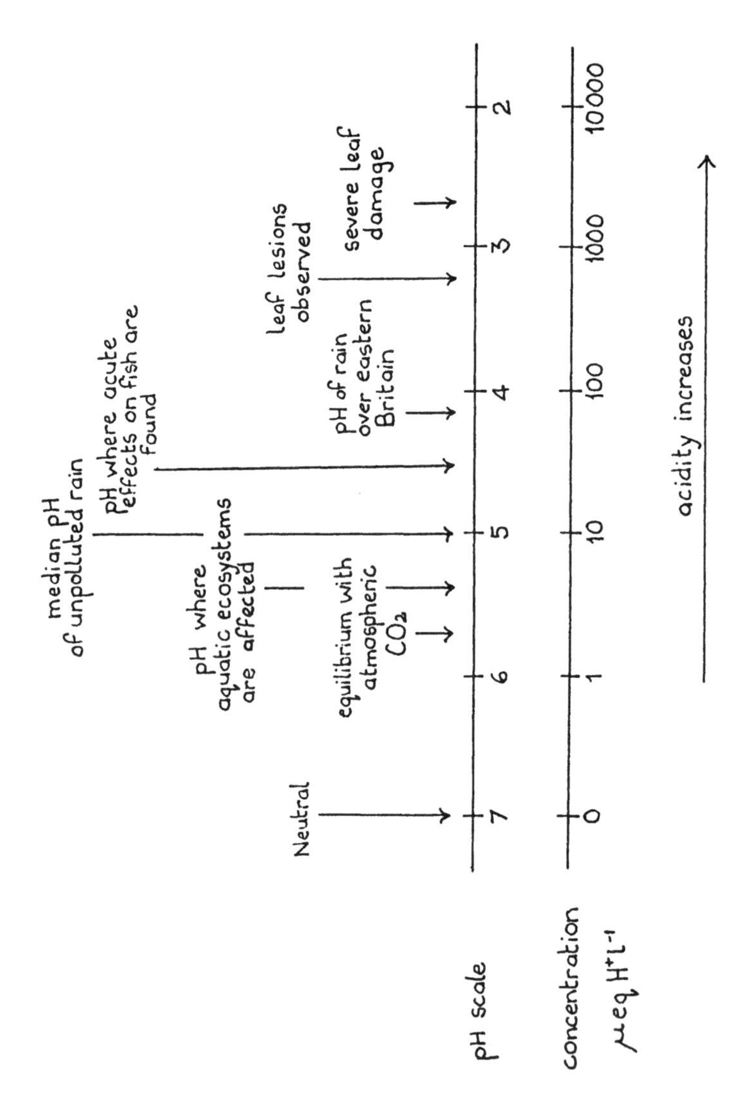

Figure 1 Scale of pH and H^+ concentration; levels of some observed effects

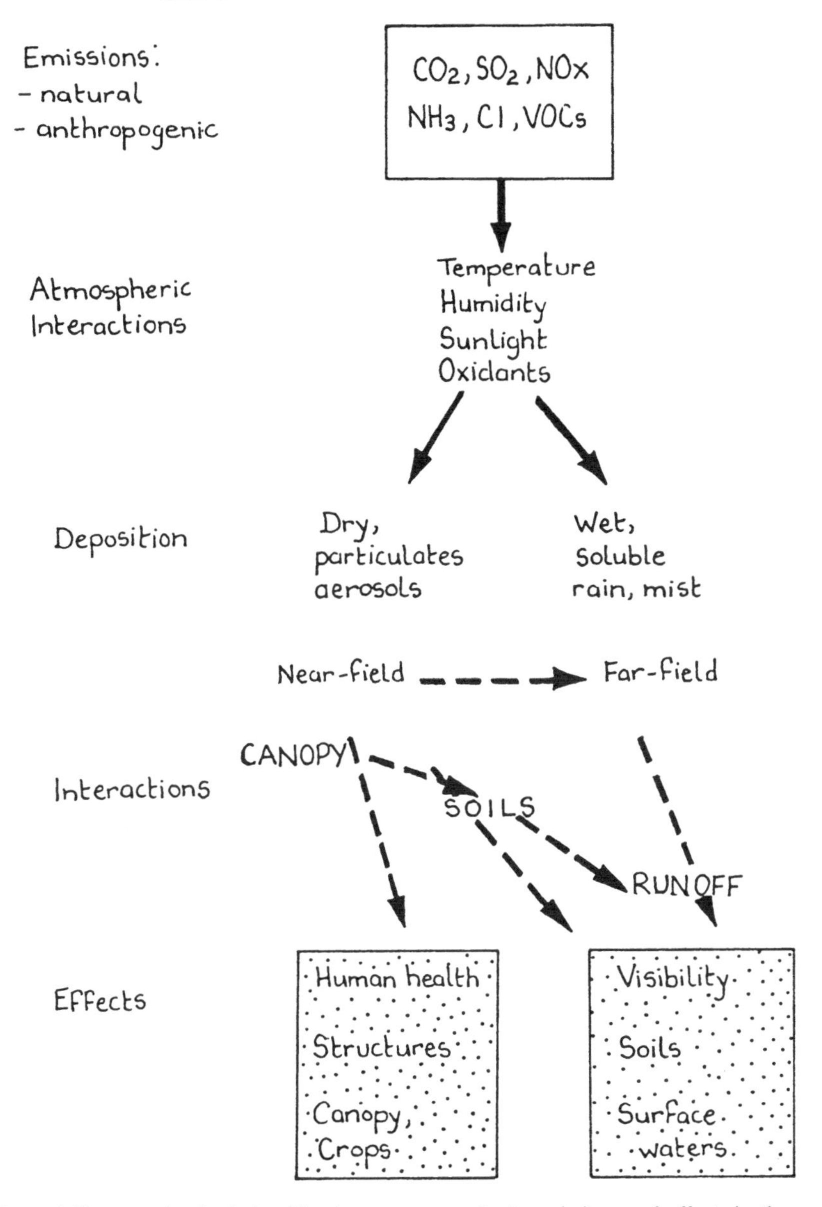

Figure 2 Framework of relationships between atmospheric emissions and effects in the near- and far-field

constant at a pH = 8.25, a consequence of its high content of solutes with substantial buffer capacity.

1.2. *Rain Chemistry*

Rain acidity reflects its chemical composition, of which carbon dioxide is a principal component. Carbon dioxide gas dissolves in water to form carbonate

(CO_3^{2-}) and bicarbonate (HCO_3^-) ions. At normal atmospheric pressure and a temperature of 25°C, this brings the pH to a value of 5.6 if no other solutes are present. In reality, the pH of rain even in pristine areas is around 5.0, since other natural acidic gases are always present, as seen in rain samples from remote areas such as oceanic islands.[1] Carbonate equilibria are critical in the transformation of deposited rain via runoff to waters of lakes and streams, due to solution of carbonate minerals, and biological respiration and root activity. Carbon dioxide pressures within soils are as much as 100-fold greater than in atmosphere[2] as a consequence. While carbon dioxide in the soil water is degassed on emergence to surface runoff (or captured in "mineral water"!), within the soil matrix it is a powerful weathering agent, providing both mineral solutes and alkalinity to the soil pore water and runoff. These interactions mean that soil water and runoff always differ from rain, and is usually less acid, even where underlain by shallow soils and rocks resistant to weathering.

In less natural conditions where industrial and agricultural activities are substantial, carbon dioxide concentrations in atmosphere are greater, as well as those of sulphur and nitrogen gases; they all increase the acidity of rain to a pH of less than 5.0. The presence of dusts, both natural and man-made, often counters this acidity by contributing base materials (Figure 3).

The importance of sulphur dioxide gas in generating acidity in rain was first recognized quite early in the last century by Adam Smith, a pollution inspector in industrial Manchester.[3] Sorenson's pH measurement system had not then been developed but it was known that sulphide minerals (vitriol) in soils or solid wastes could be oxidised to sulphate with consequent increase in acidity; early

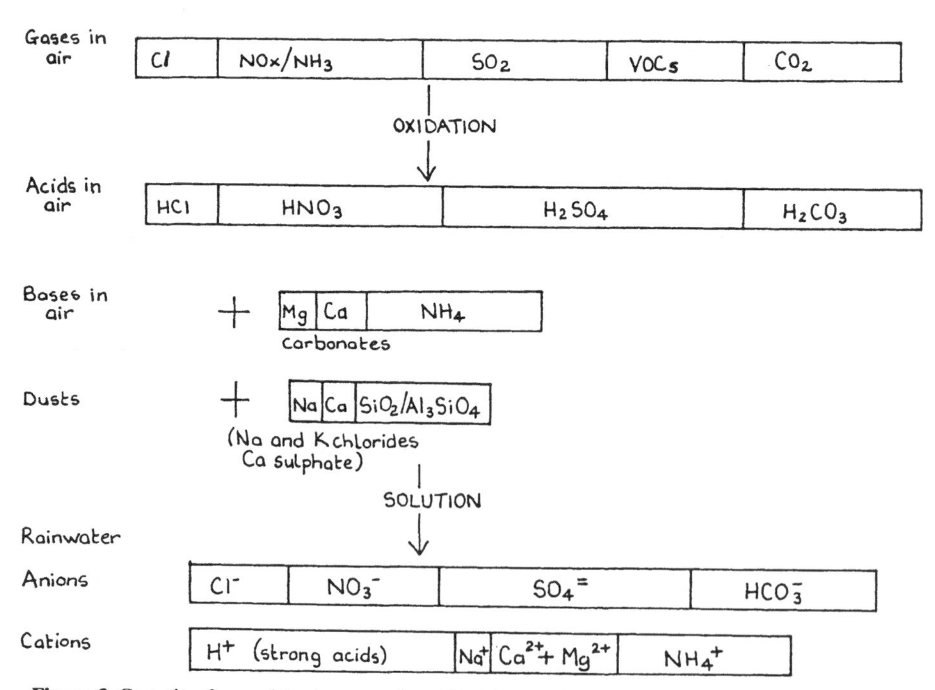

Figure 3 Genesis of an acid rain water (modified from W. Stumm and J. J. Morgan, Aquatic Chemistry, 2nd. ed. Wiley, New York, 1981)

this century, this was recognized as the cause of fish deaths in some Norwegian rivers.[4]

Recognition of the influence of nitrogen species in generating acidity came at the same time; in the last century dissolved nitrogen in rain was considered a significant benefit to agricultural crops,[5] a concept then disputed by the agricultural research station at Rothamsted, UK, where a large rain gauge was set up to measure nitrogen and chloride inputs, continuing over many decades. In general, levels similar to today's were found, but spring peaks have increased fivefold[6] as recent agricultural practices of nitrogen fertilization and intensive husbandry have led to much greater advection of nitrogen oxides and ammonia to atmosphere. While N oxides, like sulphur, are oxidised to nitrite (NO_2^-) and nitrate (NO_3^-), alkaline ammonium (NH_4^+) is formed on solution of ammonia gas (NH_3). This reduces rain acidity but on deposition, ammonium oxidizes and then, as nitrate, contributes to the acidity of the runoff.

In addition to these constituents, volatile organic compounds, many generated by natural biological processes, contribute to the acidity of rain; natural sources of VOCs are high, especially in forested areas. Although these compounds are less well-documented, estimates of man-made emissions of sulphate, NO_x derivatives, and VOCs in USA are about equivalent (as tonnes/year).[7]

The commonly expressed view that "acid rain" is synonymous with "polluted rain" is misleading. A recent dictionary entry, "rain containing pollutants, chiefly sulphur dioxide and nitrogen dioxide, released into the atmosphere by burning coal and oil",[8] ignores both the crucial influence of carbon dioxide and ammonia, as well as the greater variety of natural or man-made sources. Another atmospheric gas, ozone, often an agent of significant vegetation damage, is generated in atmosphere by the interaction between other contaminants (especially nitrogen oxides and VOCs); it does *not* contribute to the acidity of rain. In popular perception forest damage is often linked with acid rain, but direct gas fumigation, mineral deficiency, and ozone are more credible causes.

1.3. *The Changing Focus of Concern*

While the acidity of rain and surface waters has been recognized as "damaging" in a general sense during the last several decades, the focus of scientific and political concern has changed since 1970. In 1971, the Swedish "Case" distinguished the principal issues as the dual effects of sulphur dioxide gas on health, metals corrosion and vegetation, and the effects of acid deposition on soils, fresh waters and forest production.[9] The subsequent Norwegian SNSF (Forests and Fish) programme (1972–1980) was directed principally to forest effects of acid rain, but also to understanding rain, snow and surface water chemistry. While fish as a target were included, most findings were related to the extent of acidified waters and their depleted fish communities.[10] Since then, independent research in several countries has investigated acid effects on physiological, behavioural and community processes in the aquatic ecosystem and established more quantitatively, the critical and causal links between deposition, hydrology and soil reactions.

With the later Swedish "Action Programme",[11] more emphasis was given to measures to reduce emissions of sulphur dioxide, nitrogen oxides and hydrochloric acid gas, even acid discharges from mining wastes. These were seen as needed on both a national and an international basis. Other measures proposed (paradoxically) included fertilization of forests to reduce the acidification of forest soils, as well as liming of agricultural soils, and lake and river liming. The most recent governmental "programme statement" through to 1992–93, gives even greater emphasis to new questions on liming strategies, reacidification, and effects of liming on a variety of organisms and communities.[12]

In 1980, scientists became aware of adverse changes in European forests ("Waldschaden"), attributed to a novel and mysterious disease which caused loss of foliage, especially of conifers. This came to be associated with air pollution and acid rain. A similar phenomenon was soon reported for northeastern USA and Canada. Later analysis found the damage to be associated only indirectly with acid rain in some areas where mineral deficiency was exacerbated by foliage leaching and lack of soil mineral replenishment.[13] While excess nitrogen deposition and natural climatic stresses play some part in the forest effects, in Europe a prime agent was identified as ozone, possibly in association with acid mist, but it has not been possible to reproduce the symptoms of forest decline in experiments, bringing this hypothesis into question again, although attribution to direct acid rain exposure is now discounted.

North American views have also changed. In 1985, the American National Academy of Sciences reported "conclusive evidence that emissions of sulphur dioxide from power stations acidify lakes and kill fish", but little action followed.[14] A regional relationship was found between sulphur dioxide emissions, sulphate aerosol, reduced visibility and wet deposition of sulphate, consistent with sulphate concentrations and the loss of alkalinity in surface waters (Figure 4).[15] Evidence of pH decline in Adirondack lakes, New York (as much as 1 pH unit over 20 to 40 years) was linked to emissions and acidification, and to a decline of fish populations. Reduced tree growth and increased mortality of red spruce was linked to high acid deposition, but possible effects of acid deposition and of other climatic stresses on trees could not be distinguished. Other equivocal relationships were also exposed: sulphur emissions in NE America have been virtually constant over the past 60 years, and changes in lake alkalinity were greater than could be accounted for by acid deposition, which had changed little. Data for fish populations at the time were considered insufficient to make projections either in space and time.

The subsequent NAPAP review has confirmed many of these conclusions and has quantified the degree of damage and simulations of ecosystem response to the control of acid emissions is now possible. Although a suite of control measures in USA has now been ratified by a new Clean Air Act, the tangible benefits of reduced S emissions to targets of acidification are specific and limited (see section 4.2). Indeed, one critic[16] argues that "the rhetoric of . . . ecological damage from 'acid rain' . . . outstrips the reality of the health and environmental benefits to be gained from legislation." Further, it is claimed that the Clean Air Act objectives are largely unattainable, and regulation will be both economically and environmentally inefficient. A reformulated NAPAP programme will continue with overall assessment of acidification, and, significantly, it will also monitor the costs and benefits of the Clean Air Acts amendments of 1990.

2. THE ASSESSMENT OF EFFECTS: TRENDS AND TARGETS

Many types of evidence are available for assessing the effects of acidification and their time-scales; alone, any single method has evident deficiencies in providing

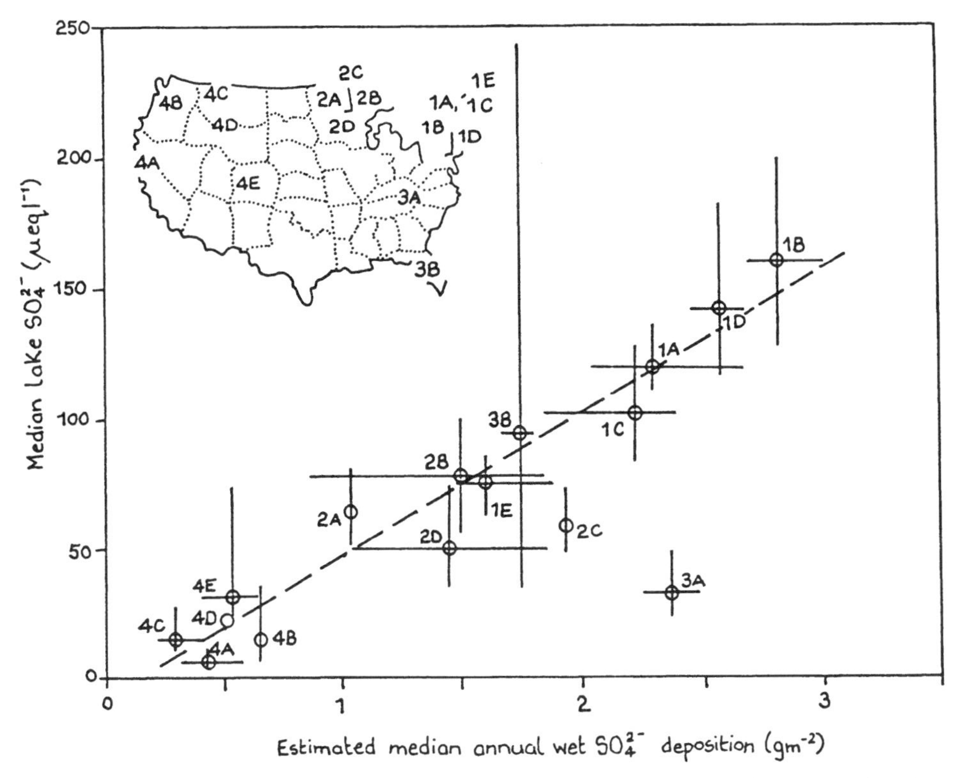

Figure 4 Relationship between estimated wet SO_4 deposited and lake/stream SO_4 concentrations for NAPAP investigated waters (after Sullivan *et al.*, 1988).[15] The linear relationship is drawn by eye

a scientific "proof." The advantages of bringing together the results of convergent enquiries are substantial if they are consistent, although there are also dangers in combining data of varied precision, and in spatial and temporal extrapolation to other conditions. In this and the following section, several approaches and data collations are reviewed. The main thrust will be to consider the relationship of emissions, deposition and surface water effects, justified by the more significant findings of surface water acidification than for other perceived effects, considered here only briefly.

2.1. *Biological Effects: Forests, Soils, Surface Waters*

During the transfer and transformation of acidic gas emissions through to deposition and ultimate neutralization several potential targets of successive impact from emission to "sink" can be identified (Figure 5). It is worth noting that "rain" is substantially modified in its transfer through to surface water (Figure 6) so that direct targets of acid rain exposure are to be distinguished from secondary effects.

Industrially emitted gases which may have direct adverse effects on vegetation (crops and forests) are threshold limited. Visible foliar injury was often seen prior to 1960 when urban concentrations of sulphur dioxide commonly exceeded 100 ppb in North America and Europe;[17] these conditions are seldom seen today

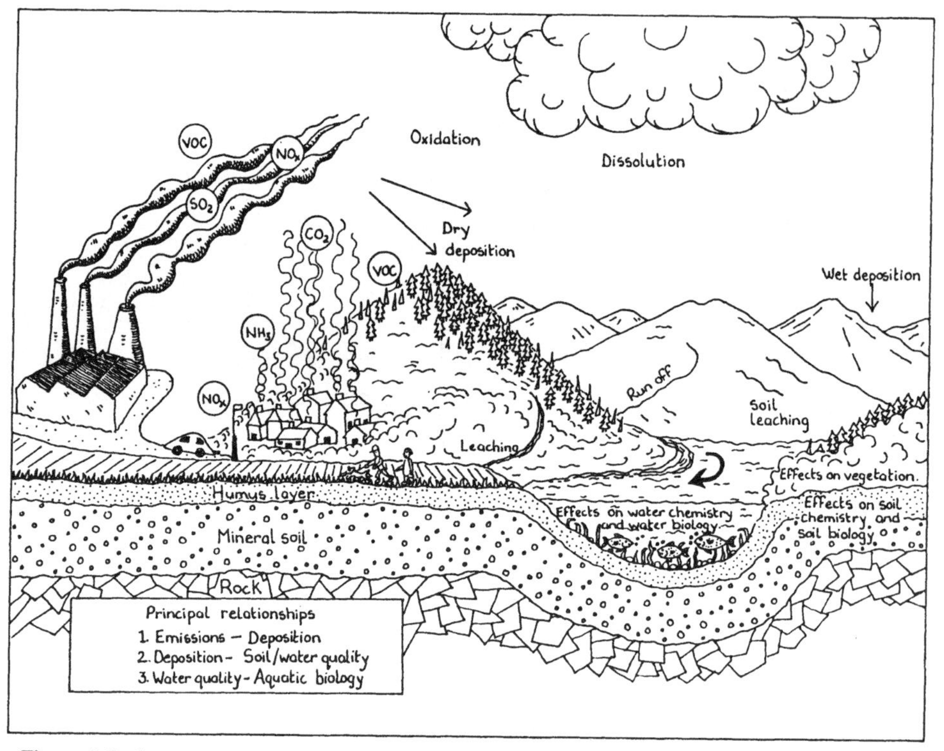

Figure 5 Pathways from emissions through to targets of acidification and principal relationships discussed (after G. Persson (ed.), *Acidification Today and Tomorrow*, Swedish Ministry of Agriculture, 1982)

and such damage is found only where there is direct fumigation in the vicinity of a source from low stack height and inadequate emission control.[13,18–20] Acid rain *per se* has little adverse effect (at pH ≥3.5) even on sensitive crop and tree species,[21–24] although increased leaching of magnesium from the canopy may induce a deficiency on magnesium-poor soils.[17,25] Increased leaching of minerals from foliage of some forest stands at high altitude, associated with climatic extremes, is also reported in USA.[7]

As acid deposits at the soil surface, however, progressive acidification of the litter and soil horizons occurs, with leaching of base cations, loss of base content and accumulation of hydrogen ion (from both natural or pollution sources), mobilizing aluminium from the soil.[26,27] Soils vary in their capacity to neutralize acidity and also in the degree to which they retain or release sulphate. In sensitive soils, acid conditions may prevail, often without damage, provided bases continue to leach, although this has important implications for surface waters if bases there are insufficient for fish health. In soils without sufficient base reserve, H^+ and Al^{3+} are leached, with toxic effects on fish. There is also concern that soil acidification leads to unacceptable changes in the quality of drinking water abstracted from shallow wells.[28] However, such changes are slow, possibly attributable to a variety of causes, and difficult to interpret. The long-term consequence of deposited acid on soils is similar (but in addition) to that resulting from natural vegetative growth in the absence of corrective management, such as lime applications.[29,30]

As acid is transferred via runoff to lakes and streams, it is accompanied by

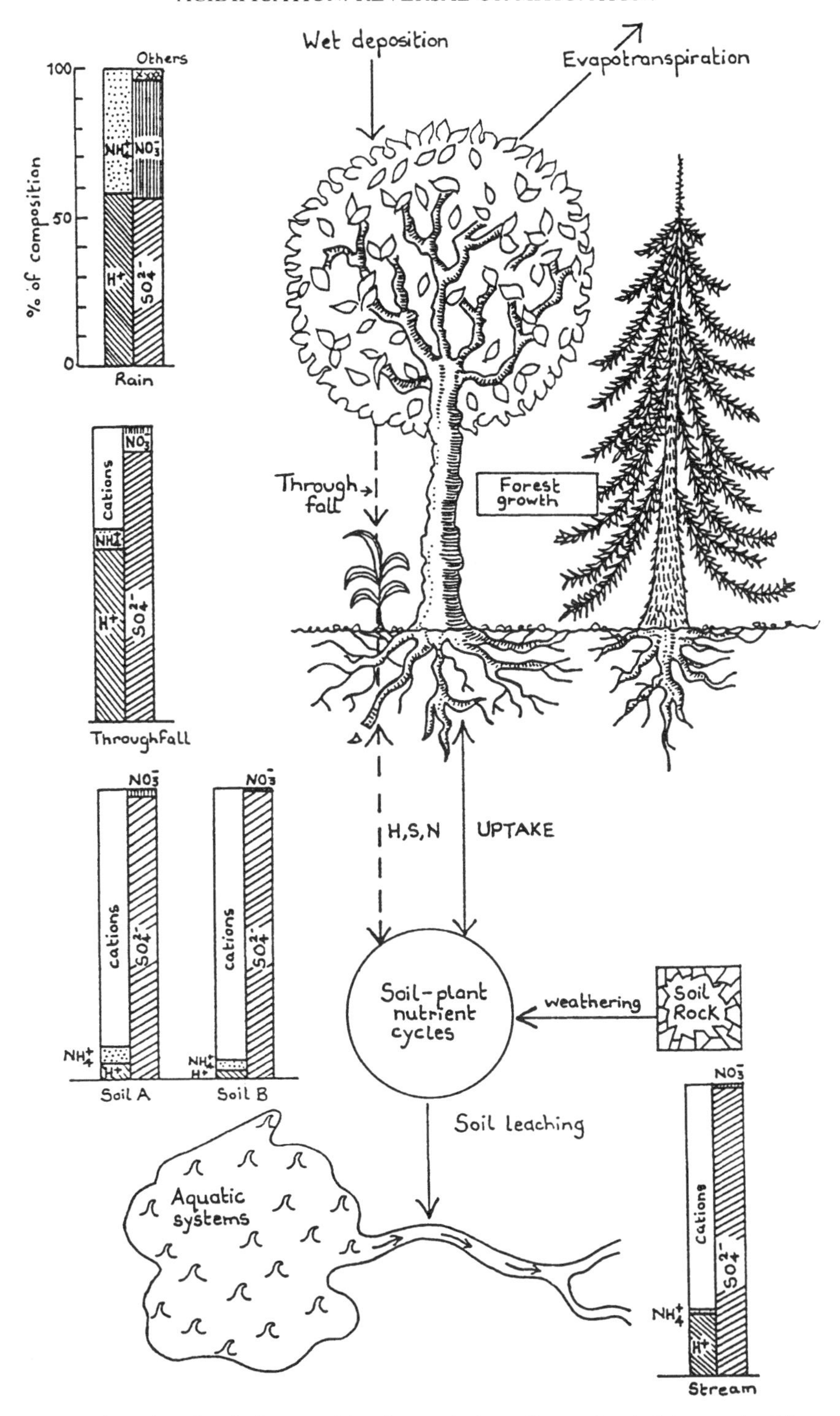

Figure 6 Changing chemical composition from rain to surface water draining a mixed forest catchment, and principal processes and pathways

labile monomeric aluminium ions (Al^{3+} and hydroxides) displaced from soil when mobile anions are in excess.[26] In acid soils, deposited sea salts (even at high pH) can also release soil acid and aluminium ions to cause acid episodes in runoff.[31] Organic materials, if present, may complex the aluminium to less toxic forms;[32] complexing of other potentially toxic minerals also occurs.[33] Within the aquatic environment, this acidification brings substantial ecological changes, notably affecting the variety, distribution and health of fish populations.[34,35] Toxic aluminium, possibly in conjunction with other trace metals, appears to be as important as acidity, especially at pH 5.0–5.5[35] and a critical threshold of calcium, leached from underlying geology, is evident.[32,36]

The most recent review of acidification effects in the USA[7] acknowledges that S and N compounds impair visibility, corrode urban structures, and can pose a threat to human health. The possible effect on trees remains uncertain and limited to some high altitude sites. The NAPAP analysis demonstrates, however, that acidification has indeed occurred in sensitive aquatic systems in USA sometime during this century, and that it cannot be ascribed to natural causes or land use changes but is due to increased deposition of S and N compounds. On a national basis, about 4% of lakes greater than 10 hectares and 3% of streams surveyed, mostly in the north east, were judged to be acidic (i.e. pH <5.5, alkalinity <0 $\mu eq\ l^{-1}$). Data records for historic conditions were found to be unreliable for most areas, but in the Adirondack region (NY State) lakes have become more sensitive by the progressive loss of acid neutralizing capacity (ANC).[37]

2.2. *Acidification: Temporal Trends*

2.2.1. *Emissions.* While more rapid acidification is associated with increasing industry since the 1960s in Europe, a longer history of both natural and man-induced acidification is evident over a world-wide scale. During the last four decades, three European programmes (EACN, OECD and EMEP) have followed changes in both atmospheric conditions and deposition. However, atmospheric concentrations of SO_2 and sulphate concentration measured in rain have not always matched changes reported in S emissions;[38] this may stem from differences in sampling and analysis, but also from lack of measured, rather than estimated emissions. While daytime atmospheric sulphate concentrations seem to be proportional to S emissions, uncertainties remain about emission/wet deposition relationships. There is evidence that while S emissions have fallen in recent decades, both emission and deposition of N species has clearly increased in Europe and North America, and overall there is little change in acid deposition.

2.2.2. *Rain.* Although rain must have always been acid (see para 1.2), it has undoubtedly become more acid in the last century or so. This trend was uncertain prior to Sorenson's pioneering techniques; although some rain constituents were measurable earlier, even quite recent (since 1950s) equipment for measuring pH of rain and surface water measurements in the field prior to the last 15 years or so has doubtful precision, particularly so where concentrated colorimetric agents were added to dilute natural waters, or where temperature compensation of potentiometric systems was ignored.[39] In a few instances, where measurements of natural waters have been compared using historic and modern methods; the earlier colorimetric methods overestimated pH to a considerable degree.[40] In addition, the transient nature of pH in natural waters means that occasional samples may not be representative and cannot be used to establish a trend.[41] These considerations limit the scope and validity of long-term records for both rain and surface water pH. A further complication, often ignored, is that "cleaner" industry has reduced the emission of dusts, potentially neutralizing.

Changes in rain pH or sulphate in UK over the past 30 years are small, with little benefit seen from the 40% reduction in national sulphur emission within this period.[42,43] Over a decade, it is judged that significant changes would be masked by year-to-year variations dependent on weather patterns. In Norway, only a slight decline in acidity and sulphate of rain has been observed.[44] In Sweden, a twofold decrease in S emissions since the early 1970s has also led to little change in rain pH and only a marginal decrease in sulphate deposition.[45] In contrast, in N America between 1955 to 1975, although S emissions in the affected northeast changed little,[14] a pH decrease in rain of about 1 unit or less (i.e. a four- to tenfold increase in acidity) is reported, concurrent in some cases with a significant increase of sulphate or nitrate concentrations. Emissions of acid generating materials (SO_2, NO_x and volatile organics) peaked there in about 1970, and remain at about twice their 1900 level, but wet deposition has shown only a statistically insignificant decrease since S emission reduction in 1979.[7] A relative change in N components in rain may help to explain this; recent data from Norway shows that nitrogen (NH_4 + NO_3) is now about 20 to 30% higher than sulphate deposition (as equivalents)[46] and at a site in northwest UK, sulphur and nitrogen wet deposition were equivalent over the period, 1985–89.[47]

No clear relationship emerges between emissions and wet deposition, and even with a total avoidance of sulphur and nitrogen emissions, the scope for reducing the acidity of rain is limited since pH cannot increase much above the natural pH 5.0. In the most affected areas rain pH is about 4.2, equivalent to an excess of about 50 μeq l^{-1} of H^+. The observed proportionality of H^+ and sulphate in rain suggests that this pH change would represent about 100 μeq l^{-1} change in rain acid anions.[7] The lack of a clear proportional relationship between sulphur emissions and wet deposition remains discouraging and an increasing world atmospheric burden of carbon dioxide will further reduce the expected benefit of S or N emission control.

3.2.3. *Surface waters.* For surface waters, the situation is undoubtedly more complex. A proportional change in surface water acidity with change in deposition is rather uncertain given the diversity of hydrological and geological conditions at any site. A searching analysis of many large data sets shows little trend even for sensitive, dilute waters, but some lakes and rivers in Scandinavia and N America have shown historic pH changes of up to 2 pH units (100-fold acidity increase) over the last three to four decades, as reported to the UN Conference in 1972.[48] This scale of change is inconsistent with reported changes in rain acidity and with the expectation of moderation through transfer via the catchment to runoff. The increased acidity is often associated with increases in deposited sulphate, but solute concentration due to evapotranspiration, and soils or wetlands retaining and/or releasing sulphate both in the long-term or during episodes, are also important. Seasonal or year-to-year climatic conditions may further confound any trend in the short term.

The most certain evidence of long-term acidification of lakes has been derived from studies of the progressive changes in diatom remnants in lake sediments from which pH can be inferred. For Adirondack lakes, a number show pH falling from >6.0 to >5.0 (SE ± 0.24–0.4); in an estimated 14% of these lakes, pH fell by an average of 0.3 pH units.[49] Similar studies in European lakes show a post-glacial decline from a pH of about 7.0 to 6.5, then stability at a fairly constant pH of 6.2 to 6.6 until this century, when some lakes showed a sharp drop of 1 to 2 pH units.[50–53] In SW Scotland, diatom studies suggest that sensitive lakes there became acid at varying times since 1850, one even as recently as 1975; changes were between 0.5 and 1.5 pH units.[54] While pH/diatom relationships have been demonstrated, however, many other water quality factors (base concentrations, ANC, aluminium, *inter alia*) must also be important. Further criticisms of the diatom studies are that they focus on small, more susceptible, lakes rather than larger ones, and reflect summer water quality conditions.

3. SOME RELEVANT CASE STUDIES

3.1. *Changes in Lakes and Streams: Some Important Data Sets*

3.1.1. *Short-term changes.* Changes of as much as 10%/year in stream and lake waters (pH and sulphate) were reported in Scandinavia during the 1960s and 1970s,[48,55] but in the longer-term, streams show little evidence of change since then.[45] Some lakes in western Sweden show reductions since 1977 of 40 to 100 $\mu eq\ l^{-1}$ in sulphate although H^+ concentrations fell to a smaller extent; rain acidity during the same period actually increased, although sulphate in rain decreased (15%) while nitrate increased (34%),[56] possibly reflected as higher nitrate in soil water.

An extensive (1000 lakes) survey in Norway shows little change in acidity between 1974/75 and 1986. In some, sulphate concentrations fell while nitrate concentrations doubled;[57] similar increases were found in rivers. This has been interpreted as evidence of nitrogen saturation of soils. In waters subject to periodic deoxygenation, this enhanced nitrate has a greater potential for nitrite formation; similarly breakthrough of ammonium to runoff has toxic potential for fish.[58] Although pH has changed little, mobilized aluminium has increased.[58]

In the UK, analysis of routine chemical data for vulnerable rivers for a limited period (1970 to 1985) showed evidence of pH decline in only 6 of 75 data sets.[59] However, diatom analysis of lake sediments gives evidence of acidification in sensitive areas.[60] Reductions in rain acidity and wet deposition of sulphate have been reported since 1978 in SW Scotland and matched to pH rise and a decrease in sulphate concentration between 1979 and 1985 of 4 lakes in the area, along with an increase in base cations in 2 lakes,[61] inconsistent with expectation (Table I). These lakes are thought to have acidified about 1850. Two other lakes nearby showed a similar recent change in pH and sulphate, but quite different changes in nitrate, chloride and decrease in base cations. Over the same six year period, some recovery of acidophobe diatom populations is noted but changes in other components of the biological community are not reported.

Sulphate deposition at 8 sites in northern Britain (range 4 to 22 kg S/ha-yr in 1988) has changed little, however, since 1978 and there is substantial variation both within and between years,[43] possibly dependent on weather patterns. However, a change since 1981 in sulphate deposition in central Scotland is reported[43] and linked with a fall in S emission in UK since the 1970s, possibly equivalent to a reduction of 1.6g S/ha-yr in the Galloway lake region, or 16g S reduction over 10 years. The USA data suggests that for each kg S/ha deposition reduction, surface water alkalinity would change by about 3–4 $\mu eq\ l^{-}$,[17] thus the Galloway

Table I Changes in chemistry of some Galloway lakes
(Data from Battarbee *et al.*, 1988,[61] Howells and Dalziel, 1988,[119] Patrick *et al.*, 1991,[87] Wright and Snekvik, 1979[55])

	Chemical Components						
Lake	$[H^+]$	$[SO_4]$	$[NO_3]$	[Cl]	[Ca]	[Mg]	[Al]
Enoch	0.60	0.78	1.50	1.47	1.80	1.25	0.86
RLGH	0.77	0.81	1.50	1.60	1.52	1.21	0.70
Fleet	0.55	0.71	0.87	1.05	0.88	0.87	0.79
Dee	0.43	0.68	0.91	1.06	0.97	0.70	0.69
(Rain)	1.79	0.71	0.75	0.88	0.56	—	—

RLGH = Round Loch of Glenhead. Values shown are the ratios of 1985/1979 concentrations.

lakes would have an insignificant increase in alkalinity over the past decade. The changes in chemistry reported for 1979 to 1985 suggest that alkalinity has risen by about 4 $\mu eq\ l^{-1}$ in one lake but 26 $\mu eq\ l^{-1}$ in the other. Consistent or explicable responses might not be expected from this extrapolation[61], given the wide range of regional and site characteristics.*

3.1.2. *Longer term changes.* In USA, palaeoecological analysis shows that most Adirondack lakes of current ANC $> 25\ \mu eq\ l^{-1}$ have not suffered a fall in pH or reduction in ANC, but for the acidic lakes ANC has fallen by about 18 $\mu eq\ l^{-1}$ since 1850,[63] less than previously believed. In this and other sensitive areas, acidification is attributed to high sulphate concentrations relative to base cations; organic acidity (DOC) is relatively low. Measured changes in chemistry in relation to shorter-term changes in deposition have seldom been reported in USA. Two Maine (USA) rivers sampled in 1969/70 and 1980/82 show no significant change in acidity although sulphate decreased and nitrate increased[64].

3.1.3. *Fish and aquatic communities.* An extensive fish survey of lakes in southern Norway[65] demonstrated that the number of lakes barren of fish has doubled since the earlier survey of 1971/75. This is associated with a decrease in calcium and sulphate and increase in nitrate and aluminium concentration but pH has not changed significantly. Periodic fish kills have also continued.[66] An association of lost fish populations and water chemistry has also been shown in USA in the Adirondacks and Upper Michigan.[35] Lakes (8.1%) in the Adirondack region lost fish communities since the 1930s; these were lakes with significantly lower recent pH levels, but fewer than the 14% shown to have acidified.

Fish populations were also lost from more than half of La Cloche (Ontario) lakes between 1960 and 1970s,[67] with a concurrent fall in pH. Salmon catches for five acid rivers of the southern uplands of Nova Scotia, Canada, show a decline over the past 50 years[68] although estimated production loss is less certain.[68] Four of these rivers show a fall in pH in the period 1955 to 1981.[69] However, catch records for 6 rivers in the adjacent state of Maine, USA, show no significant trend.[69] Salmon catches declined in western Scotland over the period 1952 to 1981 associated with regional afforestation rather than acid deposition.[59] A similar decline seen in Welsh river catches is more difficult to interpret, but there was no significant correlation with sensitivity to acidification.[60]

Many records of change in invertebrate communities are reported, with loss of species considered more sensitive to more acid conditions.[35] In a few cases, recovery of aquatic communities or sensitive species has been observed with rise in pH, specifically in lakes experimentally acidified for a time, in the Sudbury lakes after local S emissions were substantially reduced, and in limed lakes (see section 3.2.4).

3.2. *Field Experiments*

3.2.1. *Acidification of lakes.* Direct demonstration of surface water acidification has been achieved by direct acid additions to a pristine lake in Ontario, Canada, demonstrating rates of acidification and recovery.[70,71] Lake 223 is a small, sen-

*The Galloway lake chemistry shows substantial marine influence,[62] so the balance of acidic and basic ions from which alkalinity is calculated is strongly influenced by climate pattern.

sitive, seepage lake with a slow rate of renewal (up to 10 yr); it was progressively acidified by addition of sulphuric acid directly to the lake between 1976 and 1983. Initial pH was >6.5 and alkalinity was maintained by catchment sources and bacterial activity in the lake sediment even though acid "loading" in this experiment was almost 100 times greater than wet sulphate input, ten times greater than that of affected areas in Canada. These massive acid additions were less effective than first expected since sulphate reduction in the lake sediments provided a "sink" for the added sulphate, generating alkalinity. The lake was gradually acidified to pH 5.0 (increasing H^+ to 10 $\mu eq\ l^{-1}$ over 8 years); changes in chemical and biological conditions are summarized in Table II. Another lake in the same area (Lake 302)[72] was divided by a polyvinyl curtain, with sulphuric acid added to the south basin in 1982, lowering the pH from pH 6.6 to pH 5.6; the north basin was acidified over the same period by addition of nitric acid, with pH falling only to 6.2.

Acid additions to Lake 223 were gradually reduced after 1983, with pH lifted to 5.4 in 1984, 5.5 in 1985, 6.0–6.5 in 1987,[70] almost regaining its earlier chemical status over 3 years. During this period alkalinity increased and sulphate decreased (by 19 $\mu eq\ l^{-1}$) although hydrogen ion concentration decreased by only 5 $\mu eq\ l^{-1}$. Chemical recovery, overall, was thus more rapid than the acidification process, although biological recovery was slower, particularly for some fish and crustacean species.[73]

Little Rock Lake in Wisconsin was also separated by a polyvinyl barrier, one part receiving periodic sulphuric acid additions since 1983, the other a "control."[74] In the acid treated basin, pH is to be lowered over three 2-year periods from 6.1 to 4.7. The control basin maintained its natural pH through 1990. Acid "loading" is about three times that received in affected areas. As with Lake 223, acidification of Little Rock Lake was slower than predicted, with enhanced sulphate reduction in sediments of the acid basin. Over the first three years, acidification had little effect on N and P cycles although DOC decreased, trace

Table II Changes in Lake 223 during acidification and recovery

Year	pH	Changes observed
1976	6.8	Normal chemistry and biotic communities of oligotrophic lake
1977	6.13	Epilimnion sulphate increased 2.5x background; S reduction in hypolimnion; alk decreased. More green algae; more chironomid emergence.
1978	5.93	Loss of copepod *D. sicilis*; decline in *Mysis*; fathead minnow recruitment failure
1979	5.64	Changes in phytoplankton; decline of copepod *E. lacustris*; *Mougeotia* mats; minnow and sculpin decline, but more young-of-the-year
1980	5.59	More dinoflagellates and BG algae; loss of *E.lacustris*; *D.catawba* appears; crayfish recruitment failure; chironomid increase; pearl dace and suckers abundant but lake trout recruitment failure
1981	5.02	No change in P cycle. Dinoflagellates and BG algae dominate phytoplankton; Cladocera increase, zooplankton species richness reduced but not biomass; decline in crayfish; white sucker and lake trout recruitment failure
1982	5.09	Phyto- and zooplankton community stabilizes, biomass and production "normal"; *Mougeotia* mats occlude trout spawning beds; crayfish about extinct; recruitment failure of coarse fish, trout condition poor, survival reduced
1983	5.13	Plankton stable; *Mougeotia* declining; crayfish absent; mayfly *Hexagenia* absent; coarse fish recruitment poor, trout condition poor, survival reduced
1987	5.50	Phytoplankton species reappear, zooplankton reduced by fish predation; crayfish and *Mysis* still absent; minnows and sculpin still absent, but sticklebacks appear; trout condition good, but recruitment poor

metal concentrations increased, and sulphate reduction increased along with the higher S input. Two thirds of added acid was neutralized by in-lake processes, principally sulphate reduction and only one third reduced lake alkalinity.

In streams, whole-system acidification experiments are of shorter duration and within-system processes play a smaller role. Norris Brook (New Hampshire) was changed from pH 5.7–6.4 to pH 4.0 by continuous addition of sulphuric acid for 6 months.[75] Similar experiments have been made with streams in Ontario. Inevitably, such manipulations of streams require continuous acid addition to be effective; similarly, reversal of acidification by flow activated lime dosing systems,[76] or hydrological management (such as ground water addition) are also quickly effective.[77,78]

3.2.2. *Reduction of acid deposition.* The benefits of control of S emissions are more difficult to demonstrate, at least in the short-term. However, emission reduction at the metal smelter in Sudbury, Ontario, provides a classic case history of water quality improvement in lakes nearby. By 1970, the Sudbury area had many acidified and metal contaminated lakes as a result of very large industrial emissions over about 80 years. In 1972, one smelter was closed and a taller stack (381m, the highest in the world) constructed, improving dispersal and dilution of the emitted materials. In addition, a programme of emission reduction was instituted, bringing emissions down from more than 2.2 mtonnes in 1970 to about 0.65 mtonnes in 1985, a reduction of 75%. Rain acidity was reduced, with pH 2.75 rising to 3.75. Sulphate deposition declined by about a quarter and trace metal deposition also.

Lakes within 1400 km of Sudbury have been monitored through to 1988.[79,80] Sulphate concentrations decreased by about 60% in two small lakes within 1 km of the plant with acidity decreasing by 59% and 98%, and increasing in alkalinity. Thus chemical recovery of the lakes has been rapid; in about 100 lakes in the area, pH has increased by a mean of 0.37 units and alkalinity by 23 $\mu eq\ l^{-1}$ over the last 10 years. Signs of recovery of the fish community were found as early as 1982, along with improvements in water chemistry (Table III); about half the lakes were suitable for lake trout. Sulphate and toxic metals (particularly copper and nickel, as well as aluminium) were reduced, although calcium concentrations declined. Recovery of invertebrate communities was slower but an increased

Table III Changes in Sudbury lakes following emission and deposition reduction (1972) (Data from Hutchinson and Havas, 1986;[79] Gunn and Keller, 1990[80,81])

Lake	pH	$[SO_4]$	[Ca]	[Al]
Baby:				
before 1972	4.04–4.19	60	10–2	—
after 1984	5.8	30	6	0.13
Alice:				
before 1972	5.87–6.10	266	11	—
after 1984	6.7	92	18	0.08
White Pine:				
1980	5.4	237	156	0.69
1988	5.9	195	115	0.40

Baby and Alice Lakes are within 1 km of the Sudbury plant; White Pine Lake is 90 km north of the plant. Concentrations are mg/l.

diversity of invertebrates with return of mayflies and oligochaetes, and increase in leeches and crayfish, are now reported.[81]

While this reduction of point source emission has undoubtedly resulted in lake improvement, it is difficult to extrapolate these benefits to lakes acidified by long-range acid deposition because of local deposit in the Sudbury area of sulphur and trace metals. The area is also characterized by high rates of mineral weathering (unlike more sensitive areas) and the slow biological recovery may reflect a reservoir of reduced sulphur and contaminating trace metals.

The only direct experimental manipulation in the field of changed acid deposition loads is that of the Norwegian RAIN (Reversing Acidification in Norway) programme at two sites in Norway; the project was initiated in 1983 and completed in June 1991. Sogndal is a pristine site in NW Norway, and Risdalsheia in S Norway is an acidified area with a high S "loading."[82] At Sogndal, response to additional acid inputs (sulphate alone and sulphate plus nitrate) was compared with controls, while at Risdalsheia, two catchments were covered to exclude natural wet deposition, one watered with "cleaned" rain with dilute sea salt ions below a roof (KIM catchment) while another (EGIL) received ambient acid deposition by a similar sprinkler system; a third catchment served as control.

The Sogndal experiment showed that added sulphate or sulphate + nitrate (100 meq m^{-2}-yr^{-1} compared with control at 30 meq m^{-2}-yr^{-1}) increased sulphate in runoff from 20 to 60 μeq l^{-1}, or to 40 μeq l^{-1}, respectively. The yield of base cations in runoff, especially calcium, increased and labile inorganic aluminium increased. In the catchment receiving both added sulphate and nitrate (SOG4), rather less change was seen in the first 4 years of treatment, but in the 5th year nitrate concentrations in runoff were also enhanced (Table IV).

At Risdalsheia, after 6 years of acid exclusion, sulphate in runoff from KIM was reduced from about 100 to 30–40 μeq l^{-1}, rather less than predicted, presumably because soil sulphate reserves still contributed. Dissolved calcium in runoff was reduced, with improved retention of base cations in soil, leading to a ratio of Ca^{2+}:H^+ in runoff falling by about 50% over earlier conditions. Inorganic, labile, aluminium was significantly increased by about 30% (Table IV). Organic anions were substantially increased and pH little changed, although the

Table IV Changes in runoff chemistry at Sogndal and Risdalsheia (Rain Project) (Data from Wright, Lotse and Semb, 1988[82])

Component	Sogndal(C)	Sogndal(4)	Risdals(C)	Risdals(K)
pH	5.8	5.4	4.1	4.1
H^+	2	4	84	77
sulphate	21	36	106	53
nitrate	1	3	37	12
chloride	37	35	103	106
calcium	20	26	19	11
magnesium	10	10	25	18
sodium	38	37	96	86
aluminium	0	27	126	90
H/SO_4	0.1	0.1	0.8	1.45
Ca/H	10	6.5	6.5	0.14
AI	−9	0	0	24

Data for 1987; concentrations are μeq/l. Sogndal (C) had no acid additions, Sogndal (4) had both S and N added, Risdals(C) had canopy but no rain clean-up, Risdals(K) had canopy and cleaned rain. AI is an index of acidification, i.e. SO_4–(Ca + Mg).

alkalinity deficit improved, possibly because of increased organic anions.[83] Not surprisingly, in the light of lower calcium and still high aluminium concentrations in runoff, juvenile trout and salmon are unable to survive more than a few days.[84]

The results of the RAIN experiment are challenging, but the sites represent extremes of sensitivity to acidification and the short period of observations limits their value for extrapolation elsewhere.

It has been argued from field observations of acid and sulphate deposition and observed water chemistry, that there are thresholds of deposition below which surface waters are not acidified. Wright and Snekvik[55] proposed a rain pH $>$ 4.7, on the basis of 700 lakes sampled in southern Norway. More recent estimates of a "critical load" suggest that lakes in areas receiving $<$10kg S/ha-yr will have few acidified waters.[63,85,86] This regional approach ignores crucial factors such as local weathering rates, hydrological patterns, soil sulphur pools, sulphate reduction processes and sea salt inputs, all seen as significant in field and experimental studies. It also assumes that "natural" weathering at a loading below the threshold is sufficient to maintain present conditions. In Scotland, for instance, such a "critical" load seems unlikely to provide satisfactory conditions since some acid sites receive $<$10 kg S/ha-yr and rain pH $>$4.7 and neither surface water acidity nor sulphate[87] are well correlated with S or acid loading.

3.2.3. *Other field experiments.* Other small catchment manipulations have been initiated following similar design characteristics as RAIN. Many acid waters in peat rich areas are characterized by high concentrations of organic matter derived from vegetation and soils.[88–90] The presence of organic acids[91] is of interest both because of a possible reciprocal relationship between organic and strong acids, and because organic components of acid waters complex toxic aluminium compounds, reducing aluminium toxicity at low pH.[92] The RAIN project showed that as strong acid input was reduced, organic acids in runoff increased, up to two-fold.[93] To test this at a pristine site, a project ("HUMEX") will treat a pristine acid lake, divided into a control and an acidified basin, with sulphuric acid and ammonium nitrate added to catchment and drainage basin. The pH in the treated basin will be reduced to 3.0 to explore the influence of soil derived humic acids on lake acidification and its recovery.[94]

The association of acidification and humic materials (brown waters) has been tested in another field acidification of a brown water stream; acid addition did not reduce the DOC levels as expected, but it did release toxic forms of aluminum even though organically bound aluminium did not change.[95]

CLIMEX is a further experiment planned for the Risdalsheia site; enhanced carbon dioxide air concentrations and raised temperature will simulate global warming conditions.[96] This experiment will help to predict the effect of raised atmospheric carbon dioxide levels but will need to distinguish this from that generated in soil from the enhanced temperature regime planned.

3.2.4. *Alternative methods for acid mitigation.* Possible alternative approaches include a variety of liming or related land-use strategies to reverse or mitigate acidification; they include "liming" direct to waters or to catchments, application of other chemicals, forest fertilization or removal, moorland burning, and hydrological management. These are listed in brief in Table V. Some examples are given here to illustrate their relative effectiveness and possible problems.

"Liming." The provision of lime materials (carboniferous or magnesian limestone, calcium oxide or hydroxide, other calcium containing materials) has been employed for at least two millennia to correct agricultural acidity. The use of such materials to counter surface water acidity has a much more recent history. A recent review[76] provides many examples and only a few are described here.

Sweden has the most comprehensive liming programme to counter acidification of surface waters. Lake Gardsjon, in SW Sweden, provides a well-documented example.[97] The lake was treated in 1982 by the addition of 110 tonnes of limestone as a slurry direct to the lake surface. Observations of chemical and biological responses of the lake continued for the following 8 years. The chemical effectiveness of direct lake liming is governed by lake retention time, here 1.36 years. Initial changes were a pH rise to >7, an increase in DIC, a decrease in aluminium. These changes were evident within a few days after liming, with calcium and alkalinity continuing to rise over the following 8 months. Improved water quality persisted over the following two and a half years, but then declined slowly, prompting further liming in 1985. After an initial decline in species and biomass immediately after liming, biological diversity and biomass steadily increased through to 1985. Response to later liming of Gardsjon is not yet reported.

Another forest lake ecosystem, Lake Lysevatten in SW Sweden, received treatment with a limestone slag (di-Ca-silicate) in 1974; 60 tonnes were spread by boat in the shallow lake.[98] A further application (98 tonnes) of agro-limestone as a slurry was made in 1986, again direct to the lake, including open water areas. The first application brought pH to circumneutral levels and was maintained by dissolution and/or reflux of stored calcium through to about 1980, when alkalinity was depleted and pH fell abruptly to 5.0–5.5. By 1985/86 the annual variation in pH had increased and the lake reacidified in response to continued acid loading from atmosphere. Biological monitoring showed evidence of a recovery of *Asellus* populations, and decline of *Sphagnum* and *Mougeotia*; it is argued that reacidification of limed lakes should be avoided by successive treatments to maintain biological stability.

Direct lake and littoral liming was undertaken at Lake Hovvatn, in Norway.[99,100] Hovvatn and a small pond upstream were treated in 1981 with a total of 240 tonnes of powdered limestone distributed on to the ice cover and along the shores. Reacidification occurred after 6 years, and the lake was then relimed with 91 tonnes of limestone. The pH was restored to circumneutral values after ice break, but the pH declined to < 6 after one year (lake retention time < 1 year), and acid episodes were still evident during periods of snow melt and high flow.

The close dependence of liming effectiveness on lake retention time[101] suggests that catchment liming would have greater persistence. Not only would a supply of base cations and alkalinity be maintained, but the soil base reserve would be replenished, protecting against episodic release of acid and aluminium from the soil exchanger. In the less acidic soil after liming, mobilized aluminium is reduced and present in less toxic forms. Some well documented examples of catchment liming[47,99] have demonstrated the longevity and stability of water quality using this approach.

The catchment of a high-elevation fishless lake, Tjonnstrand, in Norway was limed in 1983.[100] The lake retention time was only 2 months, so catchment liming was selected. An application of 75 tonnes (3 tonnes/ha) was made, with a quick response expected since the soils are thin, with many rocky outcrops. In a rain event after liming, pH rose from 4.5 to 7.1, but later storm events brought acid pulses although pH >5.8 was generally maintained 4 years after liming. Calcium levels initially increased to about 4 mg l^{-1} but then declined, although still >2 mg l^{-1} after 4 years. Aluminium was also reduced to about 100 μg l^{-1} (total).

At Loch Fleet in SW Scotland, lime applications to the catchment in 1986 at about 3 tonnes/ha overall, selectively applied to some areas at up to 20 tonnes ha^{-1}, improved water chemistry within a few days of liming. Water quality continues 5 years post-liming to meet threshold requirements of pH, calcium and labile aluminium (Table V). Extrapolation from calcium budgets predicts that it will remain suitable for fish until the end of this century, even though lake retention time is less than 6 months.[102] While invertebrate communities were initially unstable after liming and restocking, no major changes were observed, although a few "lost" species have returned. Fish were reintroduced to the lake 18 months after the first applications and show good growth and fecundity.[103]

Prediction of runoff chemistry from the limed Loch Fleet catchment was made using the SLiM model.[104,105] The model includes data on runoff chemistry from acidified and limed areas, deriving the resulting stream water chemistry. The model requires an input of adequate baseline chemistry and dissolution equations, and a good quantitative description of flow pathways through soils of the catchment from which it calculates the soil and soil water chemistry, and hence the contribution of base cations from the limed areas. The soil chemistry component includes dissolution and deactivation of limestone, cation exchange between calcium and exchangeable acidity, leaching and accumulation of dissolved species, and aqueous equilibria for carbonic acid, DIC and aluminium. Some processes not considered are silicate weathering, sulphate adsorption and nitrogen transformations. SLiM predictions for Loch Fleet water quality through 10 years post-liming are shown in Figure 7. The effective time-scale of liming has also been developed from calcium budgets.[102]

Table V Rain and water quality (annual means) at Loch Fleet prior to liming (1985/86) and five years after liming (1990/91)

Component	Rain	Sector IV	Sector VI	Sector VII	Loch
Chloride					
1985/86	129	213	170	174	181
1990/91	180	341	231	242	282
Sulphate					
1985/86	48	98	86	94	97
1990/91	63	104	97	92	98
Nitrate					
1985/86	12	11	15	15	16
1990/91	15	11	14	16	16
Calcium					
1985/86	8	77	36	50	46
1990/91	14	698	428	182	153
Aluminium					
1985/86	n.d.	260	250	260	200
1988	n.d.	20	60	80	80
Acidity					
1985/86	17	50	47	34	29
1990/91	19	0.07	0.05	1.9	0.8

Notes: Concentrations are μeq l^{-1}, except aluminium, μg l^{-1}, n.d. not detected. Sector IV, forest, limed 23.9 t ha^{-1}. Sector VI, moorland, limed 21.3 t ha^{-1}. Sector VII, moorland with wetland source area, 10 t ha^{-1}, or 3.3 t ha^{-1} overall

3.2.5. *Some other mitigation treatments.* Several other possible strategies for reversing acidification have been proposed and a few tested in the field. They include the addition of a nutrient (phosphorus), promoting algal and bacterial growth to consume protons and generate alkalinity.[106,107] In some cases, phosphorus addition has been coupled with lime treatment and/or addition of waste water.[108,109] However, more complex responses may occur, possibly involving the N cycle, and it is not known whether phosphorus alone is beneficial. The addition of organic carbon is also proposed to buffer acidity and to promote complexing of toxic metals, including aluminium, thus improving conditions for fish survival. A possible adverse effect of organic carbon addition may arise due to the oxygen demand of the added organic material. A further chemical strategy is to use sodium or calcium solutions to enhance the neutralizing capacity of natural or acidified water.[110] At a coastal site in southern Norway, sea water was mixed with upstream acidic water received through a hydroelectric diversion and salmon smolts were then able to survive.[111,112]

Hydrological manipulations are also a possible treatment if a suitable supply of ground water is available. Temporary neutralization of an acid stream (Pennsylvania) was achieved by pumping alkaline ground water from a well, reducing acidity and aluminium; the pH was held at >6.0, with conditions suitable for fish.[113] At Loch Fleet, ground water of high calcium carbonate content (pH 7.2, alkalinity 2320 μeq l^{-1}, Ca 1790 μeq l^{-1}) could be used to improve water quality; only an estimated additional 0.06 l sec^{-1} (less than 2% of estimated natural groundwater flow to the lake, 3.5 l sec^{-1}) could buffer the lake water at about pH 6.[114]

Some biological alternatives are also possible; these include the selection and stocking of acid tolerant fish species, since sensitivity varies widely, even between varieties of a single species.[35] It is possible that some sport fishing strains, stocked in the past when waters were less acid, are not now suitable for present conditions. However, this option obviously does not improve the status of other biological groups. Certainly, the stocking of acid tolerant strains is useful in conjunction with other techniques, such as liming, to protect against short term episodes. Some success in Norway has been reported for reintroduction of selected trout stocks (REFISH project).[115]

Other options include land-use and management changes. Observations that runoff from conifer afforested areas is often more acid and higher in aluminium than from adjacent moorland areas[116,117] suggest that forest control (species replacement, bankside clearing, felling) might be effective. It had also been observed that runoff chemistry changed after forest fires.[118] A few experiments of catchment burning, forest felling and other land management strategies have been tested in the field. At Loch Fleet, burning of a rough moorland sector of the catchment provided only a transient and slight change in water quality, possibly because the depth to which the soil was burnt was limited by wet conditions prevailing there.[119] Bankside clearance of conifers, with or without added lime, proved ineffective at a Welsh acidified site.[120] In the same area, agricultural improvement of moorland (ploughing, liming, fertilizing) led to improved soil chemistry (pH rise, calcium increase) but the response in runoff chemistry was slight and acid episodes continued. Moorland liming at 10 tonnes/ha alone increased calcium levels in runoff and pH rose, but the improvements were not long sustained there.[121]

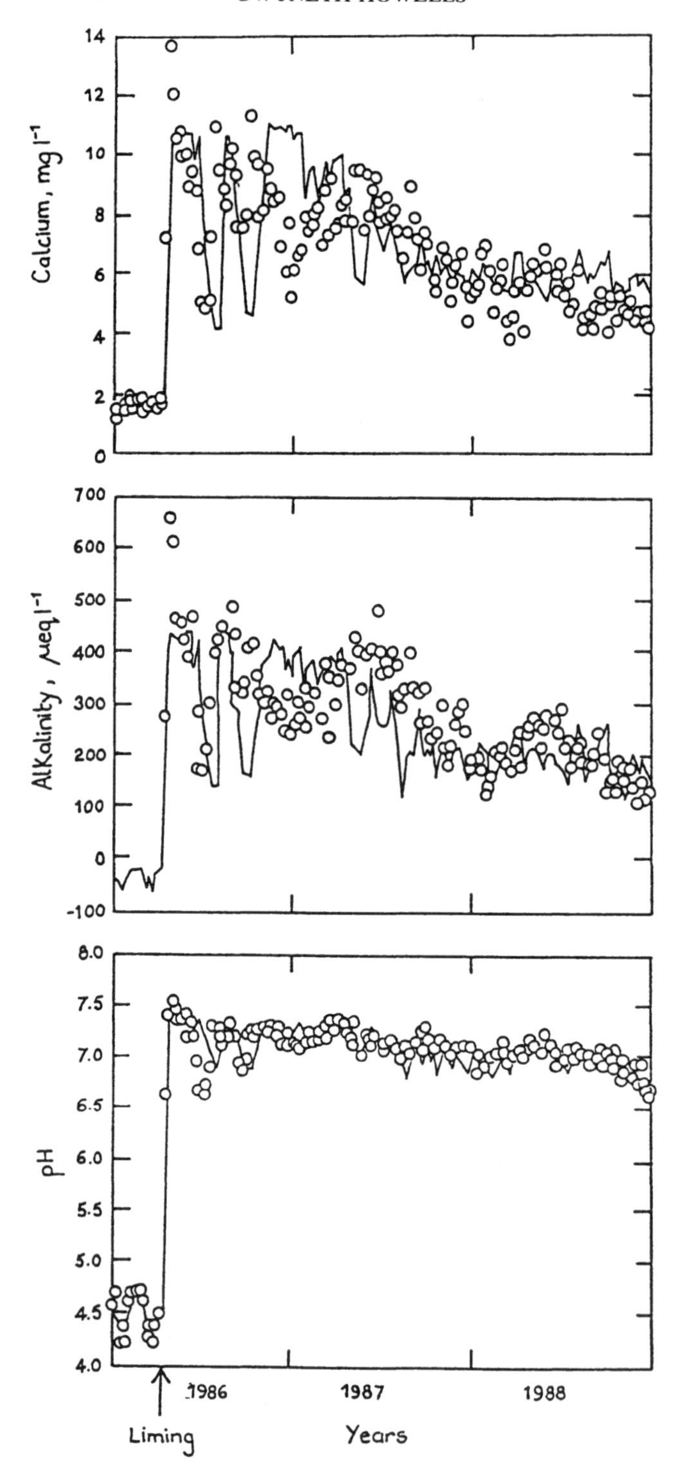

Figure 7 SLiM model predictions (—) of water quality at Loch Fleet through 3 years post-liming of a wetland area (o = measured values) (after Dalziel *et al.*, 1992)[102]

4. MODELLING

4.1. *Development of Models*

4.1.1. *Chemical models.* A number of chemical models have been formulated to estimate past, present and future chemical conditions in surface waters with chemical and physical equations to represent catchment and lake processes affecting water chemistry. They include common features but vary in complexity and detail, and in their spatial and temporal resolution; they all contain simplifying assumptions. The most reliable models used to forecast the long-term benefits of emission reduction strategies are MAGIC, ILWAS and ESSA/DFO.

Steady state models based on empirical dose-response relationships and chemical mass balances have been used quite extensively, e.g. Henriksen,[122,123] although applied to UK data sets with only limited success.[60] The hydrological BIRKENES model,[124] first formulated to account for short-term variations in stream chemistry, was later modified to simulate long-term scenarios.[125] However, these models do not predict conditions over time, and being based on specific site conditions, can misrepresent the capacity of catchment neutralization. All are simplifications, attempting to represent a complex mosaic of conditions on a catchment or regional basis including diverse vegetation, land management, geology and soil conditions, climate and "loading." Their strength is dependent on the site data from which they are developed; some critical components, such as soil chemistry, are often poorly documented. Rather little effort has been given to validation and verification of models, so that confidence in their application may not be strong. However, they have helped to develop understanding of the interactive nature of acidification and its potential for reversal.

4.1.2. *Predictive models.* The expected outcome of widespread emission control on a regional basis over time requires different models. The ESSA/DFO (Environmental and Social Systems Analysis/Department of Fisheries and Oceans) model has been used in the NAPAP assessment,[7] along with dynamic conceptual models, to predict future pH, ANC and aluminium status, and steady-state conditions. Using current surface water chemistry and deposition, and assumptions about the neutralization capacity of a lake and its catchment, the model calculates surface water ANC, pH, calcium and sulphate. The more dynamic MAGIC and ILWAS models have been used to project the effects of sulphur deposition scenarios on future water quality conditions in sample catchments and extrapolation to a target population.

MAGIC (Model of Acidification of Ground water in Catchments)[126] includes critical processes and dynamic responses—hydrological and chemical mass balances, anion (sulphate) retention in soils, adsorption and exchange of base cations and aluminium in soils, alkalinity generation by dissociation of carbonic acid in soils (and exchange of H^+ for base cations), and weathering of minerals to provide base cations. In calculating future water chemistry, it is assumed that Al^{3+} is controlled by equilibrium with $Al(OH)_3$.

Atmospheric deposition of sulphate in MAGIC is assumed on the basis of current emission estimates and deposition of base cations and acidic anions is included. Weathering rates are uncertain even for researched sites, yet the supply of bases from weathering controls the long-term response and recovery of catchments. The ability of the model to hindcast past conditions has provided a comparison with palaeoecological pH reconstruction for a few sites (Figure 8).[127] Regional MAGIC models have been used for a variety of conditions in Scandinavia, UK and USA sites.

The MAGIC model has been extended to include predictions of biological community responses to various emission scenarios.[60,128] Trout populations have been predicted on the basis of relevant stream chemistry (pH, Ca, Al) and invertebrate communities from observations of species composition and relative abundance, again relative to stream chemistry. However, invertebrate communities of Welsh waters showed no evidence of change in 1940 to 1984, in spite of other evidence of acidification during that period.[128] A similar approach, linking MAGIC predictions of future water quality to response of diverse fish communities, was applied in the NAPAP assessment,[35] also with some anomalous observations.

The ILWAS (Integrated Lake-Watershed Acidification Study) is more complex[129] including more chemical processes and more complex soil structures, but requiring more information or alternative assumptions that add uncertainty to the results. Hydrological (dynamic) processes are also included to a greater degree than in MAGIC, simulating transfers from atmosphere, canopy throughfall, snowpack release and surface runoff, soil percolation and "sink" to the deep ground water.[130] The model was developed and tested on an intensive four year data set for three forested watersheds in the Adirondacks, quantifying the relationship of atmospheric deposition, surface water acidity and fisheries response; it has not been widely applied because of its more extensive data needs.

Although designed for a more limited purpose, the SLiM (liming) model (see 3.3.4, Fig. 7) can also be used to predict surface water chemistry over reasonably long periods.

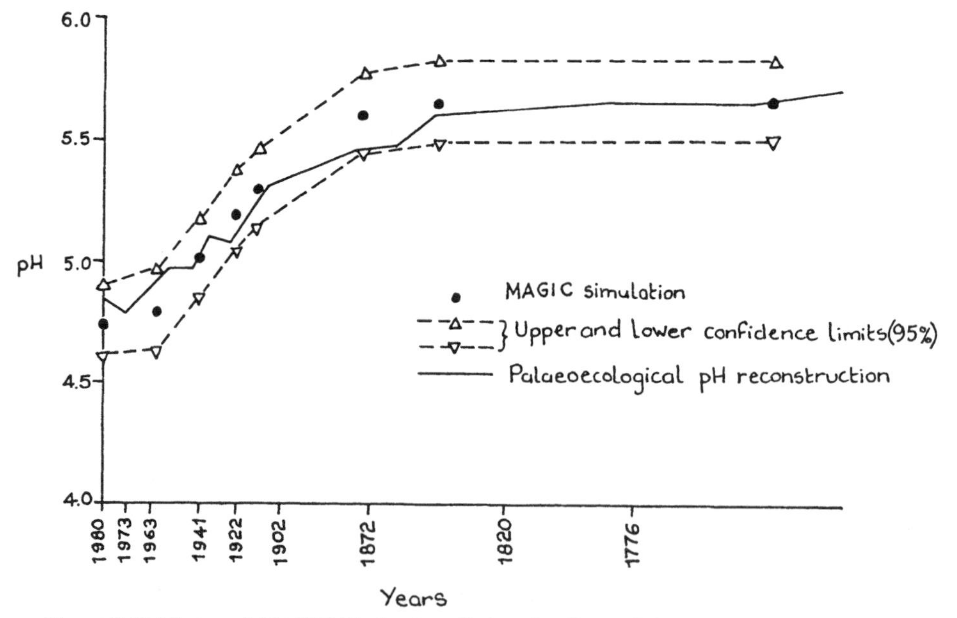

Figure 8 Validation of MAGIC hindcast predictions by diatom inferred pH at Round Loch of Glenhead (after Whitehead, 1989)[127]

4.2. *Some Results of Modelling*

The NAPAP assessment of costs and benefits (ESSA/DFO) of various emission reductions is based on the extensive 10-year assembly of data, coupled with a variety of emission scenarios.[7] These range from no new controls, to 5 to 12 million tonnes reduction in SO_2 emissions, or a combination of S and N emission reductions.

For aquatic effects of acidification, assuming that sulphate reduction is matched by ANC increase (Fig. 9), MAGIC simulations indicate an increase of 3–4 μeq l^{-1} for each 1 kg/ha reduction in S deposition. Without new controls, a further 8% of Adirondack lakes would become acidic by the end of the century and 4–9% of lakes and streams would become unsuitable for sensitive fish species. But up to a 50% reduction (10m tons), at least, would be needed for any discernible change in alkalinity of acid waters in the regions studied. In the Adirondacks, this level of emission control would reduce the percentage of lakes and tarns of pH <5.5 by about 10%, i.e. from 38% to 34%, 60 lakes.

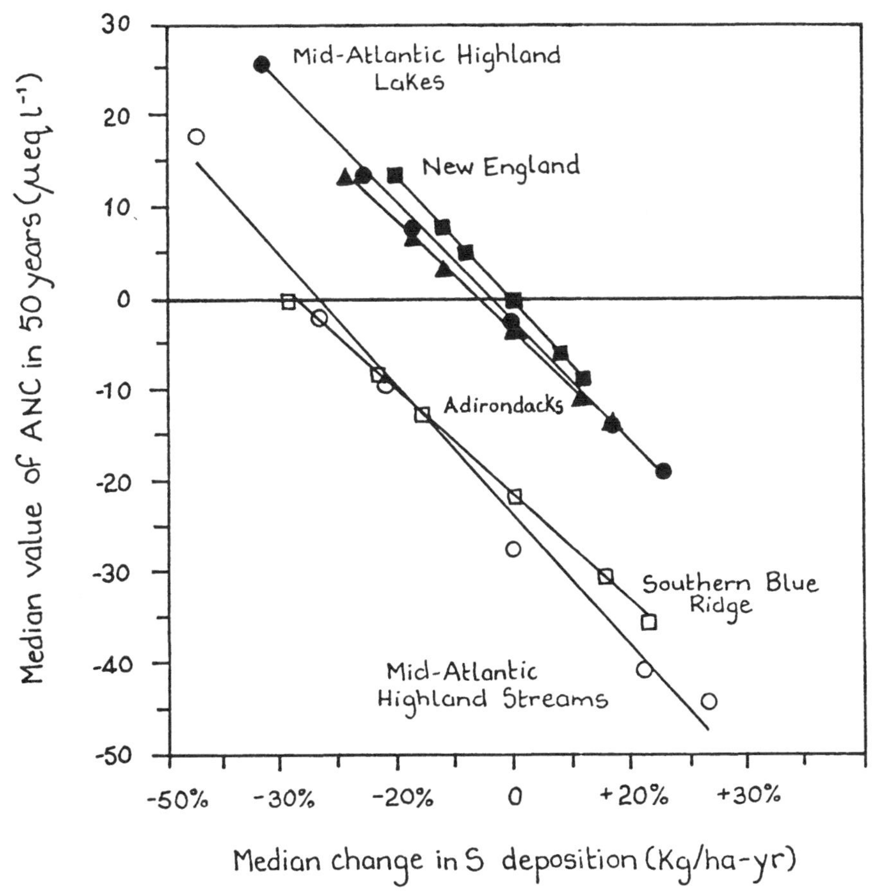

Figure 9 Changes in ANC levels predicted by MAGIC simulation over 50 years for a range of S deposition scenarios, applied to the NAPAP investigated waters

The ILWAS model, based on Adirondack sites, predicts that halving the current S "loading" in the Adirondack area would result in a site specific improvement in lake water pH, but would not significantly improve conditions during acid episodes in the most sensitive lakes.[130]

MAGIC has been applied to prediction of long-term changes in response to future deposition scenarios in UK, and as a diagnostic aid to assess the relative influences of emission control, sea salt acidification, and land-use change (Figure 10).[60] Application to a sensitive area of SW Scotland shows that with constant (1984) sulphur emissions, stream pH is predicted to decline over the next century, faster in afforested areas than in moorland, attributed to increased scavenging of aerosols. Further, the sea salt input influences the response, so that distance from the coast and prevailing airstreams are significant. Reduction of sulphur deposition by as much as 90% would be needed to restore "pristine" water quality in such sensitive areas in the UK, a control that would have to be shared by adjacent source countries. Even a 50% reduction would be required to prevent further biological changes in moorland streams. The time-scale would be long, even with early action to control emissions–probably over several decades. Since forest cover increases acid deposition by about 30%, land management could offset a policy of reducing emissions by about 25% (or for UK alone, by 75%).[84] MAGIC modelling of changes in runoff with reduced emissions can also be compared with predicted benefits from lime application (3 tonnes $CaCO_3$ ha^{-1} every 100 years), equivalent to 80% reduction in emission of S.[84]

5. SOME CONCLUSIONS AND IMPLICATIONS

5.1. *Causes*

The primary gases that generate acid aerosols are sulphur dioxide, oxides of nitrogen and volatile organic acids (VOCs). These are all products of natural

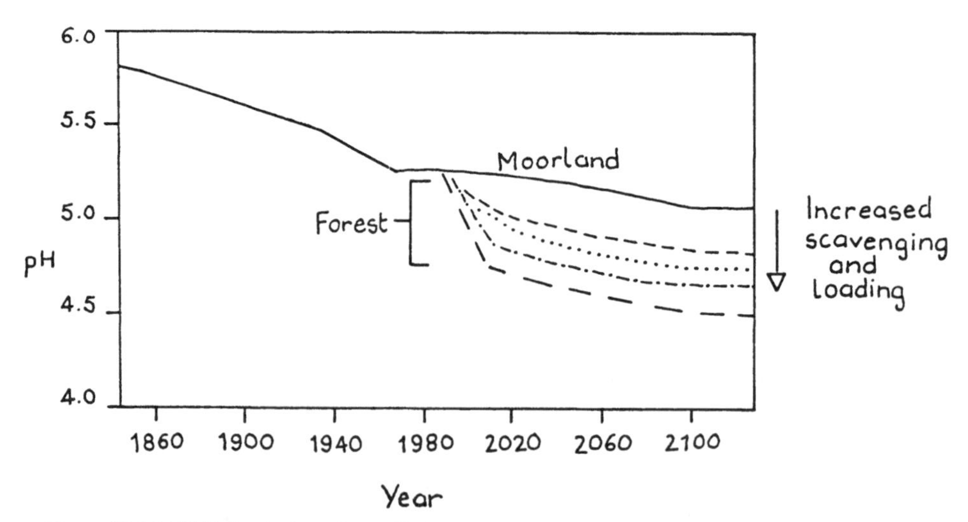

Figure 10 MAGIC simulation of runoff changes in pH over > 100 years at a sensitive UK site with different levels of S deposition (or emissions) and relative influence of afforestation vs moorland land-use (after Whitehead, 1989)[127]

processes, but anthropogenic activities have increased their emission, especially in the northern hemisphere over the last century. In the USA, since 1900 emissions of SO_2 have increased threefold, VOCs twofold, and NO_x tenfold. Natural biological sources of VOCs, NO_x and reduced S are also quite significant, and natural emissions of SO_2 from volcanic activity and disasters such as oil and gas fires are often much greater although they occur over limited periods.

5.2. *Acid Deposition*

Rain is naturally acid, about pH 5.0 in pristine areas due to the influence of these natural gases and carbon dioxide. This background acidity (about 10 μeq l^{-1} of H^+) can be compared with that of rain in industrial areas (about pH 4.2, or 60 μeq l^{-1} of H^+); reducing acid deposition to pristine conditions calls for a change of about 50 μeq l^{-1} of acidity. Although acidifying emissions in northern Europe and USA have decreased by 20 to 40% in recent decades, rain pH has changed only insignificantly, possibly due to increase in N species.

Dry deposition of sulphate aerosol is now reasonably predictable from atmospheric conditions, but wet deposition processes are complex and non-linear with atmospheric aerosol levels; thus a 50% S emission reduction has a lesser effect on wet deposition. Furthermore, since impacted areas (e.g. western areas of UK and Scandinavia) have a greater wet than dry deposition, the relationship is less favourable there.

Wet sulphate deposition (flux) has a reasonably linear relationship with surface water sulphate concentrations; in USA, regional studies indicate that a deposition of 10 kg SO_4/ha-y contributes about 50 μeq l^{-1} of sulphate in surface waters. These sulphate concentrations, however, do not necessarily match acid or ANC in surface waters, partly because natural organic acids, as well as nitrate, are present. Catchment processes provide some neutralization of acidity through weathering of soils and rocks, by chemical transformations and by bacterial or other biological activity. This acid neutralizing capacity is low in areas sensitive to acidification. Land-use also influences water quality—agricultural activities contribute excessive N from fertilizers and/or intensive husbandry, hydrological diversions may reduce the volume of more neutral ground water, and afforestation enhances canopy scavenging of atmosphere and mobilizes bases from soil, transferring and "locking" them into above-ground biomass.

5.3. *Effects*

Some surface waters in sensitive areas have fallen by as much as 2 pH units (about 100 μeq l^{-1} of H^+) through this century, even within the past few decades. On the basis of observed relationships, for example in Norwegian lakes, this could be associated with $>$400 μeq l^{-1} of sulphate, assuming other acid anions do not contribute. Recent reductions of S emissions in northern Europe have resulted in only a small decrease in sulphate concentrations, but little or no change in pH, possibly due to a greater contribution of nitrate. Accumulation of past S deposits in soils provides a pool, possibly accessible.

Fish and invertebrate communities in acid waters are sensitive to a combination of acidity, low calcium and high aluminium concentrations. Fish populations have been lost from acid lakes and streams, a phenomenon that continues even though sulphate has fallen and pH little changed. This may reflect the continued release of toxic aluminium from acid soils, mobilized by excess nitrate. Invertebrate communities in acid surface waters are less diverse than elsewhere.

Field experiments with acid additions to pristine lakes show that acidification is a slow process, even with an experimental "loading" greatly in excess of current levels in impacted areas; this reflects the ability of catchments and lake sediments to generate alkalinity. In contrast, once "loading" ceases, recovery is rapid, possibly because the capacity of these sites is high, in contrast to more sensitive acidified sites. In an exclusion experiment at a highly sensitive site, however, the initial response was not sustained, possibly due to the influence of S reserves in the soils. In such conditions, characterized by low weathering rates, chemical and dynamic models predict that a return to "pristine" conditions will require a 90% reduction of current S emissions, and that even 1900 conditions could only be achieved in >100 years. Modelling of data from affected areas in USA indicates that 50% reduction of emissions would be needed for any significant benefit in the most sensitive area.

Other perceived damages due to acid deposition include some injury to high altitude spruce forests in North America, and reduction of light transmission with loss of visibility. In the former case, it is considered unlikely that S emission control will significantly improve tree health, although a 50% reduction over 10 years is expected to bring some increase in the base saturation of poor soils. On the other hand, the loss of S deposition as a passive fertilizer for agricultural crops will bring some additional costs to farmers. In the case of diminished visibility where sulphate aerosols are responsible for a loss of about 60%, an expected fall to 37% is likely to follow USA S emission control proposals, increasing visibility by about 20%.

5.4. *Actions and Strategies*

Emission reductions in Europe over the last three decades have not improved water quality in affected areas even though in Canada massive reduction of emissions at Sudbury and ceasing acid additions to trial lakes were rapidly followed by restoration of water quality. From this, we might expect that a strategy of emission reduction would soon reduce and reverse acidification, but modelling suggests that recovery will be very slow, and might not be effective or even measurable in the short term. There are also likely to be some unwelcome effects, such as lower calcium concentrations in fresh waters, and sulphur deficiency in a few agricultural areas.

Liming and other chemical manipulations to surface waters or catchments has provided suitable water quality for fish and invertebrate communities within a few days of treatment. But its longevity is dependent on lake retention or river flow so that treatment has to be repeated or continuous to maintain target conditions. An alternative is to lime catchment or water source areas, providing suitable conditions for one or two decades.

There is some potential for other strategies. Hydrological management is promising in concept, but has not been demonstrated in the field and a sufficient resource is likely to be limited to a few sites. Other possible approaches through land-use or management have not proved successful in the field. The option to fell forest, reducing scavenging and base retention, would bring substantial loss of revenue. A further option to select and stock tolerant species for sport fishing has been successful, but cannot re-establish natural aquatic communities.

5.5. *Costs and Futures*

The costs of S emission in the USA has been estimated for a low level of control at \$150–200/tonne of SO_2 removed; and, at a greater level of control at \$800–1200/tonne. Full implementation of such controls could not be accomplished until 2015 to 2020, a time-scale of 25 to 30 years.[7] In Europe, the cost of implementing EC large plant Directive for SO_2 (reduction of 80–95%) is estimated to be about £2000m/y by the year 2003; for NO_x, control to 50% would cost £1000m (capital) to 1998.[131]

The time-scale of aquatic ecosystem recovery is likely to be much greater—some decades for water quality improvement, even longer for biological recovery. Models suggest that, even with a high degree of S emission reduction, improvements in natural ecosystems would not be seen until the 22nd century, possibly longer in highly sensitive and previously damaged sites.

A "no action" scenario indicates a loss of value of the Adirondack recreational fishery of 1 to 11 million dollars. The possible benefit of a 10 million tonnes reduction in S emissions is estimated at 7 to 13.6 million dollars in 2010; but most of the economic benefits would be achieved by then and little more by 2030. (This reflects travel-cost changes, alternative opportunities, and perceived value, rather than improved catch.) This might be judged a reasonable cost, even by those not actively participating in the sport. Other benefits are much less evident; although S emission reduction would improve visibility it is difficult to put a value on this resource, while possible improvement in spruce health is doubtful since independent factors like frost and soil deficiencies are more important.

5.6. *Overall*

While control of anthropogenic emissions suggests intuitively that it is possible to "put the clock back," present analysis is not encouraging; undoubtedly the problem is also politically, administratively and scientifically complex. On the other hand, few would condone an increasing atmospheric burden and an inevitable further decline of surface water quality. The real and quantifiable effects of acid deposition are now evident and more specific, even though judged by some as "totally irrelevant."[132] While claims in the past have often been overstated, there is a case for damage limitation, by whatever means shown to be effective, calling for a combination of strategies, including emission control along with energy conservation and positive remedial measures such as liming or other chemical treatments, land-use and hydrological management. The choice of an effective strategy will be determined by site conditions and the acceptable costs.

There are still wider perspectives in the light of political and economic changes, particularly in countries less sensitive to acidification than northern Europe and North America, and where resource exploitation has a high priority. Will control of S emissions pale into insignificance if atmospheric carbon dioxide levels increase substantially? Should we pursue a goal of ecosystem stability, rather against nature, and at what cost? It is clear, however, that any policy should not be pursued on the basis of preconceptions, rather than science or even pragmatism. After three decades of scientific endeavour, we can now at least judge the "disaster" of acidification on its real effects and economic implications.

REFERENCES

1. J. N. Galloway, G. E. Likens, W. C. Keene and J. M. Miller, The composition of precipitation in remote areas of the world. *J. Geophys. Res.*, **87**, 8771–8786 (1982).
2. U. Skiba and M. S. Cresser, Seasonal changes in soil atmospheric CO_2 concentrations and associated changes in river water chemistry. *Chemistry and Ecology*, **5**, 217–225 (1991).
3. R. A. Smith, *Air and Rain–the Beginnings of a Chemical Climatology* (Longmans and Green, London, 1872).
4. K. Dahl, Investigations into the extinction of trout in the mountain lakes of south-west Norway. *NJ & FF Tidskrift* **50**, 249–267 (1921).
5. J. Liebig, *Chemistry in its Applications to Agriculture and Physiology* (in 4th ed. 1849, eds. L. Playfair and M. Gregory) (Wiley, New York, 1849).
6. P. Brimblecombe and D. H. Stedman, Historical evidence for a dramatic increase in the nitrogen component of acid rain. *Nature*, **298**; 460–462 (1982).
7. NAPAP/IA, *National Acid Precipitation Assessment Program; 1990 Integrated Assessment Report* (NAPAP, Washington, DC, March, 1991).
8. Collins, *Collins Concise Dictionary Plus* (The Bath Press, Glasgow, 1989).
9. Royal Ministry of Foreign Affairs, Royal Ministry of Agriculture, *Air Pollution Across National Boundaries* (Stockholm, Sweden, 1971).
10. L. N. Overrein, H. M. Seip, and A. Tollan, Acid precipitation–effects on forests and fish. *Final Report of SNSF Project*, FR 19/80, (1980).
11. SNV (National Swedish Environmental Protection Board), *Action program against air pollution and acidification*. (Solna, Sweden, 1985).
12. S. Fleischer, G. Andersson, W. Dickson, I. Muniz, and J. Herrmann, Surface water acidification: effects and remedial measures. Report 3685 Swedish Environmental Protection Agency (1990).
13. T. M. Roberts, R. A. Skeffington and L. W. Blank, Causes of type 1 spruce decline in Europe. *Forestry* **62**, 179–222 (1989).
14. NAS (National Academy of Sciences, USA), *Acid Deposition–Long Term Trends* (Nat. Acad. Press, Washington, DC, 1986.)
15. T. J. Sullivan, *et al.*, Atmospheric wet sulphate deposition and lakewater chemistry. *Nature* **331**, 607–609 (1988).
16. F. Blake, Viewpoints: A skeptical observer. *EPA Journal* **17**, 59–60 (1991).
17. T. M. Roberts, *et al.*, Effects of gaseous air pollutants on agriculture and forestry in the UK. *Advances in Applied Biology* **9**, 2–142 (1983).
18. N. M. Darrall, The effect of air pollutants on physiological processes in plants. *Plant Cell Environ.* **12**, 1–30 (1989).
19. L. Blank, A new type of forest decline in Germany. *Nature* **314**, 311–314 (1985).
20. L. Blank, *et al.*, New perspectives on forest decline. *Nature* **336**, 27–30 (1988).
21. H. G. Miller, Deposition-plant-soil interactions. *Phil. Trans. Roy. Soc. Lond.* **B 305**, 339–352 (1984).
22. G. Abrahamsen, Effects of acidic deposition on forest soil and vegetation. *Phil. Trans. Roy. Soc.* (London) **B 305**, 369–382 (1984).
23. J. S. Jacobson, Effects of acid aerosol, fog, mist and rain on crops and trees. *Phil. Trans. Roy. Soc.* (London) **B 305**, 327–338 (1984).
24. A. R. Wellburn and N. M. Darrall, Physiological effects of the direct impact of gaseous pollutants on foliage (Abstract p. 132). *Internat. Conf. Acidic Deposition: its Nature and Impacts*, Glasgow (16–21 September, 1990).
25. H. W. Zoettl and R. F. Huettl, Liming as a mitigation tool in declining forests. (Abstract pp. 183–185.) In: *American Chemical Society Meeting, Symposium on Environmental Chemistry of Lakes and Reservoirs, April 14–19, Atlanta* (Lewis Publishers, Chelsea, Michigan, 1991).
26. D. W. Johnson and J. O. Reuss, Soil mediated effects of atmospherically deposited sulphur and nitrogen. *Phil. Trans. Roy. Soc.* (London) **B 305**, 383–392 (1984).
27. J. O. Reuss and P. M. Walthall, Soil reaction and acidic deposition. In: *Acidic Precipitation Vol. 4: Soils, Aquatic Processes and Lake Acidification* (ed. S. A. Norton, S. E. Lindberg and A. L. Page) (Springer-Verlag, New York, Berlin and Heidelberg, 1989) pp. 1–33.
28. U. von Bromssen, Acidification trends in Swedish groundwaters. SNV Report 3547 (1989).
29. L. Hallbacken and C. O. Tamm, Changes in soil acidity from 1927 to 1982–1984 in a forest area of southwest Sweden. *Scand. J. For. Res.* **1**, 219–232 (1986).
30. D. S. Jenkinson, The accumulation of organic matter in soil left undisturbed. *Rothamsted Report 1970* (2), 113–137 (1970).
31. S. R. Langan, Sea salt induced streamwater acidification. *Hydrol. Processes* **3**, 25–41 (1989).

32. G. Howells, T. R. K. Dalziel, J. P. Reader, and J. F. Solbe, EIFAC water quality criteria for European freshwater fish: report on aluminium. *Chemistry and Ecology* **4**, 117–173 (1990).
33. J. A. Leenheer, Chemistry of dissolved organic matter in rivers, lakes and reservoirs. (Abstract p. 285.) *ACS Symposium Environmental Chemistry of Lakes and Reservoirs* (ed. L. A. Baker) (Lewis Publishers, Chelsea, Michigan, 1991).
34. I. Muniz, and L. Walloe, The influence of water quality and catchment characteristics on the survival of fish populations. In: *The Surface Water Acidification Programme* (ed. B. J. Mason) (Cambridge Univ. Press, Cambridge 522 pp. 1980) pp. 327–340.
35. J. P. Baker, *et al.*, Biological effects of changes in surface water acid-base chemistry. (NAPAP SOS/T report 13) In: National Acid Precipitation Assessment Program, Acidic Deposition: State of Science and Technology, Vol. II (US Govt. Printing Office, Washington, DC, 1991).
36. D. J. A. Brown and K. Sadler, The chemistry and fishery status of acid lakes in Norway and their relationship to European sulphur emissions. *J. Appl. Ecol.* **18**, 433–441 (1981).
37. L. A. Baker, *et al.*, Current status of surface water acid-base chemistry. NAPAP SOS/T Report 9 (US Govt. Printing Office, Washington, DC, 4 appendices, 1990).
38. M. M. Popovics, *et al.*, Historical data of atmospheric sulfur in three monitoring periods for Europe. *Water Air Soil Poll.* **36**, 47–59 (1987).
39. J. C. Ellis and D. T. E. Hunt, Surface water acidification: an assessment of historic water quality records. *WRC Environment Report TR 240* (1986).
40. M. H. Pfeiffer and P. J. Festa, *Acidity status of lakes in the Adirondack region of New York in relation to fish resources* (New York Dept. Environmental Conservation, FW-P 168, 10/80, 1980).
41. J. T. Turk, Natural variance in pH as a complication in detecting acidification of lakes. *Water Air Soil Poll.* **37**, 171–176 (1988).
42. C. F. Barrett, *et al.*, Acid Deposition in the United Kingdom. *Report of the UK Review Group on Acid Rain* (Dept. Environ., 1983).
43. J. G. Irwin, *et al.*, Acid Deposition in the United Kingdom 1986–1988. *Third Report of the UK Review Group on Acid Rain* (Dept. Environ., London, 1990).
44. H. M. Seip, *et al.*, Predicted changes in streamwater chemistry following decreased/increased sulphur deposition using the "Birkness" model. *Water Air Soil Poll.* **30**, 909–930 (1986).
45. P. Sanden *et al.*, Acidification trends in Sweden. *Water Air Soil Poll.* **36**, 259–270 (1987).
46. D. J. A. Brown, Effect of atmospheric N deposition on surface water chemistry and the implications for fisheries. *Env. Poll.* **54**, 275–284 (1988).
47. G. Howells, *et al.*, Loch Fleet: liming to restore a brown trout fishery. *Env. Poll.* **78**, 131–139 (1992).
48. S. Oden and T. Ahl, The longterm changes in the pH of lakes and rivers in Sweden. In: *Sweden's Case Study for the United Nations Conference on the Human Environment* (supporting studies) (Roy. Min. For. Affairs, Roy. Min. Agric. Stockholm, 1972) pp. 1–13.
49. D. F. Charles and J. P. Smol, The PIRLA II project: regional assessment of lake acidification trends. *Verh. Internat. Verein. Limnol.* **24**, 474–480 (1990).
50. R. W. Battarbee, The causes of lake acidification with special reference to the role of acid deposition. *Phil. Trans. Roy. Soc.* (London) **B 327**, 113–121 (1990).
51. I. Renberg, *et al.*, Recent acidification and biological changes in Lilla Oresjon, southwest Sweden, and the relation to atmospheric pollution and land-use history. *Phil. Trans. Roy. Soc.* (London) **B 327**, 165–170 (1990).
52. J. Merilainen and P. Huttunen, Lake acidification in Finland. *Phil. Trans. Roy. Soc.* (London) **B 327**, 197–199 (1990).
53. F. Berge, *et al.*, Palaeolimnological changes related to acid deposition and land-use in the catchments of two Norwegian soft-water lakes. *Phil. Trans. Roy. Soc.* (London) **B 327**, 159–163 (1990).
54. R. W. Battarbee, *et al.*, *Lake acidification in the UK, 1800–1986; evidence from analysis of lake sediments* (Dept. Environ., Ensis, London, 1988).
55. R. F. Wright and A. Snekvik, Acid precipitation: chemistry and fish populations in 700 lakes in southernmost Norway. *Verh. Internat. Verein. Limnol.* **20**, 765–775 (1979).
56. C. Forsberg, *et al.*, Indications of the capacity for rapid reversibility of lake acidification. *Ambio* **14**, 164–166 (1985).
57. A. Henriksen, *et al.*, *1000 lake survey, 1986, Norway* (Norwegian Institute for Water Research, State Poll. Control Authority, 1987).
58. K. Brown, *et al.*, Rapporteurs' report on discussion at the Workshop on Excess Nitrogen Deposition, Leatherhead, September, 1987. *Env. Poll.* **54**, 285–295 (1988).
59. S. C. Warren, *et al.*, *Acidity in United Kingdom Fresh Waters, Interim Report of the UK Acid Waters Review Group* (Dept. Environ., HMSO, London, appendices, 1986).

60. S. C. Warren, *et al.*, *Acidity in United Kingdom Fresh Waters, Second Report of the UK Acid Water Review Group* (Dept. Environ., HMSO, London, 1988).
61. R. W. Battarbee, *et al.*, Diatom and chemical evidence for reversibility of acidification in Scottish lochs. *Nature* **332**, 530–532 (1988).
62. G. Howells, *et al.*, Nitrate and sulphate budgets at Loch Fleet, SW Scotland. In: *Acid Mitigation, Symposium ACS, Atlanta GA, April 14–19* (1991) pp. 144–147.
63. T. J. Sullivan, Historical changes in surface water acid-base chemistry in response to acidic deposition. NAPAP SOS/T Rept. 11. In: *Acidic Deposition: State of Science and Technology*, Vol. 11 (Washington, DC, 1990).
64. T. A. Haines, Atlantic salmon resources in the northeastern United States and the potential effects of acidification from atmospheric deposition. *Water Air Soil Poll.* **35**, 37–48, (1987).
65. A. Henriksen, *et al.*, *1000 lake fish status, survey 1986, Norway* (Norwegian Institute for Water Research, State Poll. Control Authority, 1988).
66. B. O. Rosseland, *et al.*, Acid deposition and effects in nordic Europe. Damage to fish populations in Scandinavia continues apace. *Water Air Soil Poll.* **30**, 65–74 (1986).
67. R. J. Beamish and H. H. Harvey, Acidification of the La Cloche mountain lakes, Ontario, and resulting fish mortalities. *J. Fish. Res. Bd. Canada* **29**, 1131–1143 (1972).
68. ICES (International Council for Exploration of the Sea) (1988) *Report of the acid rain study group* (ICES, Copenhagen, 1988).
69. W. D. Watt, *et al.*, Evidence of acidification of some Nova Scotian rivers and its impact on Atlantic salmon, *Salmo salar. Can. J. Fish. Aquat. Sci.* **40**, 462–473 (1983).
70. D. W. Schindler, The significance of in-lake production of alkalinity. *Water Air Soil Poll.* **30**, 259–271 (1986).
71. D. W. Schindler, *et al.*, Long-range ecosystem stress: the effects of years of experimental acidification of a small lake. *Science* **228**, 1395–1401 (1985).
72. M. A. Turner, *et al.*, Early responses of periphyton to experimental lake acidification. *Can. J. Fish. Aquat. Sci.* **44 (suppl)**, 135–149 (1987).
73. D. W. Schindler, Recovery of Canadian lakes from acidification. In: *Reversibility of Acidification* (ed. H. Barth) (Elsevier, New York, 1987) pp. 2–13.
74. P. L. Brezonik, *et al.*, Effects of acidification on chemical composition and chemical cycles in a seepage lake: mechanistic inferences from a whole-lake experiment. Abstract, pp 251–253. In: *Symposium on Environmental Chemistry of Lakes and Reservoirs* (ACS meeting, April 14–19, 1991, Atlanta GA): pp. 251–253 (1991).
75. R. Hall, *et al.*, Experimental acidification of a stream in the Hubbard Brook Experimental Forest, New Hampshire. *Ecology* **61**, 986–989 (1980).
76. H. Olem, *Liming Acid Surface Waters* (Lewis Publishers, Chelsea, Michigan, 1991).
77. K. R. Simmons, *et al.*, Use of a water-powered, automatic, doser to lime a remote Massachusetts stream. (Abstract p. 158.) In: *ACS Symposium Acid Rain Mitigation* (Lewis Publishers, Chelsea, Michigan, 1991).
78. P. J. Garrison, *et al.*, Mitigation of acid rain by ground-water addition: an alternative to liming. (Abstract p. 169.) *ACS Symposium Acid Rain Mitigation, April 14–19, Atlanta, GA* (Lewis Publishers, Chelsea, Michigan, 1991).
79. T. C. Hutchinson and M. Havas, Recovery of previously acidified lakes near Coniston, Canada, following reductions in atmospheric sulphur and metal emissions. *Water Air Soil Poll.* **28**, 319–333 (1986).
80. J. M. Gunn and W. Keller, Recovery of acid-stressed trout lakes following reduction of industrial emissions of sulphur. (Abstract, p. 151.) In: *Acidic Deposition: its Nature and Impacts, 16–21 September, 1990, Glasgow.*
81. J. M. Gunn and W. Keller, Biological recovery of an acid lake after reductions in industrial emissions of sulphur. *Nature* **345**, 431–432 (1990).
82. R. F. Wright, *et al.*, Reversibility of acidification shown by whole catchment experiments *Nature* **334**, 670–675 (1988).
83. R. F. Wright and T. Fragner, RAIN Project: 6 years of whole catchment manipulation of acid deposition. (Abstract p. 567.) *Internat. Conference on Acidic Deposition: Its Nature and Impacts, Glasgow, 16–21 September, 1990.*
84. R. A. Skeffington and D. J. A. Brown, Timescales of recovery from acidification: implications of current knowledge for aquatic organisms. *Env. Poll.*, **77**, 227–234 (1992).
85. J. Nilsson and P. Grennfelt, *Critical loads for sulphur and nitrogen.* Nordic Council of Ministers Report, 1988: 5 (Nordic Council, Copenhagen, 1988).
86. D. W. Schindler, Effects of acid rain on freshwater ecosystems. *Science* **239**, 149–157 (1988).
87. S. Patrick, *et al.*, UK Acid Waters Monitoring Network: site descriptions and methodology report. *Report to Dept. of Environ. London* (Ensis, 1991).

88. E. Gorham, *et al.*, Natural and anthropogenic causes of lake acidification in Nova Scotia. *Nature* **324**, 451–453 (1986).
89. P. Kortelainen and J. Mannio, Natural and anthropogenic acidity sources for Finnish lakes. *Water Air Soil Poll.* **42**, 341–352 (1988).
90. C. T. Driscoll, *et al.*, The role of organic acids in the acidification of surface waters in the eastern U.S. *Water Air Soil Poll.* **43**, 21–40 (1989).
91. E. C. Krug and C. R. Frink, Acid rain on acid soil: a new perspective. *Science* **221**, 520–525 (1983).
92. G. L. Lacroix and K. T. Kan, Speciation of aluminium in acidic rivers of Nova Scotia supporting Atlantic salmon: a methodological evaluation. *Can. Tech. Rep. Fish. Aquat. Sci.* **501**, (1986).
93. R. F. Wright, RAIN Project: role of organic acids in moderating pH change following reduction in acid deposition. *Water Air Soil Poll.* **46**, 251–259 (1989).
94. E. T. Gjessing, *et al.*, Humic lake acidification experiment, HUMEX. (Abstract, p. 342.) *Internat. Conference Acidic Deposition: Its Nature and Impacts, Glasgow 16–21 September, 1990.*
95. L. O. Hedin, *et al.*, Role of organic acids in buffering inputs of mineral acidity to surface waters: a field experiment. (Abstract p. 343.) *Internat. Conference Acidic Deposition: Its Nature and Impacts, Glasgow 16–21 September, 1990.*
96. A. Jenkins, *et al.*, CLIMEX—Climate change experiment. (Abstract p. 523.) *Internat. Conference Acidic Deposition: Its Nature and Impacts, Glasgow 16–21 September, 1990.*
97. W. Dickson, *Liming of Lake Gardsjon: an acidified lake in SW Sweden.* Report 3426 (Natl. Swedish Environmental Protection Board, 1988).
98. I. Alenas, *et al.*, Liming and reacidification reactions of a forest lake ecosystem, Lake Lysevatten, in SW Sweden. *Water Air Soil Poll.* **59**, 55–77 (1991).
99. R. F. Wright, Chemistry of Lake Hovvatn, Norway, following liming and reacidification. *Can. J. Fish. Aquat. Sci.* **42**, 1103–1113 (1985).
100. B. O. Rosseland and A. Hindar, Liming of lakes, rivers and catchments in Norway. *Water Air Soil Poll.* **41**, 165–188 (1988).
101. O. Lessmark and E. Thornelof, Liming in Sweden. *Water Air Soil. Poll.* **30**, 809–816 (1986).
102. T. R. K. Dalziel, *et al.*, Targets and timescales for liming treatments. Chap. 16 in: *Restoring Acid Waters* (eds., G. Howells and T. R. K. Dalziel) (Elsevier, London, pp. 365–391, 1992).
103. A. W. H. Turnpenny, Alternatives to catchment liming. Chapter 10 in: *Restoring Acid Waters* (eds., G. Howells and T. R. K. Dalziel) (Elsevier, London, pp. 173–196, 1992).
104. P. Warfvinge, *et al.*, Modelling acidification reversal at Loch Fleet, UK. (Abstract p. 589.) *International Conference Acidic Deposition: its Nature and Impacts, 16–21 September, 1990, Glasgow.*
105. P. Warfvinge and H. U. Sverdrup, Modeling limestone dissolution in soils. *Soil Sci. Soc. Amer.* **53**, 44–51 (1989).
106. A. O. Altshuller and R. A. Linthurst (eds.), The acidic deposition phenomenon and its effects: Critical assessment review papers. *Vol. II, Effects Sciences* (EPA-600/8–83–016BF, US EPA, Washington, DC, 1984).
107. W. Davison, Internal element cycles affecting the long-term alkalinity status of lakes: implications for lake restoration. *Schweiz. Z. Hydrol.* **49**, 186–201 (1987).
108. B. Hultman, *et al.*, Can acidification of lakes and rivers be cured by treated sewage? *Vatten* **39**, 3–14 (1983).
109. W. Scheider, *et al.*, Reclamation of acidified lakes near Sudbury, Ontario, by neutralization and fertilization. *Ont. Min. Environ. Report* (1975).
110. W. Davison and W. A. House, Neutralizing strategies for acid waters: sodium and calcium products generate different acid neutralizing properties. *Wat. Res.* TB22; 577–583 (1988).
111. B. O. Rosseland and O. K. Skogheim, Neutralization of acidic brook-water using a shell-sand filter or sea-water: effects on eggs, alevins and smolts of salmonids. *Aquaculture* **58**, 99–110 (1986).
112. O. K. Skogheim, *et al.*, Base addition to flowing acidic water: effects on smolts of the Atlantic salmon (*Salmo salar L.*). *Water Air Soil Poll.* **30**, 587–592 (1986).
113. C. J. Gagen, *et al.*, Pumping alkaline groundwater to restore a put-and-take trout fishery in a stream acidified by atmospheric deposition. *N. Amer. J. Fish. Manage.* **9**, 92–100 (1989).
114. J. M. Cook, *et al*, Groundwater contribution to an acid upland lake (Loch Fleet, Scotland) and the possibilities for amelioration. *J. Hydrol* **125**, 111–128 (1991).
115. B. O. Rosseland, *et al.*, Strains of brown trout (*S. trutta L.*) stocking and test fishing 1988 and 1989; field and laboratory toxicity testing experiments. *NIVA report 0–87178* (1990).
116. A. S. Gee and J. H. Stoner, The effects of afforestation and acid deposition on the water quality and ecology of upland Wales. In: *Ecological Change in the Uplands* (ed. M. B. Usher and D. B. A. Thompson) (Brit. Ecol. Soc., Blackwell, Oxford, 1988) pp. 273–287.

117. S. C. Bird, *et al.*, The influence of land management on stream water chemistry. In: *Acid Waters in Wales* (eds., R. W. Edwards, A. S. Gee, J. H. Stoner) (Kluwer, Dordrecht, 1990) 241–254.
118. I. T. Rosenqvist, Pre-industrial acid-water periods in Norway. In: *The Surface Waters Acidification Programme* (ed. B. J. Mason) (Cambridge Univ. Press, Cambridge, 1990) pp. 315–320.
119. G. Howells and T. R. K. Dalziel, The Loch Fleet Project. *Report of the Intervention Phase (2) 1986–1987* (CEGB, 1988).
120. A. S. Gee and J. H. Stoner, A review of the causes and effects of acidification of surface waters in Wales and potential mitigation techniques. *Arch. Environ. Contamin. Toxicol.* **18**, 121–130 (1989).
121. WWA (Welsh Water Authority), *Llyn Brianne Acid Waters Project. Second Tech. Summary* (WWA, Brynymor Press, Swansea, 1988).
122. A. Henriksen, Acidification of fresh waters—a large scale titration. In: *Ecological Impact of Acid Precipitation* (ed. D. Drablos and A. Tollan) (SNSF, Oslo, 1980) pp. 68–74.
123. A. Henriksen, Changes in base cation concentrations due to freshwater acidification. *Verh. Internat. Verein. Limnol.* **22**, 692–698 (1984).
124. N. Christopherson, H. M. Seip and R. F. Wright, A model for streamwater chemistry at Birkenes, Norway. *Wat. Air Soil Poll.* **18**, 977–996 (1982).
125. H. M. Seip, *et al.*, Hydrochemical studies in Scandinavian catchments. In: *The Surface Waters Acidification Programme* (ed. B.J. Mason) (Cambridge Univ. Press, 1990) pp. 19–29.
126. B. J. Cosby, *et al.*, Freshwater acidification from atmospheric deposition of sulphuric acid: a quantitative model. *Environ. Sci. Technol.* **19**, 1144–1149 (1985).
127. P. G. Whitehead, Future trends in acidification. In: *Acidification in Scotland, Edinburgh 8 Nov. 1988* (Scott. Devel. Dept., Edinburgh, 1989) pp. 114–121.
128. S. J. Ormerod, *et al.*, Preliminary empirical models of the historical and future impact of acidification on the ecology of Welsh streams. *Freshwater Biology* **20**, 127–140 (1988).
129. R. A. Goldstein, *et al.*, Integrated acidification study (ILWAS): a mechanistic ecosystem analysis. *Phil. Trans. Roy. Soc.* (London) **B 305**, 409–425 (1984).
130. S. A. Gherini, *et al.*, The ILWAS model: formulation and application. *Water Air Soil Poll.* **26**, 425–459 (1985).
131. P. F. Chester, Strategic planning for the control of emissions from combustion processes—the latest information. *IBC Tech. Services*, Memo (1989).
132. L. Roberts, Learning from an acid rain program. *Science* **251**, 1302–1305 (1991).

3. THE ROLE OF SCIENCE AND TECHNOLOGY: AN ANGLO-GERMAN COMPARISON

SONJA ANITA BOEHMER-CHRISTIANSEN

Emission reduction (by containment, fuel substitution or technology change in the production and use of energy) is at the core of acid rain policy. Britain and Germany responded with very different strategies and timetables of implementation. Germany has completed its abatement programme for stationary sources, and its effort for the transport sector, while less satisfactory, is well advanced. Britain has hardly began to implement its reluctant policy response, which was largely forced by external pressures. These differences reflect complex cultural, institutional and geographic differences. Three factors are isolated as being of explanatory power and general significance: the role allocated to scientific advice in environmental policy, energy policy objectives and technological aspirations. The potential of environmental policy for realising broader national goals is more readily recognised through open and competitive policy-making processes. In Britain, ideology and the nature of the political process encouraged short-sightedness.

1. THE REGULATION OF ACID EMISSIONS IN EUROPE

From 1982 onward, Britain and Germany took opposing positions in 'acid rain' negotiations in Europe. Uneasy compromises were reached only towards the end of the 1990s.[1] These meant that the UK has finally began to implement acid abatement plans, while the implementation of West German programmes is either completed or, for mobile sources, well advanced. What were the impacts of these different responses and what lessons might be learnt from them?

Since the mid-1970s, acid rain abatement has been negotiated at the intergovernmental level at the United Nations Economic Commission for Europe (UNECE) and the European Community (EC). UNECE reached acid rain agreements under the 1979 Geneva Convention on Long-range Transport of Air Pollutants (LRTAP) and its two protocols (a third on VOC, volatile organic compounds has recently been adopted) and the EC did so under a number of directives. The most important ones are the Large Combustion Plant Directive (LCPD) of 1988, for which negotiations began in 1983, and the Consolidated Directive on Vehicle Emissions from passenger cars, negotiated since 1984 but not agreed in final form, subject to future revision, until June 1991. The LCPD may also be tightened further under its 1994/95 review.

Under these agreements, member countries are significantly but differentially reducing total sulphur dioxide (SO_2) emissions by early next century and, hopefully, stabilising nitrogen oxides (NO_x) somewhat earlier. While the LRTAP protocols define only overall national reduction targets expressed a percentages of national outputs for an agreed base year, the EC directives combine similar measures with emission standards for new plant.

The LRTAP protocols for aggregate SO_2 and NO_x reductions are now in force. The former commits signatories to a 30% reduction of emissions by 1993 and the latter to their stabilisation by 1997. Only Germany has signed both protocols and, in addition, has committed itself to a 30% reduction of NO_x. Britain has not signed for sulphur, and uncertainty about its ability to comply remains given current emission trends and legal uncertainty.[2]

The LCPD applies primarily to fossil fuelled power stations and oil refineries. Agreement on common emission standards was reached only for new plants authorized after July 1987, (see Table I). However, few of these are expected to be built in the near future, especially in Britain. Reductions in the industrial sector, especially from power stations, will therefore have to be achieved through other choices decided at the national level. As for LRTAP, the 'teeth' of the LCPD are requirements for national percentage reductions. German regulations are considerably more stringent, (see Table I).

As is appropriate for vehicles, major trading goods and transfrontier pollution sources, new vehicles registered from 1992/93 will be regulated uniformly. Until 1983 vehicle emission standards had been negotiated for the whole of Europe at UNECE. After 1983, negotiations moved to Brussels and increasingly into the

Table I The large combustion plant directive

a. Selected proposed and final SO_2 reduction limits (ktonnes or % of 1980 output)

	Actual in 1980		LCPD 1983	1987			**1988**		
	Total	LDP	by 1995	1993	98	2010	**1993**	**98**	**2003**
FRG	3200	2225	60	40	60	80	**40**	**60**	**70%**
UK	4670	3883	60	22	33	80	**20**	**40**	**60%**
EC	21153	14430	60	27	42	74	**23**	**42**	**57%**

b. Emission limits for new plants (in milligrammes per cubic meter)

	SO_2 1983 Proposal	**Final**
Solid fuels	2000–400	2000–400 but plants less than 100 MW excluded*
Liquid fuels	2000–400	400 for above 300MW
Gaseous fuels	800–35	100–5**

*400 for plants over 500 MW; in Germany the 400 standard applies above 300 MW and 2000 below 100 MW; the same standards apply to existing plants.

**5 applies to liquified gas, as a rule for all plant sizes.

	NO_x 1983 proposal	**Final**	(German)
Solid Fuels (as a rule)	800	**650***	200 (>300MW)
Liquid fuels	450	**450**	150
Gaseous fuels	350	**350**	100

*This allowed the UK to not require SCR under its batneec principle—best available technology not entailing excessive cost.

NB: The German Large Combustion Plant Regulation of 1983 (GFAVo) also covers heavy metals, HLC, HF and ozone; and applies, slightly altered and with timetables for closure, to existing plants.

Source: Boehmer-Christiansen and Skea,[1] (1991).

hands of EC environment ministers, but with the decisive participation of the European Parliament after 1986.[3] (See Table II.)

The standards finally agreed to in 1991 will require fully regulated (three way) catalytic converters (CC) for all cars. Both countries, in the end accepted mandatory standards, although Germany debated the possibility of unilateralism vigorously during the mid-1980s and then moved ahead faster with a voluntary scheme based on 'economic incentives'. Controls are gradually being extended from passenger cars to goods vehicles and, at some future date, aeroplanes.

West German 'acid rain' abatement in the energy sector was completed by about 1990 (Figure 1); in the East this has to be achieved by 1997.

The influence of scientific argumentation dominated at the UNECE with emphasis on the use of distribution models, air quality monitoring[4] and the use of concepts such as critical load for potential regulatory purposes. The more scientific concept of 'critical load' is intended, in the interest of cost effectiveness, to define reduction requirements which relate abatement to the capacity of different environments to tolerate acidification. Britain has since felt it necessary to add the less scientific concept of 'target load', probably to avoid having to reduce emissions 'excessively'.[5] Germany thinks little of such measures and would find it legally and politically difficult to implement them. In contrast and more typical of the German approach, technological and macro-economic argumentation prevailed at the EC, hence the adoption of uniform emission standards as well as 'bubbles' for industrial sectors.

It is of course true that percentage reductions and uniform emission standards have little scientific rationale in the sense that they predictably relate to a reduction in pollution effects in the environment. However, the more technological

Table II The evolution of european emission standards (gm/test cycle)
(For passenger cars and NO_x only, timetable omitted)

Regulation/Agreement	NO_x Standard	Comment
1976 ECE 15–02	10	no observable impact on air quality or technology
1979 ECE 15–03*	8.5	no observable impact on air quality or technology
1984 ECE 15–04*	19	combined with HC: allowed development of leanburn for all engine sizes
1984 EC proposal	6	threat to large leanburn engines
1985 Luxembourg >21	3.5	fully regulated CAT only
Agreement ** 1.4–21	8	leanburn and simple oxidation CAT
<1.41	6	leanburn possible
1988 Small Car Agreement (rejected)	8	technology impact uncertain ***
1988 Small Car Agreement (forced by EP)	5	no leanburn, but still possible for medium-sized engines
1991 Consolidated Directive	**5**	**fully regulated catalysts for all cars**

*standards dominating on UK roads.

**emission standards were now to be related to engine size (instead of weight as under ECE)—this equalised relative costs and had technology implications. The Agreement was celebrated as a British success as the time.

***once a new driving cycle including a high speed component was also agreed, leanburn could no longer comply.

NB: Germany began introducing low emission cars (including diesels) with aid of economic incentives in the mid-1980s, by end of decade about 30% of cars on the road and most new registrations were in this category.

Source: Boehmer-Christiansen,[3] (1990).

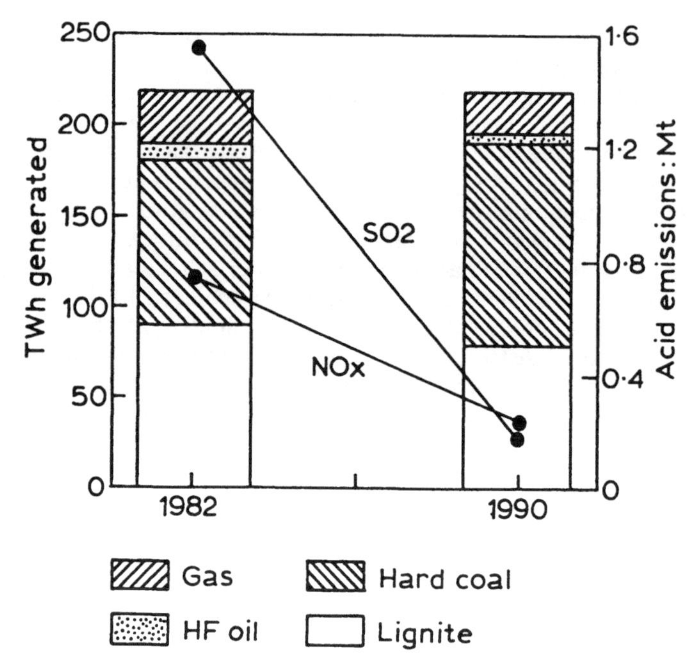

Figure 1 West German acid emissions record

approach has generally been shown to be more effective in producing actual reductions of emissions.[6] It is agreed that such reductions have some impact on pollution effects, even if these cannot be observed immediately, the goal of environmental protection has been served. Broader, non-environmental impacts are also possible, and it is with these broader, non-environmental effects (and 'causes' of policy) that this paper is primarily concerned.

The research reported here suggests that one of the reasons for preferring emission standards is both their effectiveness in reducing emission quickly and their more direct impact on technology. This in turn relates to the environmental principle of 'precaution,' which broadens the discretionary powers of policy-makers. If these powers are used wisely, precaution may be applied to increase the 'opportunity structure' of environmental policy to allow for convergence with other policy objectives. Abatement policy can then be implemented, by some of the actors at least, for non-environmental purposes. The political 'will' to act against a particular environmental problem may be strengthened and rhetoric turn into action.

2. DOMESTIC POLICIES DETERMINE INTERNATIONAL POSITIONS

Until the early 1980s, international acid rain control initiatives had come mainly from Sweden and Norway. Stimulated by this pressure and the general environmental concern of the time, domestic German policy concepts and processes began to encourage the convergence of environmental goals with other policy objectives. This in turn helped to create a 'winning of alliance' against acid emissions before the European Commission took up the issue, something it largely did at the initiative of the German federal government. International agreement allowed Germany to implement unilateral mandatory response for its own combustion plants, i.e. uniform national emission standards for new and, more importantly, for existing plants of virtually all sizes.

In Britain, policy divergence leading to paralysis dominated until external pressures forced the positive response outlined above and meant that Britain acted, for some time, as a brake on European aspiration and timetables. It excelled in the use of delaying tactics and the choice of arguments based on science (as too uncertain) and costs (as too high). In the summer of 1991, a new Environment Minister (the fifth in a decade compared to two in Germany) summed up this chapter of British environmental history by arguing that:

"One of the reasons it took so long to reach agreement of acid rain was that heroic attempts to find solutions were launched in advance of broad international agreement on the scientific nature of the problem and the priorities for action."[7]

Other reasons are explored below, but German 'heroism' is probably not quite the appropriate description for a major investment programme which coincided with an economic 'downturn'. The underlying German motivation was as self-interested as the British refusal to act quickly. The main difference, as will be demonstrated below, was that the German interest was a resultant of many forces and pressures which were able to participate in a relatively open process of policy formation dating back to the late 1960s.[8]

Fear by industry of a loss in 'competitiveness' caused by higher energy costs (and brought about both by more nuclear safety and cleaner coal-burn) then led to demands that other EC members should follow suit. This German pressure could become effective through the legal and institutional machinery of the European Community. Bonn could appeal not only to the 'pollution knows no frontiers' argument, but also to the aims of the EC as a supranational body set up to prevent unfair competition and promote technological change.

3. THE SINGLE MARKET

The prospect of the completion of the internal market and the Single European Act of 1986 benefitted German 'acid rain' initiatives in Brussels. Overarching EC goals tend to be as much economic as they are environmental, other targets being the promotion of specific technologies, investment by private industry, and managed competition. The Single Market represented a determined move towards the harmonisation of technical standards and the eradication of trade barriers, arguments which have probably done more for the harmonisation of environmental standards than those alleging transfrontier pollution or responsibility for the common environment. As stated by one Commissioner closely associated with 'acid rain':

". . . anticipatory standardisation (is) essential for the development of new technologies which have infra-structural characteristics so that companies will not invest against one another until one or more backs down."[9]

4. POLICY DEVELOPMENTS AT THE NATIONAL LEVEL

4.1. *In Germany*

Senior German administrators (supported by Willy Brandt and Hans-Dietrich Genscher) had shown a strong interest in air pollution at the beginning of the 1970s, but could not muster sufficient political support for the implementation of the environmental programme they prepared and for which legislation was passed. Without more powerful allies the small if growing environmental lobby in government remained too weak.[10] In particular, the resistance of industry to tough regulations for coal (but not foreign oil) and car emissions proved successful until the early 1980s. The public did not yet care enough. Germany did, however, press for tougher vehicle emission standards in Europe already during the 1970s and succeeded in unilaterally introducing low lead petrol in 1976. (In Britain this battle was won by the oil companies with support from the motor industry; low lead petrol did not become mandatory until the mid-1980s).

When 'Waldsterben' (forest decline, lit. dying) became a national issue in 1981, acid rain abatement policies were suddenly strongly promoted by all political parties including, after October 1982, the right-of-centre Cabinet of Helmut Kohl. Dying forests provided a chance for gaining widespread public support for air pollution measures which by their nature were directed at fossil fuels. Major cost increases related to environmental protection had already 'burdened' nuclear power. Raising the cost of coal-burn, especially of brown coal, 'balanced' energy policy, (see Figure 1). Germany's semi-privatised members of the electricity 'cartel' depend on quite different fuel mixes and therefore have to pay attention to the relative costs of the fuels they use.

In 1982, in a bold move not yet fully agreed to at home, Genscher changed Germany's position inside UNECE. Bonn itself stood to gain in legal competence (at the expense of the Länder) once air pollution had become a national crisis and subject to international regulation. This change in policy allowed acid rain to increase in domestic significance and become part of a broad, proactive environmental programme which was adopted by the Social Democrats and their Liberal coalition partner just before the fall of the Schmidt Government in the autumn of 1982. Britain was at a loss for an explanation: a major ally had deserted it. Why did the German political system accept that recently 'discovered' forest damage was caused by emissions from coal-fired power stations mainly situated in the Ruhr region? Hard scientific proof was indeed still lacking, and is likely to remain so for some time to come.

During 1982/83 a bargain between the parties in the energy sector was worked out and the Large Combustion Plant Regulation (GFAVo), much tougher than that originally proposed entered into force in June 1983—it included existing plants as well as new ones. The Green Party could not complain. The new Coalition in Bonn of Christian Democrats (CDU), Christian Social Union (CSU) and Liberals (FDP) had much stronger links with the forested Southern Länder (especially Bavaria) than the previous government. Much of this forest is privately owned.

In the South were also the Republic's firmest supporters of nuclear power and the homes of the high performance car industry. The still growing, but primarily anti-nuclear green vote needed to be directed to an issue which government was able and willing to solve. Given much pressure not only from the Greens, but also from public opinion, forest owners and industries suddenly concerned about their image, the most significant pro-acid rain abatement decision—to extend tough emission standards from new to existing plants above 50 MW and regulate for auto-catalysts—was taken.

The Federal Council, voice of the regions, was a powerful influence in favour of rapid abatement. The rising Green party was feared as competitor by both of the small coalition partners, the CSU of Franz-Josef Strauss and the FDP of Genscher. Waldsterben thus allowed the pro-nuclear, conservative South to blame the social-democratic North-West, the coal states of Saarland and North Rhine Westphalia, for the increasingly visible and measured problems of the forests. Nuclear power was proclaimed, in the South at least, as the saviour of the *Wald*. And the prime target of the Greens, nuclear power was indeed in need of a more environment-friendly image. The construction of a new batch of 'convoy' reactors was due to begin in the autumn of 1982. The amount of political energy generated by the debate to abate acid rain from power stations and cars was prodigious.

For vehicles, bargaining had also begun, but the whole issue proved to be much more difficult than power stations. The solution, fully regulated catalytic converters and a voluntary approach until Europe had caught up, was probably suggested by BMW to the Bavarian government, and a Bavarian minister responsible for both air pollution and nuclear safety, promoted the idea in Bonn. It took until 1985 before the car industry was fully persuaded, and longer for the rest of Europe,[11] (see Table II).

The arguments for acid abatement did not, primarily, come from scientists, although these played their part by suggesting the direction of policy and arousing public concern. Rather, the available knowledge base was interpreted by the mass media and philosophers in an alarmist or pessimist manner—catastrophe was just around the corner.[12] Social theorists and the churches entered a debate which, in the end, was not only about environmental threats to industrial society, but also about social cohesion, even political stability. A fairly fragmented society frightened by what seemed to be an intensification of the Cold War (e.g. Pershing missiles on German soil), was being united by ecological 'Angst'.[13] This allowed a political system which is usually slow to reach consensus to close ranks against 'vested' interests in the 'common' interest.

In 1983/84 Germany took the 'acid rain' debate to the EC to become a major subject of negotiations, which in turn strengthened the fledgling environmental lobby in Brussels. Germany now took the existence of severe damage to buildings, soils, freshwater and forests more or less for granted, although funding for forestry research increased and continues. The price paid for its acid rain policy was, in part at least, more expensive electricity, but without reduced profits to the utilities, profits which are however shared with local government. Expensive electricity and congestion on the roads, to the decision-makers, were not so much threats than incentives for technological progress. Solutions which are researched under the 'climate change' label relate to energy efficiency and information technology.

Germany experienced low inflation, a healthy trade balance and continued economic growth throughout the 1980s, inspite of a deep recession at the start of the decade. At the beginning of the 1990s only the German car industry was still 'booming'[14] and the alliance of coal and nuclear power, while still under attack, had maintained itself as the mainstay of electricity generation and is likely to remain so until and if 'naturally' replaced by new fuels and technologies. A few years later unification had altered much of this rosy picture, but the underlying commitments, though not current capacity, remain unaltered.

4.2. *In Britain*

The first (modern) fluegas desulphurisation (FGD) plant in Britain was under construction in 1991, just as FGD in general appeared increasingly unattractive to industry. In the past, opposition to measures similar to those adopted in Germany had taken the form of perfectly correct statements about scientific uncertainty: transboundary fluxes were not yet known empirically; the nature of the relationship between emission reduction and declines in deposition remained unclear; the mechanisms of how fish kills were related to acidification of fresh waters was not yet fully understood (when they were, the Central Electricity Generating Board (CEGB) announced its acid abatement programme); damage due to higher emissions in the past might not be reversible by abatement today, the causality of forest damage in particular remained poorly understood and the idea that air pollution was the main culprit an untested theory.

Considerable efforts were made in Britain to answer these scientific questions with funds provided by government or the largest alleged polluter, the state utility CEGB.[15] Some of this scientific work was done collaboratively with Norwegian scientists who apparently felt that Britain, as the alleged transfrontier polluter, approached the subject with excessive scepticism, looking for evidence which would weaken British responsibility. While this cautious approach angered environmentalists, it was by no means of disadvantage to scientific progress. Collaboration with German groups appears to have been less successful. Science therefore became an instrument of politics for defending British policies based on a range of other subjects which also influenced it, but could do so only implicitly.

These other policies ranged from political efforts to destroy the power of the British coalminers union, to deregulate the economy (a process which in fact turned into a 'torrent of radical legislation' for almost everything including the environment) and, probably most important here, to privatise nationalised industries as quickly as possible. State debts were to be reduced at all cost. Major acid polluters, however, happened to belong to this category and were protected by government from approaching costs which might reduce their value in the eyes of shareholders. For electricity, it has been argued, this meant the 'abdication of responsibility for strategic issues' such as security of supply, the future of coal and of the power plant industries.[16] It also meant that nuclear power could not be privatised—the City would not buy it.

Energy policy and macro-economic considerations therefore entered as much on the negative side as they did on the positive side in Germany. The CEGB, hoping to be privatised as a single monopoly, did not wish to publicise the true costs of nuclear power as compared to coal. The idea that the future of coal was in fact tied to that of nuclear (each in need of government support) appears not to have occurred to the coalminers, British Coal or the CEBG. No alliance between the two was arranged in the environmental and national interest.

Environmental scientists consider the soils of at least 8% of the British land surface subject to sulphur deposition above the 'critical' load level and geologists have warned about the acidification of ground water. The impact of pollution on forests is still met with official scepticism, however. It is admitted that forests are best not planted where they may scavenge air pollutants and increase acidification of water.

Was science in Britain a mere fig-leaf of policy, or is the 'scientific approach' to pollution control deeply ingrained in the British policy process? The answer is both, hence the powerful resistance to UNECE and EC proposals. British regulatory policy and the policy process itself—through the advice which is sought and the institutions which are trusted as sources of information—can be shown to direct policy outcomes away from rapid technological adjustments and towards more research and understanding. This, it is argued on the basis of the acid rain case study, may account, in part, for industrial underperformance. Insufficient attention to the technological consequences of regulation have produced the alleged but poorly understood reputation of Britain as the 'dirty man of Europe'.[17] This amounts to suggesting that Britain is 'dirty' not because it is poorer, but that it is poorer because it did not clean industrial processes early enough. How did Britain become trapped in a cycle of industrial decline initiated by a style of regulation that discouraged investment in technological innovation producing higher quality rather than cheaper products? What is the evidence?

5. ABATEMENT TECHNOLOGY AND FUEL COMPETITION

The technology required for acid emission abatement obviously depends on the fuel used and affects production costs. Fuel switching normally requires major investments. Acid emission reduction, especially from existing plants, makes coal-burn considerably more expensive, thus benefitting its competitors. Two competitors are important in the market place: nuclear power and natural gas. Nuclear power is not a significant source of acid emissions, and natural gas has several advantages, (see below). Fuel competition thus becomes part and parcel of environmental policy formation. New actors (as lobbies or directly as government or party clients) may thus enter the policy formation process, provided institutional and political processes allow for this.

The technical means for reducing, virtually eliminating the emission of SO_2 and NO_x were available in the early 1980s: fluegas desulphurisation, improved combustion technology and catalytic processes were all known about and either already commercially available or in the early stages of being demonstrated as feasible. Clean-coal technology, such as fluidized bed combustion, was also on the commercial horizon and strongly supported in Germany, but was poorly supported in the UK during the 1980s. Clean-coal technology has not yet reached commercial application, but R&D continues world wide. In the UK it was recently stopped.

All these technologies could be improved further, especially with the help of new materials and electronic monitoring. FGD as such is not 'high tech', but employs a lot of people who may previously have been building ships (one German 'actor' ceased its opposition when this was recognised). It is a scrubbing process which removes sulphur gases before they leave the stack of coal and oil fired power stations, trapping them chemically using limestone to form solid or liquid residuals. Low NO_x burners regulate the temperature of combustion so as to minimise the oxidation of nitrogen in the air. The catalytic reduction of NO_x in the fluegases of power stations (SRC) and, together with several other exhaust gases, in the exhaust system of vehicles using fully regulated catalytic converters (CC), similarly prevents NO_x emissions. Unleaded petrol is essential for engines linked to catalytic systems, thus bringing another lobby into the 'acid rain' bargaining process.

FGD and catalytic converters for vehicles had been invented in Britain(!),[1] but were rejected in the 1980s largely on costs grounds and after furious lobbying by the motor industry, even though the official explanation related to the lack of 'sound science' and loss of fuel efficiency.[18] These so-called 'add-on' technologies had been adopted in Japan and, for cars, in the USA and Japan during the 1970s. Only Japan has so far achieved a major reduction in total acid emissions, including of NO_x. In Britain and Germany NO_x emissions from most mobile sources are still rising.[19] (In Japan, the associated technological developments had been stimulated not on 'acid rain' grounds, but as measures to protect human health in cities or to reduce photochemical smog.)

FGD may lead to important subsequent disposal or recycling problems, although more expensive technologies do not generate waste sludge, but sulphur and usable gypsum instead. Major environmental impacts are most likely to be associated with the mining, transportation or disposal of large volumes of raw material or waste.[20] The problems associated with nuclear safety and waste disposal are well enough known, and clearly enter the overall cost-benefit balance of acid rain abatement if nuclear power is relied upon to reduce acid emissions.

The least costly way of reducing acid rain (assuming that no existing investments are required to be written off prematurely) are, however, gas turbines (CCGT) and Combined Heat and Power plants.[21] CCGT produce two streams of electricity instead of one and hence increase efficiency of conversion from around 35% in conventional electricity plant to 50%, with 80 to 90% of useful energy conversion if CCGT is used for combined heat and power. Gas (methane) promises three environmental advantages: significant reductions in acid emissions without FGD or SCR, significant reductions of CO_2 emissions (because of the lower carbon content and improved efficiency) and suitability for small scale production (and hence lower investments costs and quicker pay-back periods).[22]

Faced with the choice between fuel switching and technological 'fixes', Germany chose fuel security (and clean air), while Britain during the 1980s stumbled slowly and then dashed (only to be slowed down again), market-led, towards fuel switching to gas, probably at the expense of nuclear power and domestic coal. This choice has been described as one between equity and cost effectiveness.[23] The argument here is that this 'equity' strategy also had broader technological and investment implications of considerable significance for a range of industrial sectors. This is especially true for coal combustion, which will undoubtedly continue to be the major fuel world wide for decades to come.

Using FGD and SRC meant that Germany did not need to alter its fuel mix for electricity generation significantly until this was decided desirable, although there was an energy penalty.[24] This was what the utilities wanted above all else, as did the coal miners and the nuclear industry. The strategy adopted benefitted existing policy in that it improved the competiveness of nuclear power *vis-à-vis* domestic coal and developed a potential export industry for cleaner coal-burn. It also advanced fluegas cleaning technology significantly: the first commercial combined denoxification/desulphurisation (DESONOX) plant was opened in 1991. It produces no 'waste' or gypsum (of which Germany now has more than enough), but commercial sulphuric acid.[25] The capacity of 'clean' power stations to incinerate all sorts of wastes, including domestic rubbish, explains the rapid expansion of German utilities into the waste management business, e.g. RWE. This also created closed ties with their main clients, local and regional governments and communities on whose political protection they may have to depend as their 'cartel' is threatened by the policy for a common market in energy.

Britain, in pronounced contrast, intends to abate acid emissions from the energy sector by both switching massively to natural gas in the coming decades and replacing domestic with less sulphurous imported coal. This market-led strategy not only promises the mentioned major environmental and economic benefits, but does so only in future and not without a new set of broader impacts and risks. While the strategy is the cheapest from the perspective of industry (and recalls an earlier, zero cost strategy to improve air quality, the substitution of 'premium' gas for coal in the domestic sector during the 1960s), it has already led to a 30% increases in the cost of gas to large consumers, litigation against British Gas and parliamentary concern about energy security and the balance of trade. On the other hand, a slow-down in current plans to abate acid emissions by fuel switching is likely to endanger Britain's ability to comply with international obligations.

The restructuring of industry and the energy sector was undertaken with little concern for the consequences to either the national economy or technological innovation. Centralisation prevented the interests of regions most seriously affected by acidification to be well represented in the decision-making process. Acid rain was not, therefore, in Britain, a step in what Germans now call 'ecological modernisation of industry',[26] a process which may only now be beginning in Britain thanks to EC standard-setting and new liability rules, but is severely hampered by a deep recession.

Lack of co-ordination or strategic planning by government and industry meant that the technology needed for the sudden fuel switching was not available in Britain. Gas turbines for CCGT plants are being imported, for example, from Siemens. In the meantime, coal-burn will remain important and NO_x emissions a problem.[27]

This process of fuel switching now under way, if allowed to continue, will go some way towards having environmental effects similar or better than those due to 'add-on' technical fixes, but is not without other negative side effects and longer term risks.

6. 'VORSPRUNG DURCH TECHNIK' AND INVESTMENT?

"Heightened concern about the environmental impacts of energy activities has created not only a mandate but more importantly an opportunity for governments, producers and consumers to implement optimal solutions supporting goals of economic growth, energy security and environmental protection." (OECD)[28]

What are the links between economic performance and environmental regulation? There can be no simple answers, for as this chapter hopes to demonstrate, policy-making and implementation are societal processes in which many forces must reinforce each other before results can be achieved. Does the 'ecological modernisation' of an economy promote growth and development? Does it really protect the global environment?

No definitive general answers can as yet been given to these questions, especially to the last two, but it is argued from the research here summarised, that there is sufficient evidence to show that industrial pollution control can initiate a new cycle of investment and innovation which does lead to growth provided:

- the political system has the capacity to regulate in support of technological innovation and its dissemination;
- the technological capacity is available and can be mobilised to accomplish technological change in industry.

These capacities do involve science directly, but of a rather different kind than that which is required to prove the causality of environmental damage or predict emission scenarios of the future. Rather than linked to 'basic science', science on the solution side is found in applied fields and engineering. The relative status between 'doers' and 'thinkers', between those concerned with 'why' and those more interested in 'how' also becomes of fundamental importance in how society responds to environmental threats.

Environment-friendly technological change also tends to promote political stability in market-orientated economic systems, but may take place without positive impacts on less technologically advanced economies whose dependence may even be increased.[29] Environmental technology may become an economic threat to others, and thus a subject of bitter international negotiations aimed mainly at gaining time for adjustment.

The technologies adopted in Germany were not new.[30] Given the short time available for compliance (and the need for investment in an economy which had slowed down), patents or equipment were bought from Japan or the USA, but developed further. Catalytic converters were developed to allow continued emphasis on the development of powerful engines and the potential for high speed travel, as well as incentives to push along the development of lower emission technology for diesel engines. Both strategies were to help against the growing strength of Japanese manufacturers.

The investments involved, in Germany, amounted to an estimated DM 22 billion for FGD alone and somewhat less for SRC. DM 50 billion is the official figure for the total amount spent by private industry on air pollution abatement between 1983 and 1989. Investments in environmental protection in general (public and private) more than doubled between 1978 and 1990, but rose most steeply after 1984 for air pollution control, which now covers all economic sectors, including domestic boilers.[31]

The German car industry (led in the emission control area by the GM-owned firm Opel) has not revealed what it spent on changing to the electronically controlled low emission car, but the electronics component and precious metal suppliers, e.g. Bosch and Johnson Matthey, have been major beneficiaries. Catalyst cars are not of the 'old' type with something added on, but represent a new car design involving considerably more 'added value' and chemical engineering.

Equally important, and perhaps from the political perspective decisive, the German strategy of acid rain abatement by 'add-on' technologies, rather than fuel switching or waiting for clean-coal technology to become commercially available, protected past and current investments into nuclear power and German coal-mining. Some of these investments were recent and, on purely economic grounds, should not have been written off as yet; others were social rather than economic in nature. Their protection was, however, essential for the political feasibility of energy policy in Germany.[32] Environmental regulation thus came to serve rather than hinder energy policy.

In Britain, each of the 'add-on' technologies was rejected by industry and government either on cost grounds or as being untimely. The British-owned car industry fought catalysts and unleaded petrol for a number of reasons vigorously for over a decade. It is still in serious difficulties and has virtually disappeared from the world market because it could not keep up with technological change. A major British-based manufacturer, Ford Europe, had to abandon its alternative, the 'leanburn' engine, to the conventional engine combined with fuel injection and catalysts, and with it a lot of invested money and effort.

7. WHY THE DIFFERENCE?

Spending 'too much' on pollution control was not only perceived as unsound science, but also as in breach of a major British environmental principle, that of cost effectiveness. Pollution abatement was generally viewed as an unproductive cost rather than investment. Polluters were innocent until proven guilty and the industry (or the Treasury) felt it could afford to pay. Technology-forcing was also challenged by the view, in the 1980s at least, that government should not interfere with the market.

Since 1991, the British economy is, once again, in severe recession and many of the economic achievements of the Thatcher era have been revealed as wishful thinking rather than substantial improvements in productivity and innovative technology.[33] In Britain the search for explanations continues, one review concluding that:

> ". . . it is the archaism and inefficiency of Britain's social institutions that seem to lie at the root of so many of its industrial weaknesses."[34]

The analysis presented here tends to refine this view, selecting the political decision-making process as one of the weakest points in the institutional set-up. Three factors in particular are suggested as particularly relevant in explaining Anglo-German policy differences, as well as being of broader interest. Social institutions play a major role in deciding whether and to what extent these factors become effective during policy formation:

- the respective roles played by scientific, economic and technology advice in the formulation of environmental policy, with the order of significance reversed in the two countries;
- the political incentives available to the political system to promote general environmental 'awareness' or threat perception;
- the capacity of government to co-ordinate different policies in the broader national interest, e.g. of combining environmental aspirations with social, industrial and technology policy.

For air pollution at least, the fragmented, semi-privatised nature of the German energy sector (or entirely privately owned car industry) combined with a decentralised, coalition-dominated political system in such a way that the OECD prescription cited above was recognised intellectually and could be realised in practice. This was only possible because energy and environment had become closely linked in public perception and hence a politicised and widely debated subject during the 1970s. The legal base for technology forcing and stimulating economic activity through environmental regulation on the basis of the precautionary principle had been developed during the 1970s and was operationalised through the 'Stand der Technik', state of the art technology, criteria. State of the art was decided not by the administration, but largely by independent professional bodies, such as the Association of German Engineers (VDI).

In Britain, on the other hand, in part because of good energy endowments, institutional structures, underlying concepts and political objectives combined to defeat pragmatism and foresight. The potential of environmental policy for stimulating technological and economic activity, as well as improving social consensus, was not recognised by the policy system. In addition, a science-led environmental policy assisted in the process of preventing an early alliance between industry and official environmentalism. Even British academic discussions of environmental instruments has paid little attention to technical change as an instrument in its own right.[35] Lacking the wider enabling conditions prevailing in Germany, the British Government could not muster sufficient 'political will' for a proactive policy.

By neither 'forcing' technology through regulation nor helping industry with the creation of green markets (through R&D support, subsidisation of new products), government probably promoted industrial decline.[36]

8. PROACTIVE POLICY: AN OUTCOME OF COMPLEX BARGAINING

In the end, governments will only change existing policies on the basis of a calculus which includes political costs and benefits, as well as practical feasibility. Although 'acid rain' policy outcomes in each country were determined by a similar range of factors: environmental objectives, macro-economic and political/ideological considerations, energy policy commitments and goals, political concepts and attitudes, the trade-offs made during the political process were clearly very different.

In all democracies, however imperfect, policy-making is a societal process which is guided and channelled by institutions. Yet the degree of imperfection seems to matter. Flexibility and the capacity to recognise and co-ordinate interrelated policies have been more observable in Germany than Britain, in part at least for structural reasons, although individuals were also important.

Bargaining was restrained in Britain by politics, but encouraged in Germany. It was bargaining (not technocratic calculations of costs and benefits), which allowed policy convergence in favour of acid rain to take place, while in Britain a small number of people, many of whom found scientific uncertainty useful as argument against policy change, decided on the basis of more limited objectives.

9. THE LIMITATIONS OF A SCIENCE-LED POLICY

Isolating scientific evidence of unacceptable harm as the major factor determining environmental policy means that policy tends to be based on future research findings and the monitoring of extremely complex natural systems. This in turn may encourage (or justify) a wait-and-see attitude as a better understanding of the causes of undesired impacts of emissions and discharges is gained. Environmental science has little to do with abatement—but it may serve as justification for doing something on other grounds: aesthetics, sustainability, ethics, economics, health etc. The link between science and environmental policy is therefore indirect and ambivalent; natural science is rarely directly prescriptive.[37]

Basic environmental science approaches pollution in terms of understanding causation and measuring impacts. It alone is able to suggest criteria for measuring damage and hence, hopefully, allocating legal blame. It alone can provide the essential basis for economic evaluation and the allocation of liability. It is an excellent servant, should this be needed, for a defence against allegations of pollution damage. The route from pollution as proven effects in the environment to abatement policies is long and difficult. In contrast, the route from pollution as knowledge of the presence of potentially harmful wastes in the environment to technological innovation may be much shorter, looks more to the applied and engineering sciences, and may be of greater political and commercial advantage.

The German definition of pollution is not very scientific, but refers to the act of discharging 'nasty' substances into the environment. As the environment is *per se* considered over-burdened and stressed, the reduction of quantities emitted or discharged in itself becomes pollution control. If this is the aim, then it is logical for the decision-maker to call less for more knowledge about the details of an environmental threat—a generalised threat image is sufficient—but to go elsewhere to discover ways for improving technology and/or behaviour. Since political systems are much better at stimulating economic activity than changing human behaviour, engineering solutions tend to be preferred.

For stimulating innovation, the legal base may also be highly important because it allows the translation of attitudes and goals into normative and enforceable rules. As mentioned, environmental activism in Germany relied on the transformation of the precautionary principle into state of art technology. The precautionary principle is not a scientific, but an insurance concept. It legitimises discretion which may be used to take non-environmental consideration into account during policy-making. The origin of scientific advice to government also deserves closer study, but this is not attempted here. Self-censorship by advisors, in the name of departmental or national interest, would directly limit and contain the range of information and interest which government considers legitimate in policy-making, i.e. a tendency for ideology to dominate is likely to prevail.

10. CONCLUSIONS

Faith in unregulated markets, institutional inertia and deeply held ideas about how policy ought to be made, in particular the failure of government departments to organise 'systems' rather than sectors, meant that Britain failed to perceive non-environmental advantages from acid rain abatement. Recognised benefits alone remained far too small to justify action. Because the issue was perceived largely in terms of 'sound science' and affordability, only grudging policy adjustments were made to external pressures. Blessed with rather too much energy, Britain worried less about energy than economic decline, the reasons for which were not understood but which government hoped to arrest with the help of cheap energy. When adjustment were made, the losers of 'acid rain' turned out to be domestic car, coal and nuclear power industries, the balance of trade and, associated with this, energy security and employment. The winners have been British Gas and gas-based combustion technology, although this too is now in some doubt.

In Germany, acid rain, nuclear power and forest death became strongly intertwined political and electoral themes during the early 1980s, when major public policy decisions on emission reduction from power stations and passenger cars were taken. These produced investment strategies which both protected existing technology and forced technology change and dissemination. Germany protected its coal- and nuclear-based electricity sector and strengthened its car industry by actively creating markets for lower emission, higher value-added cars. Given the scale of German investments and associated risks, an active international environmental policy was a subsequent protective device which Britain, given its own commitments and perceptions of interest, could not but resist for as long as possible. Will the 'global warming' debate become a replay? If it were, then the fault would surely not lie with industry, trade unions or senior public servants, but with the weak decision-making system at the very top.

References

1. For the full story and analysis, see S. Boehmer-Christiansen and J. Skea, *Acid Politics: Energy and Environmental Policies in Britain and Germany* (Belhaven, London, 1991).
2. For example, report by Agren in *Acid News,* June 1991; ENDS Report *op. cit.*, and press reports about the pollution impact of burning orimulsion fuel in power stations, a high sulphur fuel which industry hopes to burn without FGD. The 1992/93 coal crisis further added to this uncertainty.
3. For details, see S. Boehmer-Christiansen, "European Vehicle Regulation: The Demise of the Leanburn Engine, the Polluter Pays Principle and the Small Car?" *Energy and Environment* **1**, No. 1, 1–25 (February, 1990).
4. The British monitoring system was, however, reduced from 59 to 32 stations in 1989, with the collection of rain water taking place at intervals of about 7 days.
5. Department of the Environment, *Acid Rain: Critical and Target Load Maps for the United Kingdom* (London, 1991) p. 4, states that it is not yet possible to map critical loads for trees. An 80% reduction is probably needed for restoring all surface waters their pristine state.
6. J. Wettestad, and S. Andresen, *The Effectiveness of International Resource Cooperation* (Fridtjof Nansens Institute, Oslo, 1991).
7. The Rt. Hon. Michael Heseltine, "The Politics of Energy and the Environment: Progress and Prospects" *Energy Focus* **8**, No. 2, 48–51 (July, 1991). The current strategy would appear to make British policy coincide with international developments, a process not known to lead to rapid progress in pollution control.
8. For details, see S. Boehmer-Christiansen, "The Politics of Environment and Acid Rain in FR Germany: Forests versus Fossil Fuel?" *SPRU Occasional Paper 29*, 159, Brighton (1990); also *op. cit.*, note 1.
9. Karl-Heinz Narjes, "Europe's Technical Challenge: a View from the European Commission" *Science and Public Policy*, **15**, No. 6 (December, 1988). This happened in the case of vehicle emission control, where the German supported (but British invented) catalytic/electronics-based technology won out over leanburn engine technology supported by Britain. The latter could not compete with autocatalysts for NO_x abatement.
10. E. Müller, *Innenwelt der Umweltpolitik* (Westdeutscher Verlag, Opladen, 1989).
11. *Op. cit.*, note 3.
12. E. Wiedemann, *Die Ängste der Deutschen: Ein Volk in Moll* (Ullstein, 1989).
13. N. Luhmann, *Ökologische Kommunikation* (Westdeutscher Verlag, Opladen, 1986).
14. S. Boehmer-Christiansen and H. Weidner, *Catalyst versus Leanburn: A comparative Analysis of Environmental Policy in Great Britain and Germany* (Wissenschaftszentrum Berlin, 1992).
15. The CEGB is now fragmented and privatised, many scientists have been dismissed and a major laboratory closed because basic environmental research no longer serves the industry's core interests.
16. J. Roberts, D. Elliott and T. Houghton, *Privatising Electricity* (Belhaven, London, 1991).
17. As described but not analysed by C. Rose, *The Dirty Man of Europe* (Simon and Schuster, London, 1990).
18. This raises interesting questions about whether British environmental policy should be 'blamed' on the advice given by scientists or economists! Our research suggests that there is 'innocent collusion', with the last word being spoken by the Treasury.
19. IEA Coal Research, *Annual Report,* 10–18 Putney Hill, London (1991); also OECD, *op.cit.*
20. The mining of precious metals and uranium tends to take place in 'foreign' places far away from environmental activism and does not seem to have been considered in the policies here described. For a full environmental assessment of acid rain abatement, the impacts of replacement technologies should not be ignored.
21. For a summary of clean coal technology, see J. Cooke, "Innovation and Clean Energy Generation in Industry" *Energy and Environment* **1**, No. 4, 368–385 (1990).
22. Its disadvantage is the 'leakage' of methane at various stages during production, distribution and use. A major debate therefore rages about how small this leakage must be before gas can really claim to be a greener fuel.
23. H. Dowlatabadi and W. Harrington, Policies for the Mitigation of Acid Rain, *Energy Policy* **17**, No. 2, 116–122 (April, 1989).
24. H. Schilling, in *Global Energy Issues* **2**, No. 3, 151 (1990), states that FGD increased German energy consumption, by 1988, to an extra 873,000 t of coal per annum.
25. It is said that the Japanese tightened their building regulations in order to absorb their FGD gypsum production.

26. M. Jänicke, "Conditions for Environmental Policy Success" *FFU Rep,* 90–8a (Freie Universitat Berlin, Forschungstelle für Umweltpolitik, 1991).
27. Current guidance notes for large combustion plants prepared by HMPI do not require advanced SRC technology, but are satisfied with the installation of low-NO_x burners, see *ENDS Journal,* No. 195 (April, 1991).
28. OECD, *Energy and the Environment* (Paris, 1989), p. 184.
29. *Op. cit.,* note 14.
30. For details of technologies available, see OECD, *op. cit.*
31. The possibility of the EC regulating small boilers for emissions had to be abandoned. The aim of regulating their energy efficiency caused deep alarm and powerful opposition in Britain from an industry which has not been particularly innovative but largely enjoyed a captive market in the past.
32. The German nuclear programme is now considered complete by most people. Nuclear legislation is in the process of adjustment to a new state of affairs in which a 'mature' technology must fend for itself while government invests in new technologies, such as solar power.
33. For a well-supported argument here, see W. Walker, 'Britain's Dwindling Technological Aspirations,' *Science Policy Report Group Paper,* No. 1, 22 Henrietta Street, London (1991); abridged version in Nelson (ed.), *National Innovation Systems: A Comparative Study* (1991).
34. *Ibid.*
35. D. Everest and D. Cope, "Instruments Available for the Implementation of Environmental Policy" in *Energy and Environment* **1**, No. 4, 353–367 (Multi-Science, 1990), dedicate one paragraph in 16 pages to technical innovation, listing it under 'other environmental instruments'. Neither scientists nor economists have a professional interest in technology, which tends to be viewed as means for implementing policy rather than as an independent objective.
36. The statistical growth that did take place is better explained be 'demanning', cutting labour costs and even deskilling. Economic incentives were used for the introduction of unleaded petrol.
37. Such judgements cannot by their nature be prescriptive about technical and institutional innovation, but tend towards the measurement of environmental quality, monitoring and gaining understanding of damage. They make a valuable base for decisions, but should not lead policy. For a broader discussion, see Section III in S. Andresen and W. Ostreng (eds.), *International Resource Management: The Role of Science and Politics* (Belhaven, London, 1989).

4. AIR POLLUTION AND EUROPEAN FORESTS

STEN NILSSON

This chapter deals with the air pollution conditions in Europe and their possible effects on the future development of the forest resources. A timber-assessment model has been developed for the analysis of the future wood-supply in Europe. From this model, a number of scenarios have been produced. These scenarios are both with and without the effects of air pollutants. A special dose response model has been employed to estimate the decline effects at different rates of air pollutant depositions. The aggregated results presented deal with future wood supply, economic impact and policy implications.

1. BACKGROUND AND OBJECTIVE

The European forest resources are important in many ways, with forest decline attributed to air pollution becoming a major concern of the European society since the early 1980s. Forest decline caused by pollutants is not a new phenomenon. For more than a century, forest decline caused by air pollutants was identified on a local scale in continental Europe. However, the appearance of widespread visible stress symptoms in the forests of Europe during the early 1980s has raised forest decline caused by air pollutants to one of great significance. Recent trends in forest decline have resulted in a widespread concern within the scientific industrial, labor, economic, regulatory and public sectors of the European society. The ongoing decline may lead to a number of undesirable consequences such as disturbances of trade, decreased production and increased prices of forest industrial products, increased costs for management and protection of the forests, loss of protective functions, and loss of biodiversity and recreational functions.

Forest decline attributed to air pollutants can be found in each country of Europe, although the extent of this decline differs from region to region. While many forests do not display any signs of decline, some have been all but obliterated by air pollution. Strategies to combat undesirable impacts must meet local or regional conditions; in addition, they must also be set in an international context which takes into account such matters as transboundary air pollution and trade in raw materials.

Forests are, in relative terms, slowly evolving systems. Actions taken in European forests over the past several centuries have given us the forests we have today. Actions taken over the next decade will play a large part in determining the nature of the forests and the forest economy several decades into the future. Therefore, those who would design and implement solutions for the forest decline problem must face the difficulties of co-ordinating actions from the near-term to achieve longer-term objectives for the systems under management. One dilemma of trying to cope with the forest-decline problem on a broad-scale, long-term context is that the range of affected persons and decision-makers is extremely extensive, heterogeneous, and difficult to bound.

The IIASA Forest Study, began in 1986 with the task to address the question of the long-term future development of forests in Western, former Eastern Europe and of the European part of the former USSR. The detailed results from this study have been presented by Nilsson *et al.*,[1] Nilsson,[2] and Nilsson *et al.*[3,4] This chapter is a summary of these detailed results. The major objectives of the IIASA Forest Study were to:

(1) Gain an objective view of potential future developments of the forest resources of Europe.
(2) Build a number of alternative and consistent scenarios about potential future developments and their effects on the forest sector, international trade, and society in general.
(3) Illustrate the effects of:
    (a) Forest decline caused by air pollutants;
    (b) Existing and changed silvicultural strategies; and
    (c) Expansion of the forest landbase.
(4) Identify meaningful policy options, including institutional, technological and research/monitoring responses that should be pursued to deal with these effects.

2. MODELING FOREST DECLINE CAUSED BY AIR POLLUTANTS

As indicated above, the Forest Study has had as an objective to be as quantitative as possible. There are limited possibilities to quantify all functions of the European forest resources on a European-wide scale due to the lack of basic data.

The only function which can be analyzed in a quantitative way, with some relevance, for all of Europe is the potential supply of wood for industrial and energy uses. National timber assessment models exist in a number of the European countries. However, consistent and dynamic timber assessments for all of Europe are sparse. The ECE Timber Committee has carried out four European timber trend studies (ETTS), the first of which was published in 1953, and the latest in 1986.[5] The next study is scheduled to be presented in 1995. All of these studies have been, and will be, based on an approach in which the member countries of the ECE (Economic Commission for Europe) are invited to provide forest-resources forecasts for removals and growing stock based on domestic conditions and policies.

To analyze the impact of air pollutants we had to start with the development of a consistent and dynamic timber-assessment model by which we could estimate the development of the forest resources in a consistent way for all of Europe without the effects of air pollutants.

2.1. *Timber Assessment Model for Europe*

There are large variations and differences among European countries along several dimensions of forestry. These include:

- The natural potential for forest production.
- Forest-management and silvicultural traditions.
- Economic importance of forest resources.
- Ownership patterns and objectives of owners.

Throughout Europe the environmental and social benefits of forests are increasing in importance compared with industrial benefits. Unfortunately, the non-timber benefits of forests are difficult to measure and model quantitatively.

The most important factor in designing a consistent, dynamic, timber-assessment model for all of Europe is the availability of forest-resources data. Data quantity, quality, and accessibility must be added to the list of features of European forestry for which there is large variability among countries (Table I). In

Table I Distribution of basic data used in the study for key forest-inventory variables in European countries

Country	Regions within country	Owner	Species	Age or diameter class	Site Class	Standing Volume	Growth
Nordic							
Finland	X	X	X	X	X	X	X
Norway				X	X	X	X
Sweden	X	X	X	X	X	X	X
EEC-9							
Belgium	X	X	(X)	X		(X)	
Denmark			X	X	(X)		
France	X	X	X	X		X	X
FRG[a]			X	X		X	
Ireland	X		X	X	X		
Italy			X	X		X	X
Luxembourg			X	X			
Netherlands			X	X	X	(X)	(X)
UK	X	X	X	X	X	X	
Central							
Austria	X	X		X		X	X
Switzerland	X		X	X	X	X	
Southern							
Greece			X	(X)		(X)	(X)
Portugal	X	X	X	X	X	X	
Spain	X		X	(X)		X	X
Turkey			X	(X)		(X)	(X)
Yugoslavia	(X)		(X)	(X)		X	(X)
Eastern							
Bulgaria			X	X		X	
CSFR	X		X	X	X	X	
GDR[a]			X	X		X	X
Hungary			X	X	X	X	X
Poland	X	X	X	X	X	X	X
Romania			X	X	X		
Former USSR (Europe)[b]	X		X	X	X	X	X

X = satisfactorily represented in inventory.
(X) = partially represented in inventory.
[a]Prior to German unification.
[b]17 subregions.

general, most countries collect huge amounts of data about their forests. However, such data-collection programs are seldom driven by a strong link to quantitative dynamic analysis of future forest resources. These analyses make strong demands on the choice of variables that need to be measured in the forest, variables that are often different from those usually measured to obtain a simple static description of current resources. Timber-assessment analyses require structural variables and information on site quality and growth rates (for example, data on areas, growing stocks, and growth rates for different stand types, age classes, and site classes).

While some countries have rather well-suited databases for timber assessment studies (e.g. Finland and Poland), others lack basic information (e.g. Germany and Luxembourg). In general, there are some fairly widespread data problems regarding European forests and their management:

- Insufficient data on young forests and forests made up of uneven-aged stands.
- Inconsistent definitions of forest land.
- Insufficient information about the dynamics of potential forest land.
- Sparse documentation on silviculture programs.
- Outdated inventories.
- Insufficient economic information about the forests.

Clearly, The only kind of timber assessment that can be carried out for all Europe in a consistent and dynamic way is one that projects future development of growing stocks and biological harvest potentials. Thus, at this stage, it is not possible to undertake European-wide, quantitative assessments of such aspects as future economic wood supply, future non-timber benefits, and potential contribution of forests to regional development.

Besides forest-resource data quality and availability, there are other important criteria to take into account in choosing and designing a model structure:

- The model structure must be appropriate for all countries included in the assessment.
- The model structure must be equally adaptable to both high-quality and low-quality databases
- The model must be easy to calibrate and to understand.
- Results from the model must be easy to explain.

These criteria require highly aggregated forest-resources data.

The basic demand on a timber-assessment simulation model is that it must depict the dynamics of the forest under different, exogenously determined, management regimes. The model must represent forest growth in a proper way, react to changes in management programs, react to changes in the environment, and cope with changes in the landbase allocated to forest production.

Since we are dealing with different forest structures, there is a need for different model concepts, founded on different assumptions. The two model concepts we used are the unit area and its characteristics (area-based approach) and the tree and its characteristics (diameter-distribution approach).

2.1.1. *Area-based approach.* Specific forest types in the study are described by age and standing volume. A matrix defined by 10 intervals for the volume dimension and 6–15 intervals for the age dimension is defined. The forest state is then depicted by an area distribution over this matrix. Dynamics of volume increment are expressed as transitions of areas between specific fixed states in the matrix (Figure 1).

Harvest and regeneration activities are introduced through controlled transitions. Thinnings are expressed as the fraction of the area residing in a cell of the age-volume matrix that is thinned. This area is moved one step down in the volume dimension, thus simulating the harvest of the difference in volume between the cells, whereupon the area grows in a normal way. An area unit that is clear-cut is moved to a bare-land class, the transitions out of which are controlled by a "young forest" coefficient. This coefficient can then be regarded as expressing the intensity and quality of regeneration efforts. A more extensive discussion of the area-matrix model and its characteristics is found in Sallnäs.[6]

a. *Forest types.* An age-volume matrix is established for every forest type. Here, the concept of forest type is used for a stratum that can be defined by country, geography, owner, forest structure (high forest, coppice), site class, and

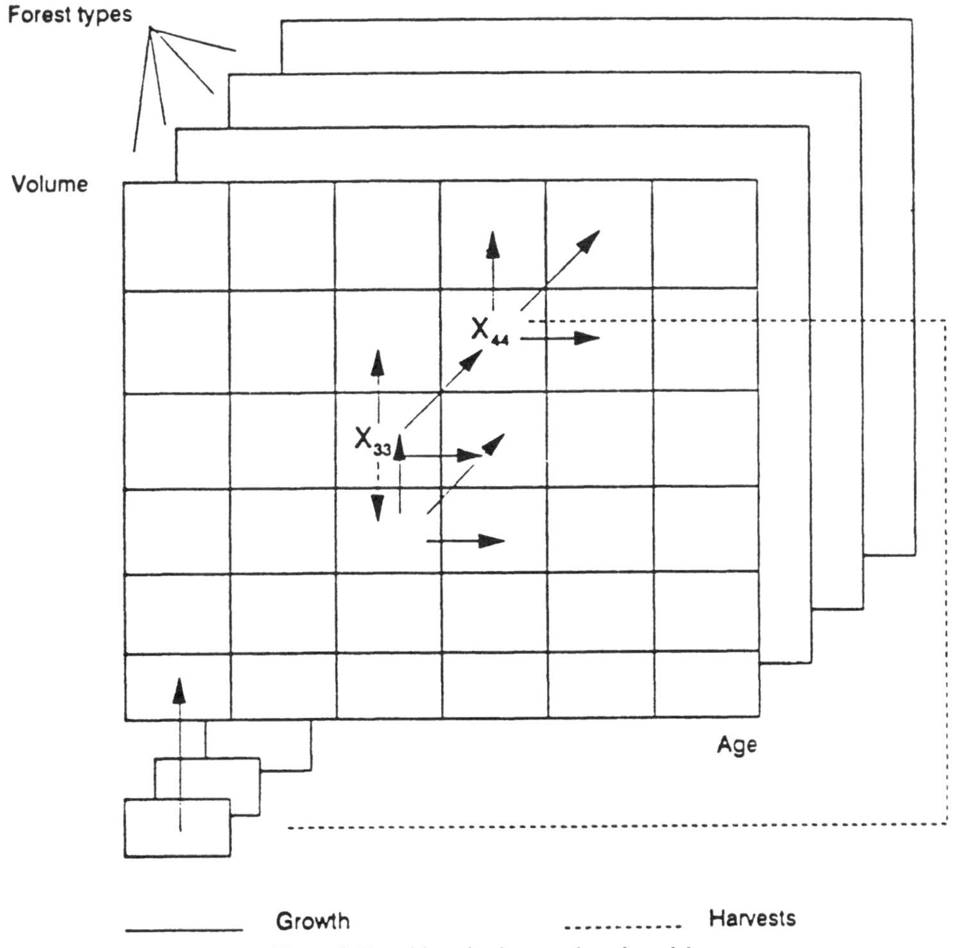

Figure 1 Transitions in the area-based model.

species. The level of aggregation into forest types is of course dependent on available data.

The number of forest types used in our study ranges from 2 to 130 for individual country. A forest type is distinguished if proper data can be found to provide the state-descriptive parameters for the forest type. A minimum demand is that the total area of the forest type must be separable into age classes for which areas and standing volumes are available.

b. *Estimation of the growth model.* Three sets of parameters must be estimated or given for the model:

(1) Parameters describing the volume distribution of the forest state, such as area and standing volume, including data about the external changes to the initial conditions, such as changes in the forest landbase.

(2) Parameters describing the biological dynamics, such as growth and site quality.

(3) Parameters describing management activities and external factors influencing dynamics.

α. *Volume distribution.* The forest-type definitions imply that only figures for area and standing volume are available at the age-class level. This means that to be able to use the model, a procedure has to be applied to produce a distribution area over volume for each age class. When advancing this procedure, two assumptions are made:

(1) The standard deviation in relation to the mean volume per hectare for different areas with similar types of forest is approximately the same. Values for the coefficient of variation that were available for some countries were extrapolated to other countries.

(2) The variance in volume per hectare increased with age. The relation between the variance and age must be established.

When the area distribution over volume classes is calculated, three variables are used: (a) the mean volume per hectare, (b) the coefficient of variation in volume per hectare, and (c) the correlation between volume per hectare and age or transformations of age. The calculation is performed in four steps. Calculate the variance in volume per hectare, using mean volume per hectare and the coefficient of variation:

$$s^2 = (mv \times cv)^2,$$

where cv is the coefficient of variation, mv is the mean volume per hectare, and s^2 is the variance in volume per hectare. Calculate the conditional variance given mean age:

$$s_{ma}^2 = (1 - r^2) \times s^2,$$

where s_{ma}^2 is the variance in volume per hectare given mean age and r is the coefficient of correlation between age and volume per hectare. Calculate the ratio of variance to age, and use this ratio to calculate the variance in each age class:

$$k = s_{ma}^2/ma.$$

The variance in age class i is then

$$s_i^2 = k \times ma_i.$$

The class limits for the volume classes are calculated using the largest volume per hectare plus three times the largest standard deviation as the class limit for the largest volume class. This span is then divided into a sequence of volume classes of increasing width. The distribution of area over the volume classes is calculated using the mean deviation and the standard deviation of volume in each age class and a modified normal distribution. After analyzing the available data it was decided to use 1n (age) as a transformation of age in the calculations involving the correlation between age and volume.

β. *Biological dynamics.* The percent volume increment is estimated with functions of the following type:

$$I_v = a_0 + \frac{a_1}{T} + \frac{a_2}{T^2},$$

where I_v is the five-year volume increment in percent of the standing volume, T is the total stand age in years, and a_0, a_1, a_2 are coefficients.

The functions are estimated from data on age and percent volume increment. The percent volume increment is calculated from data on volume increment and standing volume (cubic meters per hectare). This means that each function is associated with a series of standing volume over age. Using this method, a distribution over volume classes is created in the matrix. Consequently, the mean volume in the age-volume cell will deviate from the mean volume series. Accordingly, the percent volume increment will also deviate from the value given by the function, which means that some correction must be made. The correction is made according to

$$I_{va} = I_{vf} \times \left(\frac{V_m}{V_a}\right)^{\beta},$$

where I_{va} is the five-year percent volume increment for actual standing volume, I_{vf} is the five-year percent volume increment given by the function, I_a is the actual standing volume (cubic meters per hectare), and I_m is the standing volume (cubic meters per hectare) from the mean volume series. The relationship between the relative standing volume and the relative volume increment is described by the parameter β. From studies of this relationship in yield tables and data available for this study, the value of the parameter ranges from 0.25 to 0.45, depending on species, site classification, and the type of data used to construct the yield tables.

γ. *Management activities.* Management is controlled in two levels in the model. First, a basic management program is defined for each forest type. In this program the activities thinning, final felling, and regeneration are included. Thinnings are expressed as a percent of growth in each forecast period. This percentage of growth is then converted in the forecasting model to percentage of the area in a cell to be thinned. The thinning percentages are extracted mainly from yield tables and vary with age, species, and site.

Thinning programs can be expressed in the model by the algorithm

$$A_t = a + b \times \left(\frac{L}{c}\right)^d,$$

where A_t is the proportion of the volume increment that is thinned, L denotes the number of the age class, and a, b, c, d are parameters that can be changed.

The final felling in forests with even-aged stands is performed using stand age as a criterion for harvest. First, an age is set when final felling can occur. Then, the felling profile for the age classes above this age limit are defined. The amount of final felling in each forecast period (usually five years) is expressed as a proportion of the area in each cell. The following algorithm is used:

$$L \geq c;\ A_f = a + b \times (L\text{-}c),$$

where A_t is the portion of area that will be felled in one forecast period, L is age class number, c is the first age glass where final felling is performed, and a and b are coefficients. The coefficients a and b vary depending on species, site class, and other information about silviculture regimes.

The regeneration intensity is expressed by a coefficient controlling the transition rate from the bare-land class to the ordinary matrix. The values used range from 0.4 to 0.9 depending on region and species.

These expressions correspond to an ideal management program. If these programs are applied at an aggregated level, the resulting cutting profiles over time will be quite uncontrolled, since in many cases the present forest state does not correspond to ideal management. Therefore, a second level for defining management programs is introduced. Here, a total harvest level, differentiated by species groups and type of harvests (e.g. thinning, final felling), can be prescribed. In each period, the activity structure defined by the handbook (or ideal) program is shifted upward or downward to meet the prescribed cutting levels.

2.1.2. *Diameter-distribution approach.* In the diameter-distribution model, the basic entity on which the description of the forest is based is the individual tree instead of the forest area. The state of the forests belonging to a forest type is described by the distribution of stems over a set of diameter classes. In turn, each diameter class is associated with a figure for mean volume per stem. Dynamics are introduced via transitions of stems between the diameter classes (see Figure 2). In this case the different forest types are defined by country, region, owner, forest structure (e.g. high forest, coppice), and species. The use of site class as an additional separating variable was not possible because supporting data were not available.

The estimation of the model is generally quite straightforward. For every diameter class in the basis data sets, data on number of stems, total volume, and total growth are available. Thus, the transition probability between two subsequent classes can be calculated as

$$P_{i,i+1} = \frac{g_i}{v_{i+1} - v_i},$$

where $P_{i,i+1}$ is the probability for a stem to move from diameter class i to i + 1, and g_i and v_i are the growth and volume per stem, respectively, in diameter class i.

One major problem in estimating a model of this kind is the lack of infor-

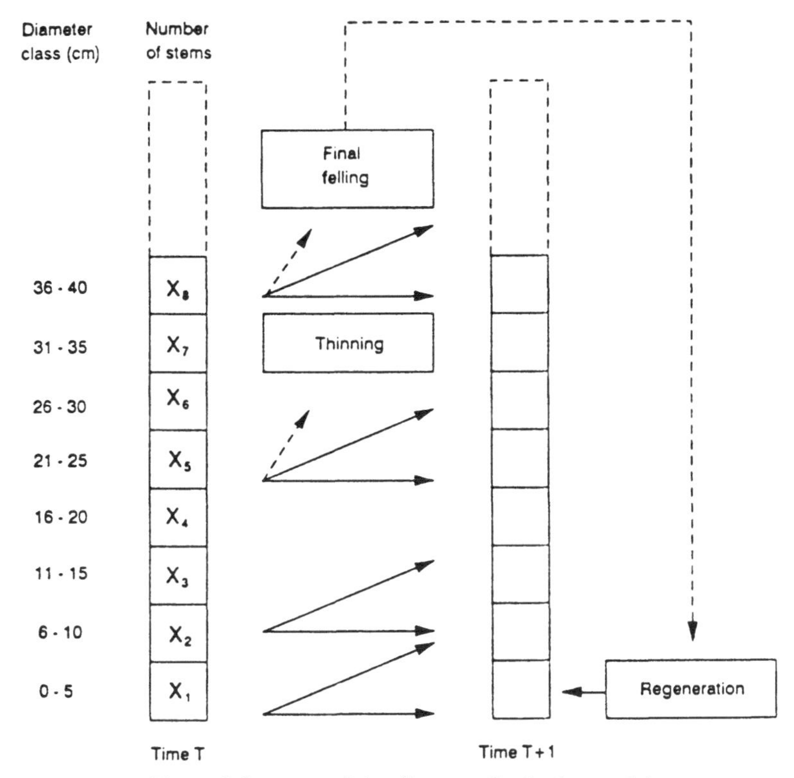

Figure 2 Structure of the diameter-distribution model.

mation about regeneration rates. Regeneration in such models must result from the final felling of one stem in a specific diameter class. Since regeneration coefficients are crucial to the properties of such a model, we decided to establish the coefficients in a way that assures stability of model results. This was accomplished by using coefficients that, together with the basic transition probabilities and the management program defined, yield a transition matrix in which the largest eigenvalue equals 1.0. For a more detailed discussion, see Houllier.[7,8]

2.1.3. *Simple approach.* We tried to use the same basic model structures for all countries in Europe. However, due to data problems this was not possible for some countries. The forest inventory of Greece and parts of Yugoslavia is described in terms of diameter-class distributions. Unfortunately, the data encompass only a narrow range of diameter classes, with no information about corresponding growth rates and standing volumes. For Turkey, standing-volume information did not match the reported forest areas. Thus, in these forests, it was not possible to use either an area-matrix approach or a proper diameter-distribution approach, so the analyses were made in a very approximate way. Potential harvests were determined as a percentage of standing volume and growth rates as initial growth percentages multiplied by a factor that depends on the relations between actual volume and initial volume, according to the formula

$$IV_t = IV_0 \left(\frac{V_t}{V_0}\right)^{\alpha},$$

where IV_t and IV_0 are growth percentages at time t and 0, respectively; V_t and V_0 are standing volumes at time t and 0, respectively, and α is a coefficient differentiating species groups and countries.

In the simulation runs using this model, two basic objectives were pursued: a steady harvest level over time and an increasing growing stock over time. As a result, for Greece, Turkey, and Yugoslavia we produced only one basic simulation (instead of three as for all the other countries) of the future wood supply as affected by silviculture and timber harvest.

2.2. *Forest Structures*

Three different forest structures can be distinguished in the forest-resource data sets: (a) even-aged high forests; (b) uneven-aged high forests; and (c) coppice. Coppice is regarded here as a forest structure more than a regeneration form. There are several different forms of coppice; for example, in the Italian national inventory[9] four different types are distinguished: simple coppice, composed coppice, coppice with standards, and transition forest which is an intermediate form between coppice and high forest.

To model the development of these different structures, the following basic criteria were used for allocating different forests to the different model concepts. In those rare cases where both a diameter-distribution approach and an area approach would have been possible, we regard simple coppice as even-aged forest, consequently modeled with the area approach, while other coppice types are regarded as uneven-aged forest, and consequently modeled with the diameter-distribution approach. As noted above, the forests of certain countries or parts thereof are described in a way that make it more or less impossible to model a more sophisticated stimulation. For these cases, the so-called simple model has been used. For most countries in Europe, however, an area-based modeling approach was used (Table II).

2.3. *Modeling Impacts of Air Pollutants*

In modeling the impact of air pollutants, a number of different factors have to be considered. These factors are illustrated in Figure 3.

The first step in the modeling process was to estimate the current and future depositions that affect European forests. The database generated by the IIASA Forest Study and discussed earlier (box 1 in Figure 3) was employed by the RAINS model[10] (box 2 in Figure 3) to calculate estimates of air pollutant deposition distribution on forests. The RAINS model emphasizes on the transboundary aspects of air pollution in Europe with the principal aim to present a spatial and temporal overview. In estimating the future depositions of Sulfur and Nitrogen, the current international agreements to reduce air pollutants were employed (The Helsinki Protocol from 1985 and the Sofia Protocol from 1988). Thus, by merging the Forest Study database (box 2 in Figure 3) we obtained the current and future (year 2000/2005) distributions of depositions over the European forest resources (box 3 in Figure 3). These distributions are not presented here, but are presented in detail by Nilsson and Posch[11] and Nilsson.[2]

The second task was to calculate each country's distribution of forests over several sensitivity classes with respect to depositions (box 4 in Figure 3). The sensitivity classes are based on the capability of forest soils to buffer against

Table II Type of model used in different European countries.

Country	Area approach	Diameter approach	Simple approach
Nordic			
Finland	x		
Norway	x		
Sweden	x		
EEC-9			
Belgium	x		
Denmark	x		
France	x	x	
FRG	x		
Ireland	x		
Italy	x	x	
Luxembourg	x		
Netherlands	x		
UK	x		
Central			
Austria	x		
Switzerland	x		
Southern			
Greece			x
Portugal	x		
Spain		x	
Turkey			x
Yugoslavia	x		x
Eastern			
Bulgaria	x		
CSFR	x		
GDR	x		
Hungary	x		
Poland	x		
Romania	x		
Former USSR 17 subregions	x		

Thus, a major effort in developing the Timber-Assessment Model has been the generation of a consistent database for all of Europe.

acidification resulting from the deposition of sulfur and nitrogen compounds. Specific critical loads and target loads of sulfur and nitrogen depositions have been assigned to the individual sensitivity classes. The critical load can be said to mean a "quantitative estimate of the exposure to one or more pollutants below which significantly harmful effects on specified sensitive elements of the environment do not occur according to present knowledge."[12] Target levels are based on the critical loads, and developed in the light of possible legal, technical, ecological, economic, and political concerns. The idea in setting target levels is that they will form the basis for negotiating internationally, acceptable emissions reduction strategies. Critical loads for such negotiations have recently been published by the UN-ECE.[13]

The target loads for sulfur and nitrogen deposition for the site sensitivity classes employed by the IIASA Forest Study are presented in Table III. The sulfur depositions are based on Chadwick and Kuylenstierna,[14] and the nitrogen on UN-ECE[15] and Nilsson and Grennfelt.[12] By combining the information on the distribution of pollutants over forest areas (box 3 in Figure 3), with the target

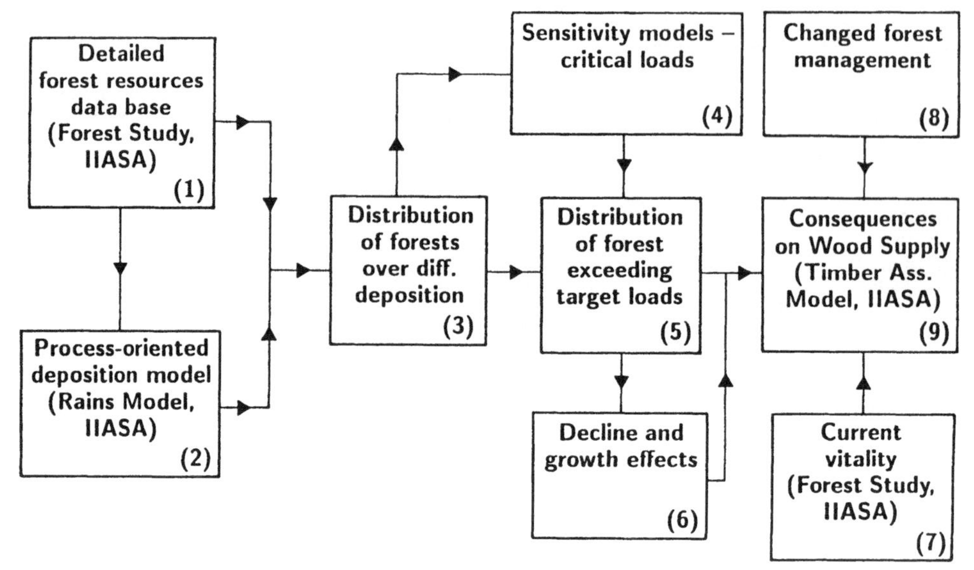

Figure 3 Components of the modeling of air pollution impacts.

Table III Target loads for sulfur and nitrogen deposition for the site sensitivity classes used in the forest decline scenarios. Data are grams of substance per square meter per year.

	Coniferous			Deciduous		
Substance	Low	Medium	High	Low	Medium	High
Sulfur[a]	2.0	1.0	0.5	4.0	2.0	1.0
Nigrogen[b]	1.5	1.0	0.3	2.0	1.2	0.5

[a]Target loads set by Chadwick and Kuylenstierna [14] based on critical loads set by UN-ECE.[15]
[b]Target loads for nitrogen are the same as critical loads set by UN-ECE.[15]
The proportions of each country's forest area in different sensitivity classes are presented in Table IV.

loads in Table III and distribution of the forest over different sensitivity classes in Table IV (covering box 4 in Figure 3), we get a distribution of forests with depositions exceeding the target loads (box 5 in Figure 3).

The detailed information on the extent of forest area with depositions exceeding target loads in individual countries and regions is presented in Nilsson[2] and Nilsson *et al.*[3] Aggregated data on the extent of forest area with depositions exceeding target loads is available in Table V. It can be seen from Table V that the current international agreements (valid end of 1991) on the reduction of emissions will improve the conditions of European forests only marginally, even if the reductions are carried out in reality. Still, with the current agreements, the situation will worsen in southern Europe. The aggregation of countries into regions is presented in Figure 4.

The key issue is the effects on forests at deposition rates exceeding the target loads (box 6 in Figure 3). To estimate the decline effects under such conditions we have used a model and field data developed, respectively, collected in the former GDR as a platform. The data and the model are described in detail by

Table IV Proportions of each country's forest area in different sensitivity classes. Data are percentages of total forest area.

	Coniferous forest area			Deciduous forest area		
Country	Low	Medium	High	Low	Medium	High
Albania	50	50	0	76	24	0
Austria	14	54	32	67	33	0
Belgium	0	100	0	23	73	4
Bulgaria	3	71	26	66	29	5
CSFR	20	45	35	71	29	0
Denmark	68	32	0	83	17	0
Finland	1	10	89	0	7	93
France	37	53	10	79	12	9
FRG	38	49	13	29	71	0
GDR	23	72	5	91	9	0
Greece	88	10	2	63	28	9
Hungary	100	0	0	100	0	0
Iceland	17	42	41	28	21	51
Italy	41	44	15	47	46	7
Luxembourg	0	100	0	47	53	0
Netherlands	79	21	0	100	0	0
Norway	0	11	89	0	9	91
Poland	37	61	2	38	62	0
Portugal	17	48	35	19	80	1
Romania	22	42	36	60	29	11
Spain	78	10	12	59	27	14
Sweden	1	16	83	2	17	81
Switzerland	18	66	16	23	20	57
Turkey	75	23	2	77	22	1
UK	20	15	65	76	12	13
Yugoslavia	24	61	15	36	57	7
USSR	58	24	18	77	16	7

Table V Exposure of European forests to significant amounts of air pollutants. Data for sulphur and nitrogen are percentages of the total forest area where target loads for the pollutants are exceeded. Data for ozone are based on diurnal concentration distributions April–September 1986.

	Region					
Pollutant/Forest type/Period	Nordic	EEC-9	Central	Southern	Eastern	European USSR
Sulphur						
Coniferous						
1985	59	88	98	62	98	27
2000	48	76	93	84	98	21
Deciduous						
1985	19	34	50	18	84	4
2000	7	24	46	40	76	3
Nitrogen						
Coniferous						
1985–2000	75	83	100	34	76	53
Deciduous						
1985–2000	52	55	86	21	47	36
Ozone						
1985–2000	1–2 × CL	1.5–2.O × CL	2.0–2.5 × CL	n.a.	1.5–2.5 × CL	?

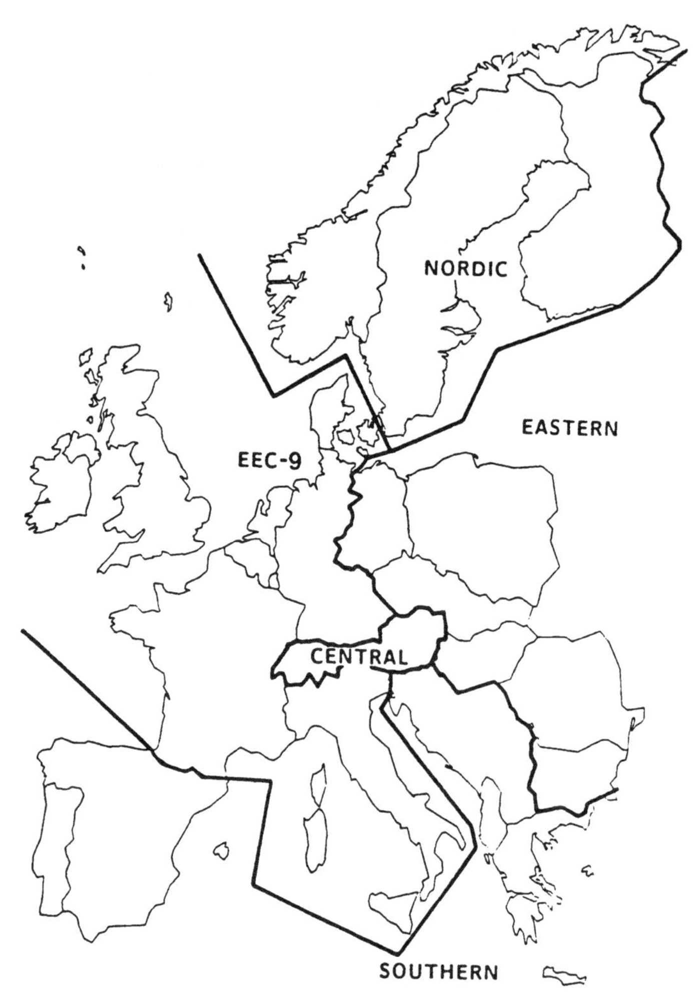

Figure 4 Aggregation of countries into regions.

Bellmann *et al.*[16] In this chapter only the structure and layout of the German system will be discussed. The basic concept for the model is a dose-response approach based on research results in cancer epidemiology. The following assumptions were considered for the development of the model concerning the loss of foliage caused by air pollutants:

- The observed loss of foliage is caused by a direct impact process (air pathway) as a cumulative process.
- The foliage decline caused via the soil pathway is a result of a cumulative and filtered dose of air-pollution concentrations. The filtered size of the dose depends on the stands capability to resist the actual doses. The degree of this dose compensation capability depends on factors such as site characteristics, mixture of pollutants, soil nutrient level and the water supply.
- The dose response is further increased by water stress and extreme temperatures.

Field data has been collected since 1947, 1962, respectively, at eight different experimental fields in former GDR. In the regions studied the loss of foliage is mainly caused by sulfur and nitrogen. The ozone concentrations have been moderate in the region. The dose response calculation for the foliage decline is schematically illustrated in Figure 5.

Thus, the model has the SO_2-concentration or S-deposition as an entry-parameter. But as the basic data—used in the analyses and described by Bellmann *et al.*[16]—illustrates, the target loads for nitrogen have also been reached or exceeded in the experimental fields. The nitrogen effects on the loss of foliage is taken into account by the changed dose compensation capability in the model.

However, the nitrogen deposition also influences the growth rate of the trees. This has been analyzed by Hofmann *et al.*[17] and is principally illustrated in Figure 6. From this study it can be concluded that:

- the nitrogen depositions compensate the individual decline effect of sulfur depositions during the enrichment and, to some extent, during the saturation stages of a stand; and
- during the damage and dissolution phase, the combined effects of sulfur and nitrogen depositions lead to an increase in decline, as compared with the individual effects of sulfur and nitrogen.

The German model (based on the field data) has been used to illustrate the combined effects on the growth rate by depositions of sulfur and nitrogen. The results are presented in Table VI. As earlier demonstrated (Table V), most of the European forests have nitrogen depositions close to, or above, the target loads for nitrogen. And it can be concluded that the dose-response calculations on the impact of air pollutants carried out by the German system takes the combined effects of sulfur and nitrogen depositions into account.

Based on the German system, there has been a possibility to estimate a damage cycle, which is presented in Table VII. Again, it should be underlined that sulfur deposition is used as an entry parameter in this table, although the effects deal with the combined effects of depositions of both sulfur and nitrogen exceeding the target loads. The damage cycle produced by the German system is supported, from a result viewpoint, with other studies discussed in Nilsson *et al.*[3]

Growth effects are linked to the loss of foliage (i.e. decline classes). We have estimated a set of growth effects, expressed in relation to undisturbed growth, according to yield tables for different decline classes (Table VIII). Again, it should be stressed that sulfur deposition is used as an entry parameter in the table, although the effects illustrate the combined effects of nitrogen and sulfur depositions exceeding the target loads.

Many studies have been conducted on the relation between growth effects and the loss of foliage. Literature reviews on these studies are presented in Nilsson and Posch[11] and Nilsson *et al.*[3] These reviews reveal support for the growth effects presented in Table VIII.

To estimate the effects of air pollutants on the future wood supply by using the timber assessment model, earlier described, information on the current vitality has to be used as a starting point (box 7 in Figure 3). As a basis for this, the results of monitoring the current vitality according to the UN-ECE methodology[5] have been employed. This methodology is based on the criteria loss of foliage. The data on defoliation can be used as a starting point because the loss of foliage is used, both as the main current vitality criterion, and as a key parameter in the quantification of the effects attributed to air pollutants.

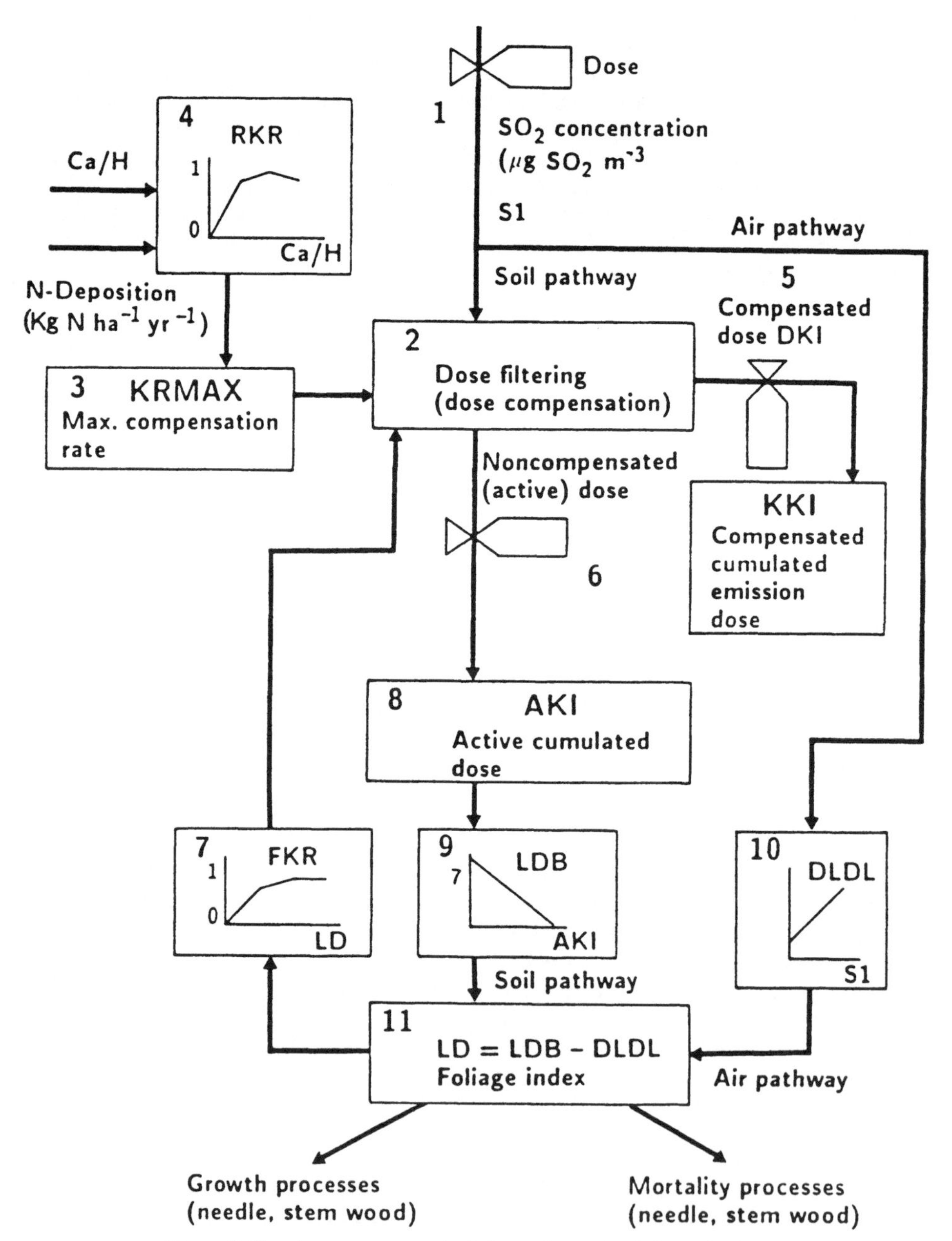

Figure 5 The dose-response calculation scheme for foliage decline.

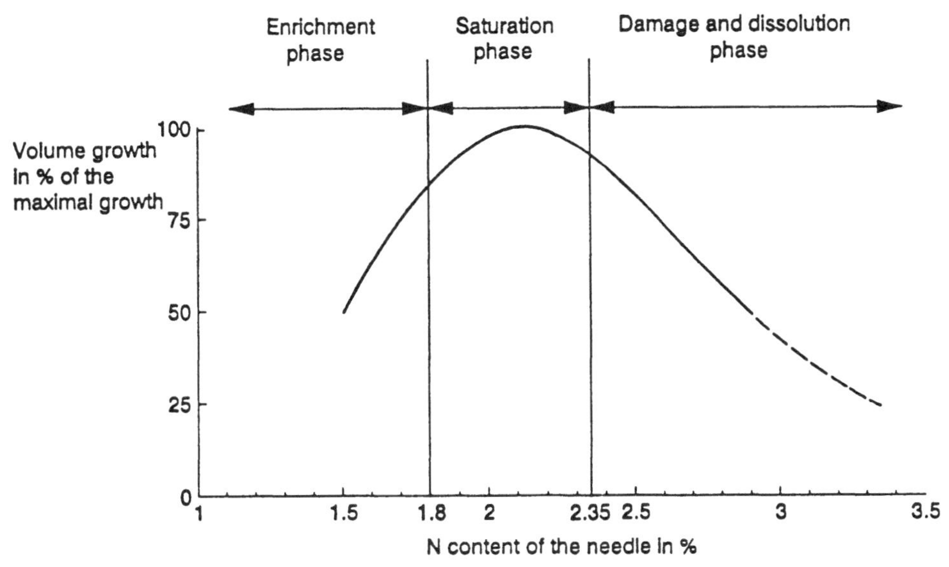

Figure 6 Volume growth in relation to the N content of the needles. Source: Hofmann *et al.*[17]

Table VI Growth responses for different test sites

Initial nitrogen content	SO_2 concentrations ($\mu g/m^3$)	Growth response during enrichment stage	Decline effects dominated by	Decline stage starts after no. of years
Low (1.2%)	72(medium)	Extremely strong effects	SO_2	15–25
Medium (1.5%)	116 (high)	Strong positive effects	SO_2	8–10
Medium (1.6%)	143 (high)	Strong effects for high N dep.	SO_2	5–10
Medium (1.6%)	18 (low)	Strong effects for high N. dep.	N	10–20
Optimal (1.8%)	69 (medium)	Limited or no effects	Combined N, SO_2	10–15
Optimal (1.8%)	41 (medium)	Limited or no effects	Combined N, SO_2	10–15

The current degree of defoliation can also be said to reflect the accumulated historical stress to forests. The detailed information on current defoliation in Western and former Eastern Europe is presented by Nilsson *et al.*[3] and for the former European USSR by Nilsson *et al.*[4]

The silviculture (forest) management has to be changed in relation to the deposition rates of air pollutants (box 8, Figure 3). The objectives of these changes are to increase the stand vitality and delay the decline process. Several research organizations have been engaged by the Forest Study to formulate explicit silvicultural responses to the decline. Examples of such measures are intensified thinnings, shortened rotation periods and changed species composi-

Table VII Damage cycles for different depositions of sulfur on middle-aged coniferous stands. Values are year-stands remaining in each decline class

Decline class[a] Sensitivity class	Deposition class (g S m^{-2} yr^{-1}) 0.55–0.99	1.0–1.99	2.0–3.99	4.0–5.99	6.0–7.99	>8
0–10%						
Low sensitivity	[b]	60	30	6	1	p.t.
Medium	50	20	10	p.t.	p.t.	p.t.
High	40	10	p.t.	p.t.	p.t.	p.t.
10–25%						
Low sensitivity	[b]	20	12	5	3	1
Medium	25	20	8	3	2	p.t.
High	20	20	8	p.t.	p.t.	p.t.
25–60%						
Low sensitivity	[b]	25	13	12	8	7
Medium	25	20	12	7	5	4
High	20	20	12	5	1	p.t.
>60%						
Low sensitivity	[b]	[b]	[b]	10	7	53
Medium	[b]	10	8	6	4	3
High	[b]	7	6	4	3	2
Number of years to death						
Low sensitivity	[b]	[b]	65	30	17	11
Medium	[b]	70	38	16	11	7
High	[b]	57	26	9	4	2

[a]Decline classes are delimited by defoliation percentages.
[b]Data do not allow any estimate.
p.t. = passed through this decline class.

Table VIII Damage cycles for different depositions of sulfur on middle-aged coniferous stands. Values are year-stands remaining in each decline class

Decline Class[a] Sensitivity class	Deposition class (g S $M^{-2}yr^{-1}$) 0.55–0.99	1.0–1.99	2.0–3.99	4.0–5.99	6.0–7.99	>8
0–10%						
Low sensitivity	-	100	100	100	100	-
Medium	100	100	100	-	-	-
High	100	100	100	-	-	-
10–25%						
Low sensitivity	-	100	100	100	100	-
Medium	100	100	100	100	100	-
High	100	100	100	-	-	-
25–60%						
Low sensitivity	-	67	60	50	63	60
Medium	-	65	63	58	69	63
High	80	63	67	64	76	-
>60%						
Low sensitivity	-	-	13	25	25	25
Medium	13	13	14	20	19	23
High	13	13	15	15	5	5

[a]Decline classes are delimited by defoliation percentages.
[b]Data do not allow any estimate.
p.t. = passed through this decline class.

tion. The required, quantified silvicultural measures are presented by Nilsson *et al.*[3] and will not be repeated here.

Thus, to adjust or adapt the timber assessment model described in Section 2.1 for decline conditions attributed to air pollutants, all of the boxes in Figure 3 and discussed above must be considered (box 9 in Figure 3). The manner in which the discussed decline information is presented in the formal timber assessment model is shown in Figure 7. Thus, the basic state description of the forests was expanded to include decline classes, sensitivity classes and deposition patterns. In addition, the transition rates were made changeable over time due to the depositions. The original management programs and the regeneration phase in the model were altered in correspondence with the detailed data presented by Nilsson *et al.*[3]

3. RESULTS

The model described in section 2 has been used to simulate the potential biological wood supply for each individual country (and subregions employed in individual countries). In total, seven different scenarios (or simulations) have been conducted. Each scenario has a time horizon of 100 years, with 1985 as the starting point for Western and former Eastern Europe, and 1988 for former European USSR. The seven detailed scenario results are presented by Nilsson *et al.*[3] for Western and former Eastern Europe, and by Nilsson *et al.*[4] for former European USSR. In this chapter, we will only present aggregated results.

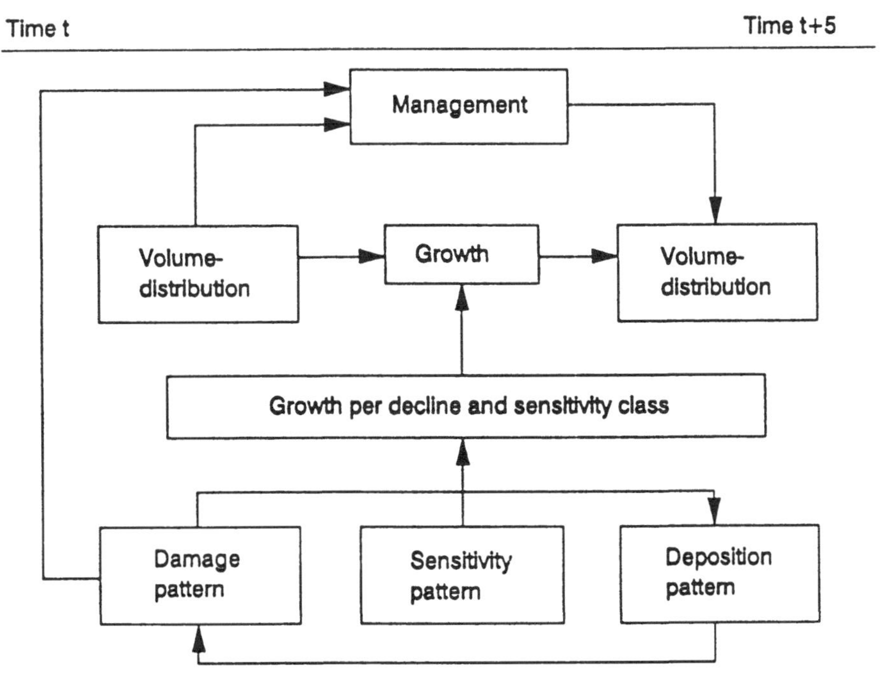

Figure 7 Schematic representation of the decline model.

Concerning the decline effects, it is important to underline the following:

- We have calculated the decline effects based mainly on sulfur and nitrogen emissions, without considering the full combination of effects from other pollutants.
- We have assumed full decline effects to the year 2005, after which growth rates will recover slowly over time. This is a result of the impossibility to estimate the emissions after 2005 in a secure quantitative way.
- The recovery rates in growth after year 2005 range from 10 to 40 years, depending on site conditions and historical deposition rates.

Therefore, we say that our analytical implementation of the effects of air pollutants on forest resources is conservative.

However, this procedure will still influence the potential biological wood supply through the entire simulation period of 100 years.

The average sustainable potential harvest per year for the next 100 years on an aggregated level is presented in Table IX. In this table, the sustainable harvest without any air pollution is compared with the effects of air pollutants, according to the current international agreements on the reduction of air pollutants, and the conditions specified in section 2 for the calculations. These harvest levels are compared with actual removals at the end of the 1980s.

The regions most affected by forest decline attributed to air pollutants are the Eastern and Central regions. The losses correspond to 27, respectively, 23 percent of the harvest potential without any pollution effects. The total loss in Western and former Eastern Europe is estimated to be about 85 million m^3 per year averaged over 100 years. This corresponds to be about 16 percent of the harvest potential under conditions of no air pollution effects. The corresponding figures for the former European USSR are 35 million m^3 per year and about 11 percent. Thus, the average conditions, from an air pollution viewpoint, are better in the former European USSR than in the rest of Europe. Although, basic data on the air pollution components in the calculations is weaker for the former European USSR.[4] In addition, this region has severe subregional forest decline problems caused by air pollutants, which are not seen in the aggregated figures.

It can also be seen from Table IX that without any air pollution effects there is the potential to increase the sustainable harvest with about 175 million m^3 per year in total Europe, in comparison with the current harvest. This potential deals

Table IX Potential average harvest levels per year for 100 years. Expressed in million cubic meters

	Potential harvest in million m^3 o.b./year for all species and average for 100 years		
	without air pollution effects	with air pollution effects	removals in 1987
Nordic	155	144	121
EEC-9	150	126	109
Central	25	19	22
Southern	78	71*	72
Eastern Europe	126	92	100
European USSR	321	286	257
Total Europe	855	738	681

*Decline effects for Spain are not included due to lack of basic data.

only with the sustainable biological potential and not the economic potential of wood supply. In reality, there are a number of factors which restrict the potential biological harvest, such as prices and costs, behavior of forest owners, and restrictions for non-wood benefits. These factors have not been possible to quantify in the above estimates.

The demand for forest products is estimated to increase strongly in Europe in the future. Even in the case of no future decline in European forests, Western and former Eastern Europe may face an annual roundwood deficit of some 40 million m^3 per year already by year 2010. If we account for the decline as we have estimated above, that deficit amounts to roughly 130 million m^3 per year. The demand/supply balance for former European USSR is more difficult to estimate for year 2010. The balance is strongly dependent on how fast the transition and reconstruction of the economy will take place. As the development looks now, there will not be any deficit in the supply for this region in year 2010, even if the decline effects of air pollutants are considered.

4. ECONOMIC IMPACT

The valuation of air pollutant impacts on European forests should aim to be a basis for international negotiations. In addition, the valuation should start from the society/politically economic or nationally economic evaluation.[2] It can also be concluded that there is no single economic theory capable of dealing with the valuations of all aspects of natural resources degeneration, although there is a consensus among the different European countries that the validation should be based on a sustainability approach. Nilsson[2,18] has argued that the standard neoclassical theory is not suitable for valuation of the effects of air pollutants on European forests due to:

- limitations to deal with sustainability;
- the odd effects of the discounting principle;
- unsatisfactory treatment of intergenerational equity;
- abuse of international laws;
- limitations to reflect the political and social debate and concerns in the valuations.

The basic approach to demonstrate the aggregate or the bottom-line impact of different policies and effects on the national economy is to study changes in the income accounts or GNP.[19] Nilsson[18] has illustrated that there is no consensus on how the current GNP-calculations should be adjusted to include sustainability in the valuations, either from a theoretical or applied point of view. The adjustments of the GNP-calculations carried out in the case of forest decline attributed to air pollutants in Europe are described in Nilsson.[2,18] Before presenting the aggregated results of the valuation, it should be underlined that no single concept can satisfactorily address all economic and policy questions connected to forest decline attributed to air pollutants. And no single system of economic thought is right or wrong.

The aggregated economic impacts based on multiple-use forestry and primary forest products industry is presented in Table X. Thus, this valuation is based on emissions reduction under current international agreements (end 1991), and their corresponding effects averaged per year during 100 years.

Table X Economic impacts on multiple-use forestry/primary forest products caused by air pollutants

Region	Billion dollars/year
Nordic	2.9
EEC-9	7.4
Central	1.6
Southern	1.8
Eastern	8.5
European USSR	7.5
Total Europe	29.7

The control expenditures required to bring all European forest resources within the target loads discussed in section 2 have been analyzed by Shaw and Nilsson.[20] With existing control technology they found it impossible to bring all the forests within the target loads. The best result was achieved by implementing the Best Available Control Technology (BACT). The corresponding control expenditures are presented in Table XI. Implementation of BACT means a reduction by 70–80 percent of SO_2 emissions, and by roughly 60 percent of NO_x emissions. The costs for ammonia reduction are not calculated on a European scale. Klaassen[21] has presented some calculations for the Netherlands and Finland.

Thus, to save the European forests from negative impacts attributed to air pollutants, control expenditures of at least 90 billion dollars per year are required. In making this investment, most of the negative effects of air pollutants influencing other sectors of society will probably also disappear.

5. POLICY IMPLICATIONS

Based on the results produced by the IIASA Forest Study a number of Policy Exercises have been carried out in the attempt to identify the most important policy implications. These Policy Exercises are documented in detail by Duinker *et al.*[22] The following can be summarized concerning the policy implications:

- Pollutant emissions are unlikely to be controlled in Europe as rapidly and comprehensively as desirable.
- Several of the major polluting countries in Europe face an uncertain political and economic future, at least over the near to medium term.
- A host of constraints will act to prevent an easier recovery of the forest resources vitality using improved management and forest policies.
- Many policy people have a pessimistic outlook about the possibility to implement expansion of the forest landbase.

Table XI Control Expenditures in billion dollars/year

	SO_2 Emissions	NO_x Emissions
Best Available Control Technology (BACT)	50.2	39.7
Current Reduction Costs (Plan to 1995)	8.0	~8.0

- Lower prices may be seen for roundwood, whose development may result in lower incentives for intensified management and afforestation.
- Strong demand for non-wood benefits can be expected in the coming decades.

References

1. S. Nilsson, O. Sallnäs and P. Duinker, "Forest Potentials and Policy Implications: A summary of a Study of Eastern and Western European Forests by the International Institute for Applied Systems Analysis" (Executive Report, IIASA, Austria, 17 Feb., 1991).
2. S. Nilsson, (ed.), *European Forest Decline: the Effects of Air Pollutants and Suggested Remedial Policies* (The Royal Swedish Academy of Agriculture and Forestry, InterAction Council and IIASA, Stockholm, Sweden, 1991).
3. S. Nilsson, O. Sallnäs and P. Duinker, *Future Forest Resources of Western and Eastern Europe* (Parthenon Publishing Group Ltd., United Kingdom, 1992a).
4. S. Nilsson, O. Sallnäs, M. Hugosson and A. Shvidenko, *Future Forest Resources of the former European USSR* (Parthenon Publishing Group Ltd., United Kingdom, 1992b).
5. U.N., *European Timber Trends and Prospects to the Year 2000 and Beyond* (ECE-FAO, New York, NY, U.S.A., 1986).
6. O. Sallnäs, "A Matrix Growth Model of the Swedish Forest" *Studia Forestalia Suecica* **183** (1990).
7. F. Houllier, "Echantillonnage et modelisation de la dynamique dépeuplements forestiers: application au cas de l'Inventaire Forestier National" (Thèse de Doctorat Univ. Cl.-Bernard, Lyon, France, 1986).
8. F. Houllier, *Data and Models used for France in the Context of the Forest Study* (Working Paper WP-92-40 IIASA, Laxenburg, Austria, 1992).
9. Anon., *Inventario Forestale Nazionale 1985: Sintesi Metodologica e Risultati* (Ministero dell'Agricoltura e delle Foreste, Rome, Italy, 1988).
10. J. Alacamo, R. Shaw and L. Hordijk (eds.), *The RAINS Model of Acidification: Science and Strategies in Europe* (Kluwer Academic Publishers, Dordrecht, Netherlands, 1990).
11. S. Nilsson and M. Posch, *Pollutant Emissions and Forest Decline in Europe* (Unpublished Working Paper: Forest Studies. International Institute for Applied Systems Analysis, Laxenburg, Austria, 1989).
12. J. Nilsson and P. Grennfelt, (eds.), "Critical Loads for Sulphur and Nitrogen" *Miljö rapport 15* (Nordic Council of Ministers, Copenhagen, Denmark, 1988).
13. UN-ECE, "Mapping Critical Loads for Europe. United Nations Economic Commission for Europe Convention on Long-Range Transboundary Air Pollution" *CCE Technical Report No. 1* (Co-ordination Center for Effects, National Institute of Public Health and Environmental Protection, Bilthoven, Netherlands, 1991).
14. M. J. Chadwick and J. Kuylenstierna, *The Relative Sensitivity of Ecosystems in Europe to Acidic Depositions: A Preliminary Assessment of the Sensitivity of Aquatic and Terrestrial Ecosystems.* (Stockholm Environment Institute, Stockholm, Sweden, 1990).
15. UN-ECE, *ECE Critical Levels Workshop Report, Bad Harzburg, Germany.* (ECE, Geneva, Switzerland, 1988).
16. K. Bellman, P. Lasch, H. Schulz, F. Suckow, S. Anders, G. Hofmann, D. Heinsdorf and B. Kalweit, "The PEMU Forest Decline Model, Cumulated Dose Response Approach to Evaluate Needle Loss in Pine Stands under Sulfur and Nitrogen Depositions in the Northeast German Lowlands" Appendix in: Nilsson *et al. Future Forest Resources of Western and Eastern Europe* (Parthenon Publishing Group Ltd., United Kingdom, 1992a).[3]
17. G. Hofmann, D. Heinsdorf and H. H. Kraub, "Wirkungen anthropogener Stickstoffeinträge auf Produktivität und Stabilität von Kiefern-Forstökosystemen" *Beitr. Forstwirtsch.* **24**(2), pp. 59–73 (1990).
18. S. Nilsson, *Valuation of Sustainability of European Forest Resources in Respect to Forest Decline* (International Institute for Applied Systems Analysis, Laxenburg, Austria, in press).
19. OECD, "The Economics of Sustainable Development: A Progress Report" (OECD, Paris, France, 1990).
20. R. Shaw and S. Nilsson, "European Commercial Wood at Risk from Sulfur and Nitrogen Deposition" **In**: (S. Nilsson, ed.) *European Forest Decline: The Effects of Air Pollutants and Suggested*

Remedial Policies (The Royal Swedish Academy of Agriculture and Forestry, InterAction Council and IIASA, Stockholm, Sweden, 1991) pp. 119–146.

21. G. Klaassen, "Cost of Controlling Ammonia Emissions in Europe". (Paper presented at IIASA workshop "Ammonia emissions in Europe". International Institute for Applied Systems Analysis, Laxenburg, Austria, Feb. 4–6, 1991).
22. P. Duinker, S. Nilsson and F. Toth, Testing the "Policy Exercise" **In**: Studies of Europe's Forest Sector: Methodological Reflections on a Bittersweet Experience. (Submitted to *Simulating & Gaming*, 1991).

5. THE NUTRITIONAL STATUS OF DECLINING FORESTS: A REVIEW AFTER A DECADE OF RESEARCH

R. F. HUETTL

Various types of forest decline prevailing for more than a decade in Europe and Eastern North America are related to nutritional disturbances. Mg-deficiency—a truly new phenomenon in high elevation forests—represents the most widespread disorder and is detailed in depth. To explain this and other far reaching changes of the nutrient supply in many forest ecosystems, direct as well as indirect damage pathways have been suggested. It is now clear that anthropogenic impacts in combination with natural stress factors affect forests in different ways that are always site and stand specific. However, all types of forest decline associated with nutritional disturbances can generally be mitigated by appropriate nutrient changes.

1. INTRODUCTION

The health status of forest trees and stands is determined by numerous site factors such as chemical, physical, and biological soil factors, water supply, climate, weather conditions, management history as well as atmospheric deposition impacts. In this context the nutrient supply is an important evaluation parameter. Forest trees well supplied with nutrients are more resistant to stresses that affect the forest ecosystem than other trees. This is true for both biotic and abiotic influences.

Therefore the investigation of the so called "new type" forest damage was aimed at the exact determination of the health status of damaged trees. When considering the complete forest ecosystem, health (= vitality) means the sustainable ability to withstand negative environmental influences and still remain stable and productive.[1] From this viewpoint an optimal nutritional status is a prerequisite for an optimal health status. Hence, it is not surprising that the widespread new type forest damage was found to be frequently associated with acute nutritional disturbances. Of particular concern are the phloem-mobile nutrient ions magnesium (Mg) and potassium (K). In addition, the elements calcium (Ca), phosphorus (P), zinc (Zn), and manganese (Mn) occasionally play an important role. With regard to atmospheric deposition the nutrient elements nitrogen (N) and sulphur (S) are also of interest.[2–4] It is the aim of this chapter to demonstrate the relationships between the new type forest damage and the nutrient supply, to briefly discuss relevant hypotheses, to identify diagnostic tools and to propose revitalization/restabilization measures.

2. NEW TYPE OF FOREST DAMAGE

Increased environmental pollution has brought about a problem of enormous importance, concern and complexity for humanity. Air, soil, and water have been affected at least at times and places by a wide variety of toxic gases, liquids, and solid substances that individually and in combination adversively affect the environment.[5] Although interest in the effects of air pollutants on trees and forests has greatly increased in recent years, apprehension about air pollution has been expressed for a long time. Formerly, however, concern was focussed on local forest damage directly related to short-distance emission sources. The main pollutants at that time were sulphur dioxide (SO_2) and particulates (mainly alkaline dust, containing nutrients such as Mg and Ca). With increasing industrialization, enhanced combustion of fossil fuels and the "high stack policy", the air pollution problem became far more serious and complex. The recently observed forest decline is therefore called "new type forest damage".

The term "new type forest damage" comprises a number of damage symptoms which have been observed in various tree species on very different sites since the mid-1970s. However, they occurred much more intensively in the early 1980s and have since then spread quickly and can nowadays be found over large forest areas. Generally, this forest damage is thought to be related to negative impacts of air pollutants.[6]

From intensive research into the new type forest damage within the recent decade it was learned that a number of damage types, for example foliar losses which have been attributed to this new problem have, in fact, been known for a long time. However, there are a number of damage types which reflect new or new type phenomena. Interestingly enough, these are always associated with nutritional disturbances.

3. NUTRITION RELATED DAMAGE TYPES

Before discussing nutrient disturbances in damaged forests of central Europe and North America some basic remarks are of interest.

3.1. *Forest History*

In general, forest sites are poorly supplied with nutrients. The historical development of commercial forests in central Europe is marked by the circumstance that the "better" sites were utilized for agriculture for a long time. The majority of sites left for forest management were agriculturally not usable. These are sites with poor nutrient supply in geo- and topographic unfavourable areas, and/or with extreme weather conditions. Also, before modern forest management based on economic and ecological principles, manifold devastations of forest sites were the case.[7]

Anthropogenic impacts on forests are as old as mankind.[8] The first people exclusively settled on fertile deciduous forest sites. With the cutting down of these stands a relative increase of conifers occurred. When the settlements were abandoned, the soils were frequently degraded. Eventually, such areas were regenerated naturally by less demanding coniferous trees. During the middle-ages and even into the 20th century forest pasture and litter-raking (for agricultural purposes) led to a further depletion of many forest soils. In numerous cases the soils were so heavily impacted that for regeneration only undemanding tree species such as Norway spruce (*Picea abies*) and Scots pine (*Pinus sylvestris*) were at hand. Furthermore, a number of uncontrolled heavy biomass extractions (for example in Germany due to so called "reparation cuttings" to compensate claims after the first World War) without sufficient reforestation had a negative impact on site quality. Regional forest devastations were caused by heavy cuttings for various industrial purposes and for fuel wood. In 1338 the sowing of seeds from coniferous trees was carried out for the first time, and this application technique favoured a fast and widespread distribution of Norway spruce.[9] Despite the fact that Norway spruce grew well in less fertile soils and with regard to the uncomplicated regeneration by sowing, the yield performance of Norway spruce was remarkable. Also with the siviculture of Norway spruce the amounts and qualities of wood could be produced that the "old forest" never could have supplied.[9]

It is therefore not surprising that Norway spruce grows on many sites which would be much more appropriate for other tree species. With regard to the presently damaged Norway spruce forests in Germany, for example in the Black Forest, in the Fichtel Mountains, in the Harz Mountains and in costal regions, we are generally dealing with off-site stands stemming from artificial regeneration and from seeds of unknown origin.

3.2. *Diagnostic Tools*

The relations between site quality and nutritional status as well as growth rates of forest trees and stands have been studied by Wittich and Laatsch, the founders of the modern forest nutrition science,[10–13] and many other authors have investigated the relationship between nutrient supply and productivity in commercial forests. To diagnose sufficient and optimal nutrient supply in forest trees a number of tools exists. In this context chemical foliar and soil analyses are of particular importance.

3.2.1. *Foliar Analysis.* The best tool to diagnose the actual nutritional status of forest trees and stands is foliar analysis.[14–20] Ingestad and many others have shown that the nutrient content in leaves and needles adequately reflects the nutritional status of forest trees when sampling and analysis techniques are standardized.

By way of multiple foliar analytical investigations of trees by hydroponic culture experiments and fertilizer trials threshold values and/or ranges of deficiencies, sufficient and optimal supply were determined for various tree species. For the interpretation of foliar data in conifers generally the current (i.e. 1-year-old) needles of the first whorl were used. In Table I, the elements contents in 1-year-old needles are presented for the evaluation of the nutritional status of Norway spruce. Strebel,[13] Huettl[21] and others have indicated that the element contents of needles of the same age vary in relation to crown position. Furthermore, element specific increases or decreases in foliar contents of various needle year classes of a specific whorl were described.[21–24] As a rule, the contents of phloem-mobile nutrients such as N, P, K, Mg and Zn decrease with needle age whereas immobile ions such as Ca and Mn are accumulated in older foliage.

An optimal nutritional status depends not only on the uptake of adequate nutrient amounts but also on a balanced ratio of nutrient ions in the plant. These

Table I Element contents in 1-year-old needles (top whorl) for evaluation of the nutritional status of Norway spruce.[21,47,63]

Element	Deficient supply	Sufficient supply	Optimal supply
mg·g^{-1} d.m.			
N	< 12–13	13–15	>15
P	< 1,1–1,2	1,2–1,5	>1,5
K	< 4,0–4,5	4,5–6,0	>6,0
Ca	< 1,0–2,0	2,0–3,0	>3,0
Mg	< 0,7–0,8	0,8–1,0	>1,0
µg·g^{-1}d.m.			
Mn	<20	20–500	>500
Zn	<13	13–25	> 25
Cu	< 2–4	4–12	> 12
B	< 8–10	10–30	> 30

relations were detailed by Ingestad.[18,25] In greenhouse studies optimal growth of Norway spruce, Scots pine and other tree species could only be achieved when the essential nutrients were taken up in certain proportions to N. Ingestad also found that the absolute concentration of nutrient ions in the soil solution does not determine the plant uptake but rather their permanent availability. The amount of uptake is related to the specific growth rate of the plant.

In addition to these relative proportions absolute ratios of various nutrients can be utilized for describing balanced nutrition. This is particularly useful when nutritional disturbances are to be determined. In Table II some of these ratios are listed.

Table II Nutrient element ratios for the evaluation of balanced nutrition of Norway spruce based on element contents of 1-year-old needles.

Ratio	Range of balanced nutrition	References
N:P	6–12*	7,13,64,65**
N:K	1–3	3,66
N:Ca	2–20***	3
N:Mg	8–30	3,52
N:S	> 8	67
P:Zn	30–150	68
K:Ca	0,8–2,4	13
K:Mg	2,2–6,4	70
Ca:Al	>****	21
Mg:Al	>****	21

*Comerford and Fisher:[71] critical N:P-Ratio: 14–15

**Ingestad:[65] for optimal nutrition: N:P = 5
N:K = 2
N:Ca = 20
N:Mg = 20

***preliminary values

****the larger the better

3.2.2. *Soil Analysis.* With regard to the heterogeneity of forest soils, it has been difficult to develop appropriate sampling and analysis methods in order to characterize the plant available content of nutrient ions present in the rooted solum. It also has to be taken into account that the nutrient uptake is not only related to the nutrient supply in the soil but it also depends on the rooting depth, fine root and mycorrhiza development, water and air supply in the soil, on the physical and biological soil status, on the transpiration rate of the trees, as well as on other factors. Due to the rather unsuccessful efforts to find relations between the nutritional status of trees and stands and the specific nutrient supply of the soil, foliar analysis was used almost exclusively to characterize the nutritional status of forest trees/stands. During the discussion of accelerated soil acidification probably related to "acid rain"[26] it was necessary to adequately determine soil parameters in order to characterize the chemical soil status. In addition to pH measurements for which Ulrich *et al.*[26] suggested ecological buffer ranges, the analysis of exchangeable cations such as Ca, Mg, K as well as aluminum (Al) and protons (H), the effective cation exchange capacity and the base saturation were the focus of interest. To determine these parameters various extraction and percolation methods were developed, utilizing dry fine soil probes from different horizons or soil depths. By using the equilibrium soil solution technique in addition to cations anions are also measured.

To define Al toxicity the ionic Al content is of interest. Therefore, a separation of Al species by way of ion exchange is applied. With regard to the so-called acid toxicity[27] critical Ca/Al and Ca/H ratios were introduced. These ratios were obtained from hydroponic culture experiments with non-mycorrhizal Norway spruce and European beech (*Fagus sylvatica*) seedlings and have since been critically discussed. There is, however, no doubt that Al may impede the Ca and Mg uptake in the sense of an antagonist. In fact, antagonisms can be found for all exchangeable cations and anions when they are present in the soil in unfavourable proportions.

First useful relationships between foliar nutrient contents and soil chemical parameters were found by relating ratios of cations in the soil such as Mg/Al-, Ca/Al- and K/Ca-ratios with Mg, Ca and K foliar contents, respectively.[28–30] These relations were verified by Matzner *et al.*[31] also with regard to the soil solution data. Anyhow, the analysis of the ion content of the soil solution seems to be a very useful method to characterize the nutrient supply of the soil; for with this method the amount of nutrient ions is determined that is readily available to the roots. Based on this approach also the questions of nutrient element replenishment due to mineralization and silicate weathering are relegated to the background even though these processes—as well as the external element input—are of crucial importance for a sustainable nutrition of forest stands.[7,32–34]

Remarkable results were obtained by chemical analysis of undisturbed soil samples.[35–37] By utilizing this method Kaupenjohann *et al.*[38] could establish significant correlations between the chemical soil status and foliar analytical data. In this case the undisturbed soil samples were percolated with solutions that mimiced the throughfall acidity of the investigated forest stand. Horn,[39] as well as Hildebrand[36,40a,b] indicated that the chemical soil status and in particular the nutrient availability of K and Mg were influenced by the soil structure. Hildebrand proved that structure related imbalances of the soil to be the cause of K and Mg disorders in damaged forest stands. The depletion of nutrient ions from the aggregate surfaces seemed to play a most important role. Finally, Liu and

Trueby[41] applying NH_4Cl extractions to very different soils stemming from the large variety of geological substrates of southwestern Germany were able to determine a threshold value for Mg deficiency in the mineral soil (2 $\mu eq \cdot g^{-1}$ finesoil ≤2mm). This value is practically the same as that put forward by Matzner *et al.*,[31] obtained from soil solution analysis. Even though new approaches have advanced soil analysis as an important diagnostic tool an exact characterization of the actual nutrient supply of forest trees and stands is not yet possible. Foliar analysis is still the more accurate tool.

3.3. *Deficiency Symptoms*

In case of acute nutrient deficiencies the impacted trees are generally marked by typic deficiency symptoms. As various types of the new type forest damage are associated with acute nutritional disturbances, the respective deficiency symptoms were frequently described. Therefore only the most important and widespread nutrient deficiency, i.e. Mg deficiency, is detailed here.

3.3.1. *Mg Deficiency in Coniferous Trees.* In needles of conifers the usually bright-yellow discoloration starts at the tip of the leaves before it eventually spreads to the base; the symptom is therefore called tip-yellowing. The chlorosis starts in the older needle year classes and in the middle to lower crown area. Discoloration symptoms occur only on those needle or twig sides that are directly exposed to radiation. After a certain time interval the length of which is not predictable the needles turn necrotic and fall off. Should the Mg supply be improved, for example due to better soil water status, a natural regreening of yellow needles may occur.[42] Except for pines, the younger shoots are normally not affected by tip-yellowing. In case of continued Mg deficiency, discoloration progresses to the next younger needle year class when the new shoots develop in the spring.[43,44] This is related to the translocation of the relatively mobile Mg from older into the younger more productive shoots. In most cases the vigorous growth of moderately yellowed conifers is not—at least at first—reduced. Analyses of volume growth, however, indicate reduced increments in discolored conifers as compared to healthy trees (e.g. Evers[45] and Aldinger[46]). The heaviest foliar losses are often found in the middle crown area leading to the impression of a "sub-top die-back." Finally, the trees may die; this occurs frequently in combination with other stress factors such as frost, drought, insect attacks or fungal infections. Mg deficiency is observed most commonly in young but nowadays also in old trees.

3.3.2. *Mg Deficiency in Deciduous Trees.* In various deciduous trees such as European beech and oak trees (*Quercus spp.*), older leaves reveal yellow flecking in the intercostal tissue. Generally, these flecks merge rapidly and subsequently cause a complete yellowing of the foliage except for the ribs which remain green. At further stages dark brown necrotic spots develop, eventually resulting in leaf mortality. Ultimately, the dead leaves fall off. However, yellow foliage may also drop, and thus a premature loss of older leaves is frequently associated with Mg deficiency in deciduous trees. Usually, the younger leaves remain green for a certain time.

4. NUTRITIONAL DISTURBANCES IN DECLINING FORESTS

Forest damage types associated with nutritional disturbances (i.e. deficiencies, imbalances) may be grouped into specific damage types based on site factors and stand conditions. As mentioned above, the most dominating nutritional disturbance is Mg deficiency in Norway spruce on poor acid base but relatively K-rich substrates. Only a brief discussion of this and other nutrient disorders in declining forest is presented here, as they have been recently discussed in great detail elsewhere.[47]

4.1. *Mg Deficiency*

Tip-yellowing of Norway spruce has been observed since the mid-70s in mountain forests of central Europe and, to a lesser degree, of those in northeastern USA; this has been termed "high elevation spruce disease" or "mountain yellowing". From the viewpoint of forest nutrition it has developed gradually. The relatively sudden and widespread appearance of visual symptoms in the early 1980s is probably best explained by extreme weather perturbations such as frequent drought periods during the end of the 1970s and the early 1980s. Stagnation in the development of yellowing since the mid-1980s as well as the occasionally natural regeneration of tip-yellowed spruces is correlated with years of more favourable precipitation regimes.[48,49] The phenomenon that healthy trees are found next to damaged individuals is probably related to micro-site differences such as differences in water supply and Mg availability as well as to genetic variations of the Mg uptake.[47] Mg deficiency is found in spruce only when it is also prevailing in the rooted solum. In case of an insufficient Mg supply high Al contents may also hinder Mg uptake. Mg deficiency is practically only found in forest stands that are marked by a low atmospheric Mg input. In addition to numerous German forest areas such as the Black Forest, Bavarian Forest, Fichtel Mountains, Harz Mountains and Thuringian Forest, mountain yellowing was also observed in Austria[50] (e.g. Bohemian Forest); France[51] (e.g. Vosges Mountains); Belgium[52] (e.g. Ardennes Mountains); in the CSFR[53] (e.g. Ore Mountains); and in Poland (e.g. Iser Mountains, personal communication). Also in North America[54] (USA, e.g. northern Appalachians) and Canada[55] (e.g. Lower Laurentians in Quebec)[55] natural forests (e.g. *Picea rubens*) and forest plantations (*Picea abies*) are nowadays affected by Mg deficiency.[56] However, the intensity of Mg deficiency in the uncommercial forests of North America is far less than in the commercial forests of Europe. More recently, Mg deficiency was also found in other tree species as well as at lower sites. There, however, Mg deficiency is frequently combined with other nutrient disturbances, such as P and/or K deficiency.

4.2. *K Deficiency*

K deficiency has been known for a long time in various tree species and occurs naturally in calcareous soils as well as in organic substrates. Since the beginning of the 1980s this syndrome has spread in e.g. southern Germany, France and Switzerland.[6,21] The reasons for this development are unclear. However, K deficiency is new in Norway spruce on acid clay-rich soils. Until recently stands on these soils have been found to be well supplied with K. This damage type, which has also been observed since the early 1980s, has supported widespread needle cast fungi epidemics. Also for this type of decline the cause-effect relationships have not yet been fully established.

4.3. *Combined Mg and K Deficiency*

In addition to single Mg and K deficiency situations in stands on poor acid base substrates K and Mg deficiency can also occur as a combined disorder. This damage type which is more frequently found in middle and northern Germany is also linked with specific site conditions. Again the chemical soil status of the rooted solum, in particular the Mg and K supply, plays a dominant role. Single Mg and K, as well as combined K-Mg deficiency, may occur in combination with optimal N nutrition as well as with N deficiency. N deficiency, however, may impede the development of typical deficiency symptoms. On the other hand optimal N nutrition stimulates the appearance of these symptoms. In forests marked by these damage types and receiving high N input rates, in contrast to the usual decrease of N contents with needle age, N may accumulate in older shoots. This is another new finding and leads to extremely unbalanced N:Mg and N:K ratios in Mg and/or K deficient trees. In these deficiency types, Ca nutrition varies. Frequently, however, Ca nutrition in Mg or K-Mg deficient stands is poor. In this context a further new finding was made. In Norway spruce stands of southern Germany, but more pronounced in stands of northern Germany, a decrease of Ca contents in older needles was found. As mentioned above (3.2.1), Ca is a typically immobile nutrient element and therefore is usually accumulated in older tissue. The physiological implications of this finding have not yet been studied.

4.4. *P Deficiency*

In coniferous stands on poor acid base parent materials, particularly when these sites were degraded by litter-raking, P deficiency is a common problem. Under such site conditions, a combined P and Mg deficiency is therefore not surprising. This complex nutrient disorder is primarily found in stands in areas with elevated N input at lower and middle altitudes. Contrarily, conifers exhibiting symptoms of montane yellowing are generally characterized by good to optimal P nutrition. At these sites the P uptake is obviously not hindered by high Al^{3+} contents in the soil or soil solution. Because of the surprisingly good P nutrition of stands marked by mountain yellowing significant root or mycorrhiza damage can be excluded as a cause of this decline type. Higher N deposition rates and improved microclimatic conditions in more heavily damage stands (higher needle loss = more light to the soil and higher temperature in the soil; lower evapotranspiration rates = more water in the soil) lead to enhanced mineralization of organic matter and thus probably to an improved P supply. This can also be seen from changes in humus forms and soil vegetation, as well as from high P needle contents of spruces on organic soils. It could be inferred that the remarkably poorer P nutrition of Mg deficient trees on sites at lower elevations might be related to a more unfavourable water supply due to much lower precipitation at these sites as compared with high elevation sites.

4.5. *Mn Deficiency*

Mn deficiency occurs in Norway spruce, Scots pine and European beech on sandy soils as well as on clay rich calcareous soils. On sandy soils as well as on base-rich organic soils, Mn deficiency can be found in combination with K and/or iron (Fe) deficiency. The visual discoloration complex of this combined nutrient disorder is called "lime chlorosis". The reason why this phenomenon has spread and/or intensified since the early 1980s is still not quite clear. On soils derived from poor acid base rocks Norway spruce may have extreme high as well as low to even insufficient Mn contents. Mn nutrition is, at least in this tree species, determined by the content of exchangeable Mn in the soil. High Mn needle values are only present in trees on soils with exchangeable Mn contents above 0.5 to 0.1 $\mu eq \cdot g^{-1}$. In reality, Norway spruce can be classified as a Mn tolerant species. This, however, seems not to be the case for douglas fir (*Pseudotsuga menziesii*).

5. CAUSAL HYPOTHESES

In order to explain the far reaching changes of the nutritional status found in many forest ecosystems exhibiting symptoms of new type forest damage, various hypotheses have been formulated. These can be separated into direct and indirect pathways.

5.1. *Direct Pathways*

Direct impacts of gaseous air pollutants such as SO_2, NO_x, NH_x or O_3 as single factors, or as factors acting additively or synergisticly, have been excluded relatively early as primary agents.[6] As O_3 has a relatively high potential of phytotoxicity, it does not seem impossible that this photooxidant may play a contributing role in bringing about forest declines in certain areas.[57]

Elevated H and NH_4 concentrations in precipitation can possibly increase the leaching of mobile and exchangeable nutrients such as Mg, Ca, K, Zn and Mn from the canopy.[58] One can therefore imagine that this mechanism may cause nutrient disorders when it is acting for a long time and when the leaching rate is higher than the annual uptake rate of the nutrient ions concerned. As this process is determined by a number of other factors it can also be excluded as a monocausal hypothesis for the nutrient deficiencies nowadays observed in declining forests.

5.2. *Indirect Pathways*

There is no doubt that indirect pathways, i.e. mechanisms and processes that cause changes of soil conditions, may lead to nutritional disturbances in forest ecosystems. In this context the "acid rain" hypothesis, i.e. enhanced soil acidification and Al toxicity, was formulated.[26] It appears likely that accelerated soil acidification due to enhanced acid deposition and other effects on poor acid base soils may cause a further depletion of alkaline cations. Consequently the forest stands that grow on these soils are not sufficiently supplied with these nutrients and acute deficiencies may develop. On the other hand it has been found that forest trees usually growing on acid soils have developed adequate defence strategies against Al toxicity and are probably Al tolerant. This is particularly true for trees which are well supplied with nutrients and which possess roots with mycorrhizae.[59] So far the Al toxicity hypothesis postulated by means of hydroponic culture experiments could not be verified under natural site conditions.

Elevated N deposition (NH_4-N:NO_3-N>1) in forest ecosystems eventually causing N saturation is another important factor with implications for forest tree nutrition.[60] An improved NH_4 supply may reduce Mg and K uptake in Norway spruce as well as in other tree species (Mg/NH_4 and K/NH_4 antagonism). This mechanism may lead to lower pH values of the rhizosphere. Anion absorption may be increased. NO_3 nutrition acts in an exactly opposite manner. In areas with high anthropogenic N loads N uptake may also occur directly through the foliage. Generally, a better N supply improves tree growth. From this, dilution effects and nutrient imbalances can result. High N input may enhance soil acidification and may cause a considerable worsening of the cation content of nutrient poor soils. However, the N hypothesis cannot serve as an overall explanation of widespread nutritional disturbances; for on the local level and even more so on the regional level very different N deposition rates are measured. Also due to site and stand specific differences in the N supply additional N input affects different forest ecosystems in very different ways. The expected effects may thus be both positive or negative.

In conclusion, the anthropogenic impact on the chemical composition of the atmosphere and the resulting deposition patterns have to be seen as a site factor. From this viewpoint the site and stand specific forest damage types associated with nutritional disturbances are considered to be caused by different sets of multiple stress factors. The frequently poor nutrient supply of forest sites predominantly related to geology and to multiple anthropogenic impacts (i.e. site specific management history) plays an important role in this assumption. Externally enhanced soil acidification (i.e. depletion of cationic nutrient elements), elevated N deposition, reduced Mg and Ca input due to drastically reduced particle emission as well as enhanced canopy leaching are factors which further worsen this situation and are viewed as accompanying factors. Extreme weather perturbations (e.g. droughts) may have either positive or negative character. Biotic diseases are considered as secondary effects. Since the factorial combination is generally complex, as these influences are highly interactive and the parameters are site and stand related, the specific causes can only be determined at the site itself. Global explanations are thus not very helpful—in fact they are misleading.

6. REVITALIZATION/RESTABILIZATION OF DECLINING FOREST ECOSYSTEMS

Since various types of the new forest decline are associated with nutritional disturbances it appeared appropriate to initiate fertilizer and liming trials in order to overcome this problem. Here a typical example of a revitalization trial in a Norway spruce stand is presented. As Mg deficiency is the most severe and widespread damage type this trial illustrates the effects of Mg fertilization on Mg deficient trees. Furthermore, the findings of recent fertilization experiments will be summarized.

In the spring of 1986 a 34-year-old Norway spruce stand exhibiting pronounced tip-yellowing and moderate needle loss was fertilized with a Mg-K-Ca trial fertilizer. The trees grow on a sandy loamy poor acid base brown earth derived from quarzite covered with moor-like raw humus. The site is located at a 560 m elevation in the Hunsrueck Mountains of western Germany, receives about 1000 mm precipitation annually and the mean annual temperature is *c.* 7°C.

The soil analysis of the control plot revealed Mg deficiency (Table III). The exchangeable Al contents were extremely high, the pH and base saturation values very low. Two and a half years after fertilization an increase of available Mg and Ca content was determined in the mineral top soil. No enhanced K adsorption was noticed. Coinciding with the improved Mg and Ca supply the Mg and Ca foliar contents were also raised as indicated by an investigation in the fall of 1987 (Table IV). This was also evident from a pronounced revitalization of the fertilized spruces. Practically all deficiency symptoms had disappeared, needle loss was stopped and height growth enhanced. Also at the end of the 4th growing period after the nutrient application the positive fertilizer effects were visually as well as analytically (Table IV) present. The Mg and Ca needle contents were still clearly improved indicating sufficient nutrition. The K content was not affected by K fertilization. However, this was not expected as it was optimal in both the control and the fertilized trees. The N and P supply was good. The somewhat reduced Mn contents in the older needles indicate a limited Mn availability. The Zn values were surprisingly high. Ca nutrition was sufficient. The Fe and Al concentrations reflect the very high soil concentrations of these ions.

This as well as many other diagnostic fertilizer trials conducted in declining forests of Europe and North America—recently summarized by Huettl[47]—clearly indicate that the widespread Mg deficiency syndrome (as well as other nutrient disorders) can be mitigated rapidly and sustainably by the application of rapidly soluble fertilizers. Mg deficiency can also be mitigated by the application of Mg containing limestone, e.g. dolomite. However, Zoettl and Huettl[61] pointed out that a revitalization is only possible as long as a critical degree of damage had not yet occurred that eventually led to tree mortality. It was shown that high exchangeable monomeric Al concentrations and low Mg as well as low Ca soil contents, occasionally resulting in extremely low Mg/Al and Ca/Al ratios, did not present an obstacle to higher Mg absorption by the trees after Mg supply in the soil has been improved via Mg fertilization. Also this technique did not lead to a long-term change of the Al concentrations in the soil solution or pH values of the mineral soil.[2] The assumption that Mg fertilizers based on sulphate would lead to root damage of conifers on acid substrates[33,36] could not be substantiated.[47]

On the contrary, an improvement of fine root development within the upper mineral soil was achieved by this treatment.[62] As the exchangeable Mg soil content was improved remarkably by Mg fertilization without negative ecological effects a sustainable improvement of forest health (and growth) can be achieved, a fact now demonstrated for almost a decade. Finally, the investigation of previous experiments, where Mg fertilizers or Mg lime had been applied prior to the development of tip-yellowing in coniferous forests in which this syndrome has manifested itself since the early 1980s, showed that the addition of Mg impeded the occurrence of Mg deficiency in the fertilized trial stands.

Table III Chemical soil status of a diagnostic fertilizer trial in a 34-yr-old strongly tip-yellowed Norway spruce stand on a podzol derived from quarzite in the Forestry District Saarburg-Ost of the Hunsrueck Mountains in western Germany, fall 1988.

Soil depth cm	C	N	P	K^+	Ca^{2+}	Mg^{2+}	Fe^{3+}	Al^{3+}	H^+	CEC	Mg/Al (mol)	Ca/Al (mol)	pH (KCl)	BS %
		$mg \cdot g^{-1}$				$\mu eq \cdot g^{-1}$								
control														
0–5	121	5.1	0.140	2.3	3.0	1.6	3.9	167	14.4	194	0.015	0.03	3.2	4.0
5–10	73	3.0	0.140	1.4	1.6	1.0	0.9	140	5.9	152	0.010	0.02	3.5	2.8
10–30	74	3.0	0.150	1.5	2.4	1.3	0.6	110	4.2	121	0.018	0.03	3.5	4.5
fertilized plot														
0–5	119	5.0	0.120	2.3	5.4	3.0	3.2	142	30.7	187	0.031	0.11	3.0	5.9
5–10	75	3.5	0.130	1.6	2.8	1.6	2.2	138	11.1	158	0.018	0.03	3.3	4.1
10–30	67	2.9	0.130	1.5	3.0	1.3	0.9	128	8.3	144	0.015	0.04	3.4	4.3

CEC: cation exchange capacity
BS: base saturation

Table IV Nutrient element contents in different needle year classes of the 1st and 4th whorls of 34-year-old spruces in the diagnostic fertilizer trial, Saarburg-Ost, fall 1987 and 1989.

whorl	needle year class	N	P	K	Ca	Mg	Mn	Zn	Cu	Fe	Al
				mg · g^{-1} d.m.					µg · g^{-1} d.m.		
control plot											
1	1987	15.7	1.6	10.4	2.1	0.41	466	44	5	101	115
4	1987	16.5	1.5	9.3	1.3	0.30	273	39	4	128	127
	1986	16.3	1.1	7.6	1.2	0.24	184	28	4	138	241
	1985	11.1	1.1	8.9	1.3	0.18	183	29	5	147	253
	1984	10.0	1.2	9.1	1.1	0.12	180	29	4	180	347
fertilized plot											
1	1987	14.9	1.6	9.0	2.8	0.72	496	42	4	102	113
4	1987	15.3	1.4	9.1	3.6	0.64	586	48	4	106	126
	1986	13.4	1.2	7.8	3.4	0.42	444	21	4	118	217
	1985	12.3	1.1	8.0	2.8	0.33	342	22	4	137	262
	1984	14.7	1.0	8.9	1.7	0.30	247	25	3	176	302
1	1989	15.2	2.0	9.2	2.6	0.71	n.a.				
4	1989	18.3	1.7	6.1	2.6	0.67					
	1988	16.2	1.1	7.6	3.2	0.36					
	1987	15.1	1.0	8.3	3.5	0.27					
	1986	13.5	0.9	6.5	3.1	0.24					

n.a.: not analyszed

7. CONCLUSIONS

The findings on the nutritional status of various types of the new type forest damage observed in central Europe (and North East America) since the mid-1970s and intensively researched for a decade clearly show that the health status of forests is determined by multiple site and stand specific factors. In this context the nutrient supply plays an important role. To determine the nutritional status of forests various diagnostic tools, such as chemical foliar (and soil) analysis, are available. In cases of acute nutrient deficiencies generally specific deficiency symptoms can be observed. "Mountain yellowing", i.e. Mg-deficiency in Norway spruce and other tree species in forests at high altitudes, is the most widespread nutritional disorder. Other nutrition related decline types involve the elements K (Mg/K), P (Mg/P) and Mn. To explain the far reaching changes of the nutrient supply found in many forest ecosystems exhibiting symptoms of new type forest damage, various hypotheses have been formulated. These can be separated into direct effects (gaseous air pollutants, acid foliar leaching) and indirect pathways (soil acidification, i.e. depletion of alkaline nutrient cations and/or Al toxicity, N saturation). However, it is now clear that these anthropogenic impacts in combination with various other stress factors (e.g. extreme climatic perturbations) affect forest ecosystems in different ways. The various effects are always site and stand specific and can therefore only be determined exactly on the site itself. Global explanations are thus not very helpful, in fact they are misleading. Furthermore, it is now also clear that all types of the new type forest damage associated with nutritional disturbances can be mitigated by appropriate nutrient changes. When the right (i.e. site and stand specific) approach is chosen, fast and lasting revitalisation/restabilisation of declining forest ecosystems can be achieved.

References

1. L. E. Liljelund and B. Nihlgard, "Nutrient balance in forests affected by air pollution" **In**: F. Andersson und T. Persson, *Liming as a Measure to Improve Soil and Tree Conditions in Areas affected by Air Pollution* (National Swedish Environmental Protection Board, Report 3518, pp 93–114, 1988).
2. H. W. Zoettl, "Ernährung und Düngung der Fichte" *Forstw. Cbl.*, **109,** 130–137 (1990).
3. R. F. Huettl, "Forest decline and nutritional disturbances" **In**: D. W. Cole and S. B. Gessel, *Forest Site Evaluation and Long-Term Productivity* (chapter 18) pp 180–186 (1988).
4. M. Kaupenjohann and W. Zech, "Ergebnisse des IMA-Querschnittseminars in Bayreuth: Waldschäden und Düngung" *Allg. Forstz.*, **37,** 1002–1008 (1989).
5. T. T. Kozlowski and J. B. Mudd, *Responses of Plants to Air Pollution.* Academic Press, New York, pp. 1–8 (1975).
6. FBW (Forschungsbeirat Waldschäden/Luftverunreinigungen des Bundes und der Länder), "Waldschadensbericht" *2. Bericht* (1986).
7. W. Wittich, *Auswertung eines forstlichen Düngungsversuches auf einem Standort mit für weite Gebiete Deutschlands typischem Nährstoffhaushalt* (Ruhr-Stickstoff A.G, Bochum, pp. 1–48, 1958).
8. H. Markl, "Wissenschaftliche Forschung und ökologische Herausforderungen" *Allg. Forstz.*, **42,** 554–556 and 582–584 (1987).
9. K. Hasel, *Forstgeschichte. Ein Grundriß für Studium und Praxis* (1985).
10. H. W. Zoettl, "Waldschäden und Nährelementversorgung" *Düsseldorfer Geobotanisches Kolloquium*, **2,** 31–41 (1985).
11. L. Leyton, "The mineral requirements of forest plants" *Handb. Pflanzenphysiologie, Bd. 4*: Die mineralische Ernährung der Pflanze, 1026–1039 (1958).
12. J. Wehrmann, "Methodische Untersuchungen zur Durchführung von Nadelanalysen in Kiefernbeständen" *Forstw. Cbl.*, **78,** 77–97 (1963).

13. O. Strebel, "Mineralstoffernährung und Wuchsleistung von Fichtenbeständen *(Picea abies)* in Bayern" *Forstw. Cbl.*, **79,** 17–42 (1960).
14. H. Lundegardh, *Die Blattanalyse* (G. Fischer, Jena, 1945).
15. C. O. Tamm, "A study of forest nutrition by means of foliar analysis" (VIII Int. Botanic. Congress, Paris, 1954).
16. L. Leyton, "The relationships between the growth and mineral nutrition of conifers" *Symposium Baumphysiologie* 323–345 (1957).
17. W. Laatsch, "Die wissenschaftlichen Grundlagen der Waldbodenmelioration" *Staatsforstverw. Bayerns,* **29,** 50–61 (1957).
18. T. Ingestad, "Macroelement nutrition of pine, spruce and birch seedlings in nutrient solutions", *Medd. f. Stat. Skogsforskningsinst.*, **51,** 1–150 (1962).
19. J. Wehrmann, "Möglichkeiten und Grenzen der Blattanalyse in der Forstwirtschaft" *Landwirtschaftl. Forschg.*, **16,** 130–145 (1963).
20. H. W. Zoettl, "Diagnosis of nutritional disturbances in forest stands" *FAO-IUFRO Symposium on forest fertilization, Paris,* 75–95 (1973).
21. R. F. Huettl, " 'Neuartige' Waldschäden und Nährelementversorgung von Fichtenbeständen (*Picea abies Karst.*) in Südwestdeutschland" *Freiburger Bodenkundl. Abh.*, **16,** 195 (1985).
22. J. B. Reemtsma and E. Ahrens, "Untersuchungen zur Interpretation der Analyse älterer Fichtennadeln" *Allgem. Forst- u. J.-Ztg.*, **143,** 54–58 (1972).
23. J. B. Reemtsma, "Untersuchungen über den Nährstoffhaushalt der Nadeln verschiedenen Alters an Fichte und anderen Nadelbaumarten" *Flora,* **156,** 105–121 (1966).
24. F. H. Evers, "Die Düngung und Kalkung, eine Maßnahme zur vorübergehenden Verhinderung von Waldschäden in Baden-Württemberg". Sana Silva-Tagungsbericht: *Düngung – eine Perspektive für den Schweizer Wald?* 35–49 (1988).
25. T. Ingestad, "Quantitative mineral nutrition of forest trees". (Lecture by the 1989 Marcus Wallenberg Prize Winner, Falon. Sept. 14, 1989.)
26. B. Ulrich, R. Mayer and P. K. Khanna, "Die Deposition von Luftverunreinigungen und ihre Auswirkungen in Waldökosystemen im Solling" *Schr. Forstl. Fak. des Univ. Göttingen und des Nieders. FVA* **58,** 291 (1979).
27. K. Rost-Siebert, "Aluminium-Toxizität und -Toleranz an Kleinpflanzen von Fichte (*Picea abies Karst.*) und Buche (*Fagus sylvatica L.*)" *Allg. Forstz.* **39,** 686–689 (1983).
28. R. F. Huettl and H. W. Zoettl, "Ernährungszustand von Tannenbeständen in Süddeutschland—ein historischer Vergleich" *Allg. Forstz.* **40,** 1011–1013 (1985).
29. H. W. Zoettl and R. F. Huettl, "Schadsymptome und Ernährungszustand von Fichtenbeständen im südwestdeutschen Alpenvorland" *Allg. Forst.* **40,** 197–199 (1985).
30. G. H. Tomlinson, *Effects of Acid Deposition on the Forests of Europe and North America* (CRC Press, Cleveland, Ohio, 1990).
31. E. Matzner, K. Blanck, G. Hartmann and R. Stock, "Needle chlorosis pattern in relation to soil chemical properties in two Norway spruce (*Picea abies Karst.*) forests of the German Harz mountains" (1988). **In**: J. B. Bucher and E. Bucher-Wallin "Air pollution and forest decline" *14th International Meeting for Specialists in Air Pollution Effects on Forest Ecosystems, IUFRO Project Group P2.05, Interlaken, Switzerland, 2–8 October 1988,* 195–199 (1989).
32. H W. Zoettl, "Die Mineralstickstoffanlieferung in Fichten- und Kiefernbeständen Bayerns" *Forstw. Cbl.*, **79,** 221–236 (1960).
33. B. Ulrich, "Die Rolle der Bodenversauerung beim Waldsterben: langfristige Konsequenzen und forstliche Möglichkeiten" *Forstw. Cbl.* **105,** 421–435 (1986).
34. G. Glatzel, "The nitrogen status of Austrian forest ecosystems as influenced by atmospheric deposition, biomass harvesting and lateral organomass exchange" *Plant and Soil,* **128,** 67–74 (1989).
35. E. E. Hildebrand, "Zustand und Entwicklung der Austauschereigenschaften von Mineralböden auf Standorten mit erkrankten Waldbeständen" *Forstw. Cbl.* **105,** 60–67 (1986).
36. E. E. Hildebrand, "Ionenbilanzen organischer Auflagen nach Neutralsalzdüngung und Kalkung" *Forst und Holz* **43,** 51–56 (1988).
37. R. Hantschel, M. Kaupenjohann, R. Horn and W. Zech, "Kationenkonzentrationen in der Gleichgewichts- und Perkolationsbodenlösung (GBL und PBL) – ein Methodenvergleich" *Z. Pflanzenernähr. Bodenk.* **149,** 136–139 (1986).
38. M. Kaupenjohann, R. Hantschel, R. Horn and W. Zech, "Ergebnisse von Düngungsversuchen mit Magnesium an vermutlich immissionsgeschädigten Fichten (*Picea abies* (*L.*) *Karst*) in Fichtelgebirge" *Forstw. Cbl.* **106,** 78–84 (1987).
39. R. Horn, "The role of structure for nutrient sorptivity of soils" *Z. Pflanzenernähr. Bodenk.* **150,** 13–16 (1987).
40a. E. E. Hildebrand, "Bedeutung der Waldbodenstruktur für die Kalium-Ernähhrung von Fichtenbeständen" *6. Statuskolloquium des PEF, Karlsruhe 6.–8. März 1990"* (in press).

40b. E. E. Hildebrand, "Der Einfluß von Forstdüngungen auf die Lösungsfracht des Makroporenwassers" *Allg. Forstz.* **45,** 604–607 (1990).
41. J. C. Liu and P. Trüby, "Bodenanalytische Diagnose von K- und Mg-Mangel in Fichtenbeständen (*Picea abies Karst*)" *Z. Pflanzenernähr. Bodenk.* **152,** 307–311 (1989).
42. O. Kandler, W. Miller and R. Ostner, "Dynamik der 'akuten Vergilbung' der Fichte" *Allg. Forstz.* **42,** 715–723 (1987).
43. E. Mies and H. W. Zoettl, "Zeitliche Änderungen der Chlorophyll- und Elementgehalte in den Nadeln eines gelb-chlorotischen Fichtenbestandes" *Forstw. Cbl.* **104,** 1–8 (1985).
44. O. L. Lange, H. Zellner, J. Gebel, P. Schramel, B. Kostner and F. C. Czygan, "Photosynthetic capacity, chloroplast pigments and mineral content of the previous year's spruce needles with and without the new flush: analysis of the forest decline phenomenon of needle bleaching" *Oecologia* **73,** 351–357 (1987).
45. F. H. Evers, "Welche Erfahrungen liegen bei Kalium- und Magnesium-Großüngungsversuchen auf verschiedenen Standorten in Baden-Württemberg vor?" *Allg. Forstz.* **39,** 767–768 (1984).
46. E. Aldinger, "Elementgehalte im Boden und in Nadeln verschieden stark geschädigter Fichten-Tannen-Bestände auf Praxiskalkungsflächen im Buntsandstein-Schwarzwald" *Freiburger Bodenkundl. Abh.,* **19,** 266 (1987).
47. R. F. Huettl, "Nährelementversorgung geschädigter Wälder in Europa und Nordamerika" (Habil. Schrift) *Freiburger Bodenkundl. Abh.* **28** (1991).
48. P. Ende, "Wirkung von Mineraldüngern in Fichten- und Buchenbeständen des Grundgebirgs-Schwarzwaldes" *Freiburger Bodenkundl. Abh.* **27** (1990).
49. T. M. Roberts, R. A. Skeffington and L. W. Blank, "Causes of type 1 spruce decline in Europe" *Forestry* **62,** 179–222 (1989).
50. G. Glatzel, M. Kazda, D. Grill, G. Halbwachs and K. Katzensteiner, "Ernährungsstörungen bei Fichte als Komplex-Wirkung von Nadelschäden und erhöhter Stickstoffdeposition – ein Wirkungsmechanismus des Waldsterbens" *Allgem. Forst- und J.-Ztg.* **158,** 61–67 (1987).
51. M. Bonneau G. Landmann and C. Nys, "The revitalization of young and adult fir and spruce stands at various stages of decline by fertilization (liming) in the Vosges and in the French Ardennes" *Water, Air, and Soil Pollution* **54,** 577–594 (1990).
52. F. Weissen, H. J. van Praag, A. Hambuckers and J. Remacle, "A decennial control of N-cycle in the Belgian Ardennes forest ecosystems" *Plant and Soil,* **128,** 59–66 (1990).
53. J. Materna, "Waldschäden in der CSSR" *Österreich. Forstz.* 17–19 (1987).
54. A. J. Friedland, G. J. Hawley and R. A. Gregory, "Red spruce (*Picea rubens Sarg.*) foliar chemistry in northern Vermont and New York, USA" *Plant and Soil* **105,** 189–193 (1988).
55. B. Bernier, D. Pare and Brazeau, "Natural stresses, nutrient imbalances and forest decline in southeastern Quebec" *Water, Air, and Soil Pollution* **48,** 239–250 (1989).
56. A. Adamovicz, "Temporal changes in the nutrition of Norway spruce (*Picea abies L. Karst.*) plantations and response to fertilisation" (MS Thesis, Pennsylvania State University, USA, 1991).
57. G. H. M. Krause and B. Prinz, "Experimentelle Untersuchungen der LIS zur Aufklärung möglicher Ursachen der neuartigen Waldschäden" *LIS-Berichte* **80,** 221 (1989).
58. R. F. Huettl, S. Fink, H.-J. Lutz, M. Poth and J. Wisniewski, "Forest decline, nutrient supply and diagnostic fertilization in southwestern Germany and in southern California" *Forest Ecology and Management* **30,** 341–350 (1990).
59. K. Kreutzer, H. Reiter, R. Schierl and A. Göttlein, "Effects of acid irrigation and liming in a Norway spruce stand (*Picea abies* (*L.*) *Karst.*)" *Water, Air, and Soil Pollution* **48,** 111–125 (1989).
60. B. Nihlgard, "The ammonia hypothesis – an additional explanation of the forest die back in Europe" *Ambio.* **14,** 2–8 (1985).
61. H. W. Zoettl and R. F. Huettl, "Nutrient supply and forest decline in southwest Germany" *Water, Air, and Soil Pollution* **31,** 257–265 (1986).
62. W. Schaaf and W. Zech, "Einfluß unterschiedlicher Löslichkeit von Düngern" *Allg. Forstz.* **46,** 766–768 (1991).
63. R. F. Huettl, "Forest fertilization: Results from Germany, France and the Nordic Countries" *The Fertiliser Society* **250,** 1–40 (1986).
64. H. J. Fiedler and H. Höhne, "Das NPK-Verhältnis in Kiefernnadeln als arteigene Erscheinung und Mittel zur Ernährungsdiagnose" *Beitr. Forstwirtschaft. Berlin* **18,** 128–1332 (1984).
65. T. Ingestad, "Versuche mit konstantem inneren Ernährungszustand in Birkenpflanzen" *Tagungsberichte* (Tharandt, DDR) **84,** 71–76 (1968).
66. W. Flückiger, "Wie präsentiert sich das Problem der Waldernährung in der Schweiz aufgrund laufender Untersuchungen heute?" Sana Silva-Tagungsbericht: *Düngung – eine Perspektive für den Schweizer Wald?* 70–73 (1988).
67. W. Zech, T. Sutter and E. Popp, "Elemental analyses and physiological responses of forest trees in SO_2-polluted areas of NE Bavaria" *Water, Air, and Soil Pollution* **25,** 175–183 (1985).

68. W. Zech and E. Popp, "Magnesiummangel, einer der Gründe für das Fichten- und Tannensterben in NO-Bayern" *Forstw. Cbl.* **102,** 50–55 (1983).
69. C. O. Tamm, "Meraomgranens Gulspetssjuka" *Sv. Skogsvardsför. Tidkr.* **51,** 1–7 (1953).
70. K. E. Rehfuss, "Standort und Ernährungszustand von Tannenbeständen *(Abies alba Mill.)* in der südwestdeutschen Schichtstufenlandschaft" *Forstw. Cbl.* **86,** 321–348 (1967).
71. N. B. Comerford and R. F. Fisher, "Using foliar analysis to classify nitrogen-deficient sites" *Soil Sc. Soc. Am. J.* **48,** 910–913 (1984).

6. ACID RAIN: AN ISSUE FOR REGIONAL CO-OPERATION

JEAN MICHEL JAKOBOWICZ

The dramatic development of the acid rain phenomenon and its effects has generated an increasing awareness among the international community to the importance of fighting this type of environmental degradation. But it was not until 1979, when the Convention on Long-range Transboundary Air Pollution was signed within the framework of the United Nations Economic Commission for Europe (UN/ECE) after three years of difficult negotiations, that an international strategy for fighting acid rain was organized both by the countries affected by acid rain and by the countries causing it.

This chapter is organized around two key ideas. Firstly, that the acid rain issue can only be addressed at a regional level; and that a convention, such as the Convention on Long-range Transboundary Air Pollution, is an ideal instrument for the international community. The second idea is that an international instrument, such as the above-mentioned Convention, should evolve both in form and in spirit. If evolution does not occur, this international instrument will very quickly become obsolete.

Therefore, this chapter also concentrates on the evolution of the various protocols to the Convention in terms of their complexity, efficiency and their overall consistency. If the Convention in itself provided the international community with a framework for dealing with the acid rain problem, the protocols elaborated within this framework have through their increasing contingency on Parties and therefore their increasing effectiveness become instrumental in solving this problem. Finally, some potential developments regarding abatement strategies for acid rain are outlined, such as the critical load approach.

1. INTRODUCTION

The adverse effects of air pollution have long been recognized.[1] Queen Elizabeth I of England promulgated laws in the seventeenth century in order to control noxious fumes from coal fires in London. A century ago the British chemist, Roger Angus Smith, observed that the rainfall in the Manchester area was becoming polluted. Noting its corrosive power, he called it "acid rain."[2] Since this period and until very recently the phenomenon of acid rain has been developing dramatically. If only one hundred years ago the phenomenon of acid rain could be localized in one region, such as the Manchester area, currently the effects of acid rain can be traced back throughout Europe. One of the best-known examples is the acidification of the Scandinavian lakes which originates in countries several hundreds or thousands of kilometres away. Nowadays, it is not only lakes or more generally fresh waters which are affected by the acidification process, but also soil and vegetation, in particular forests and monuments, as well as human health.

The dramatic increase of the acid rain phenomenon and its effects has generated an increasing awareness in the international community to the importance of fighting this type of environmental degradation. It was at the 1972 Stockholm World Conference on Human Environment that, for the first time, some delegations raised the issue of acid rain in an international forum.[3] But it was not until 1979 when the Convention on Long-range Transboundary Air Pollution elaborated within the framework of the United Nations Economic Commission for Europe (UN/ECE) was signed,[4] that an international strategy for combating acid rain was organized by the countries affected by acid rain and also by the countries causing it.

The present chapter is organized around two key ideas. Firstly, that the response to the acid rain issue can only be a regional response, and that within this framework a convention, such as the Convention on Long-range Transboundary Air Pollution, is an ideal instrument for the international community[5]. The second idea is that an instrument such as the above-mentioned Convention should evolve in form and also in spirit. If evolution does not occur, this international instrument will very quickly become obsolete.

2. ACID RAIN: A REGIONAL PROBLEM

2.1. *Emissions*

Only a few decades ago, air pollution was considered a local problem that could be solved by building tall stacks that would reduce ground-level concentrations. It was only at the end of the 1960s that this consideration was put into question by the observation that the acidification of precipitation was occurring all over Europe, including areas far from industrial centres. Since then, long-range transport of air pollutants and their effects in terms of acid rain have been extensively studied.

Sulphur dioxide and nitrogen oxides are the chief sources of acid rain. Other air pollutants of concern on a regional scale are ozone and various toxic substances, such as heavy metals and organic micro-pollutants. The main source of sulphur and nitrogen oxides is the combustion of fossil fuels (oil and coal) for heat and electricity production and transportation vehicles. It is thus to be expected that with the development of industrial societies, in particular in the northern hemisphere in the last century, this type of emission has been increasing dramatically.

It has been estimated that sulphur emissions have increased in Europe, from a level of 8 million tonnes per year at the beginning of the century to around 30 million tonnes today. The largest increase occurred after the Second World War when fuel consumption rose by about 10% per annum (between 1945 and 1970, for example, the consumption of oil increased fifteen times). The atmosphere also receives sulphur from natural sources such as volcanoes, the seas and oceans and from certain soil processes. Throughout the world, man-made sulphur emissions are, on average, equal to natural ones. However, in the more industrialized regions, such as the ECE region, man-made emissions are 90% of all emissions.

In the ECE region, the major emitters of sulphur in 1987 were the United States, followed by the USSR, Canada and the five new Länder of Germany[6] (Tables I and II).[7] Most, if not all, former centrally planned economy countries

Table I Total SO_2 Emissions 1980–2005 in the ECE Region (In thousands of tonnes S per year)

	1980	1981	1982	1983	1984	1985	1986	1987	1988a/	1989a/	1990b/	1993b/	1995b/	2000b/	2005b/
Austria	185			109		89		68	57		47		40	39	
Belgium	414	356	347	280	250	226	237	207			210		215		
Bulgaria	517			570				535	515		515	390	300	260	260
Byelorussian SSR c/	370	365	355	355	345	345	345	345	319	298	292	260	246	228	200
Canada	2322					1843					1881		1526	1535	1549
Czech and Slovak Federal Republic	1550					1575	1510	1480	1400						
Denmark	224	182	184	156	148	170	139	125	121		133		95	88	
Finland	292	268	244	187	184	191	166	164	151		126			133	
France	1669	1294	1245	1047	933	735	671	645	613	667					
German Democratic Republic	2132	2158	2775	2318	2519	2670	2679	2803	2629	2621					
Germany, Federal Republic of d/	1605	1515	1430	1355	1320	1225	1170	990	650	530				430	
Greece	200					250									
Hungary	816					702	685	646	609	542	582		571	547	
Iceland	3				3	3									
Ireland	111	96	79	71	71	70	81	87	76	81	84	87	114	120	129
Italy	1900			1575	1328	1252	1185	1205	2406						
Liechtenstein	*														
Luxembourg	12			7		8					5				
Netherlands e/	[illegible]			167		138	138	141	138	127			88	53	
Norway	71	64	56	52	48	49	45	38	33	33		35			
Poland f/	2050					2150	2100	2100	2090	1955	2250		2050	1450	1000
Portugal g/	133			153		99	117	109	102		106	135	139	152	
Romania	900									900					
Spain	1625					1095									
Sweden h/	257								107		102	95	90	91	96
Switzerland	63				48						31		29	30	32
Turkey					138	161	177								
Ukrainian SSR c/	1925					1732	1696	1632	1606	1536	1495	1350	1170	1105	870
USSR i/	6400	6309	6237	5700	5926	5555	5925	5452	5109	4682	4790	4550	4410	4110	3700
United Kingdom	2424	2194	2082	1909	1837	1838	1929	1932	1832		1916	1763	1621	1223	1009
USA	11950	11750	11000	10750	11050	10800	10600	10200							
Yugoslavia	650	650	650	700	725	750	750	775	800	775					
EEC k/						6813									

Notes: * <0.5
[a]) Preliminary data. [b]) Projected estimates. [c]) Also included under USSR. [d]) 2000 = 1998 estimates. [e]) 1995 = 1994 estimates; 2000 estimates: according to national environmental policy plan. [f]) 1990 estimates: not taking into account planned abatement measures; 2005 = 2010 estimates; 1995, 2000, 2010 emission estimates: taking into account implementation of emission control programme. [g]) For the years 1990, 1993, 1995 and 2000 emissions of small combustion plants were not calculated. [h]) Projections based on current administrative regulations. [i]) European part of USSR within EMEP. [k]) CORINAIR total, based on separate inventory of 12 EEC member States.

Table II Ranking of countries by level of Sulphur and NO_x emissions at the end of the 1980s, per capita and per unit of GDP

SO_2 Emission	NO_x Emission	SO_2/POP*	NO_x/POP*	SO_2/GDP**	NO_x/GDP**
United States	United States	Germany Dem.	United States	Poland	Poland
USSR	USSR	Canada	Canada	Hungary	Hungary
Canada	Germany,Fed.Rep.	Cszechoslovakia	Czechoslovakia	Yugoslavia	Yugoslavia
Germany Dem.	United Kingdom	Hungary	Finland	Canada	United States
Poland	Canada	Bulgaria	Norway	Spain	Canada
United Kingdom	Italy	Poland	Luxembourg	Ireland	Ireland
Ukraine	France	United States	Denmark	United Kingdom	United Kingdom
Cszechoslovakia	Poland	Yugoslavia	Iceland	United States	Luxembourg
Italy	Ukraine	United Kingdom	Germany,Fed.Rep.	Italy	Spain
Spain	Spain	Belarus	Sweden	Finland	Finland
Yugoslavia	Czechoslovakia	Finland	United Kingdom	Belgium	Norway
Germany,Fed.Rep.	Germany Dem.	Spain	Germany Dem.	Luxembourg	Netherlands
France	Netherlands	USSR	Poland	Denmark	Germany,Fed.Rep.
Hungary	Yugoslavia	Denmark	Netherlands	France	Sweden
Bulgaria	Sweden	Luxembourg	Ireland	Netherlands	Italy
Belarus	Belgium	Ireland	Switzerland	Sweden	Iceland
Belgium	Finland	Belgium	Italy	Germany,Fed.Rep.	Belgium
Turkey	Belarus	Italy	Belgium	Iceland	France
Finland	Hungary	Sweden	France	Austria	Austria
Netherlands	Denmark	Iceland	Austria	Norway	Denmark
Denmark	Norway	France	Belarus	Switzerland	Switzerland
Sweden	Switzerland	Germany,Fed.Rep.	Hungary		
Portugal	Austria	Portugal	Spain		
Ireland	Bulgaria	Netherlands	USSR		
Austria	Ireland	Norway	Ukraine		
Switzerland	Portugal	Austria	Yugoslavia		
Norway	Luxembourg	Switzerland	Bulgaria		
Luxembourg	Iceland	Turkey	Portugal		
Iceland	Turkey	Ukraine			

Notes: * Emissions per capita

** Emissions per unit of GDP.

Source: see tables I and III

were among the biggest polluters, whereas Scandinavian countries can be found at the bottom of the list. However, the picture is slightly different if the same data are presented on per capita basis or on a per unit of production basis (Table II). On a per capita and on a per unit of output basis the top ranks are occupied almost exclusively by eastern countries, while the United States and the USSR are further down on the list. The size of the population and the value of GDP does not seem to influence the ranking of smaller countries such as Switzerland, Austria and Iceland, as well as of Scandinavian countries which remain among the smallest polluters. These variations in ranking are due mainly to the fuel mix used in the various countries but also to the technology and the enforcement of environmental regulations. The "productivist" approach which prevailed in eastern Europe without proper enforcement of environmental regulation is one of the major reasons for their being among the major emitters of SO_2.

The other main cause of acidification is the emissions of nitrogen oxides, which are formed in all types of combustion. Unlike the sulphur dioxides that come entirely from the sulphur contained in the fuel or raw material, most of the nitrogen oxides are formed by the reaction of nitrogen gas in air during combustion with oxygen. Therefore the volume of emitted nitrogen oxides is much more difficult to estimate. The largest single source of nitrogen oxide is road traffic. In certain regions, such as Scandinavia, two-thirds of all nitrogen oxide emissions come from motor vehicles.

The major NO_x emitter in the ECE region is the United States, followed by the USSR, Germany and the United Kingdom (Tables II and III).[8] In terms of emission of nitrogen oxides per capita or per unit of production, the picture is different. While the United States remains the largest emitter per capita followed by Canada and Czechoslovakia and Finland, in terms of NO_x emission per unit of output Poland, Hungary and Yugoslavia come first followed by the United States and Canada. It is interesting to note that on a per capita basis eastern countries and the USSR are among the smallest emitters of NO_x.

The sources of emission of both SO_2 and NO_x vary greatly according to the countries. However, while SO_2 comes mostly from combustion or industrial sources, NO_x is emitted mostly from mobile sources (Table IV).[9]

Trends in emission of primary air pollutants in the northern hemisphere show great variations, but certain general conclusions can be drawn. Emission of sulphur dioxide increased between 1950 and 1970 by a factor of 3 in most of Europe and parts of North America, but then stabilized. In the United States, however, the increase was much greater and continued into the 1980s, while emissions markedly declined in some parts of western Europe. Between 1950 and 1980 there was also a general shift from low-level domestic emissions to high-level stack emissions (energy and other industries), reducing the local concentration of gaseous pollutants.

The data on nitrogen oxides are less comprehensive, but until 1970 the trends probably followed those of SO_2. Then, into the 1980s, emissions appear to have increased in general, with the growth in the number of internal combustion engines.

2.2. *Transportation of Acid Compounds*

Pollutants emitted from urban areas or other sources are dispersed within the lower layer of the atmosphere where, in general, their concentration decreases with their distance from the emitting source. Thereafter, these air pollutants can

Table III Total NO_x Emissions 1980–2005 in the ECE Region (In thousands of tonnes NO_2 per year)

	1980	1981	1982	1983	1984	1985	1986	1987	1988a/	1989a/	1990b/	1993b/	1995b/	2000b/	2005b/
Austria	233			228		230		218	212		201		171	155	
Belgium	442				295	281	292	297			300		250		
Bulgaria					150			150	150		150	150	150	150	150
Byelorussian SSR c/	244	235	235	231	226	220	258	287	262	263	271	269	251	187	159
Canada	1959					1887					1923		1864	1929	1995
Czech and Slovak Federal Republic	1204					1127	1060	965	950						
Denmark	241	208	224	218	225	258	266	262	249		254		224	177	
Finland d/	264	248	245	236	233	251	256	270	276				321	226	193
France	1823	1701	1688	1645	1632	1615	1618	1630	1615	1772					
German Democratic Republic								701	708	705					
Germany, Federal Republic of e/	2970	2880	2870	2880	2980	2930	2990	2940	2860					1980	
Greece						746									
Hungary	273					262	268	276	259	249	264		280	279	
Iceland	13				12	12									
Ireland	73	86	86	85	84	91	100	115	122	127	135	144	130	129	121
Italy i/	1480				1568	1595	1607	1700	1755						
Liechtenstein	*				*										
Luxembourg	23			21		19					15				
Netherlands f/	548					544	559	559	552	552			422	268	
Norway	184	178	180	190	206	203	222	232	227	226				155	
Poland						1500	1590	1590	1550	1480				1345	
Portugal g/	166			192		96	110	116	122		142	173	187	193	
Romania															
Spain	950					950									
Sweden h/	398					394			390		373		343	341	347
Switzerland	196				214						184		138	127	123
Turkey															
Ukrainian SSR c/						1059	1112	1095	1090	1065	1099	1096	1056	930	885
USSR i/	3167	3515	3560	3510	3396	3369	3330	4218	4201	4418					
United Kingdom	2418	2328	2321	2230	2162	2278	2350	2429	2480		2573	2471	2300	1822	1718
USA	20300	20300	19500	19100	19700	19700	19300	19500					19100		
Yugoslavia	350	360	370	370	380	400	420	440	480	430					
EEC k/						10428									

Notes: * <0.5

[a]) Preliminary data. [b]) Projected estimates. [c]) Also included under USSR. [d]) 2005 = 2010 estimates; 2000 and 2010 projections by Nitrogen Oxide Commission. [e]) 2000 = 1998 estimates. [f]) 1995 = 1994 estimates; 2000 estimates: according to national environmental policy plan. [g]) For the years 1990, 1993, 1995 and 2000 emissions of small combustion plants were not calculated. [h]) Projections based on current environmental regulations. [i]) European part of USSR within EMEP. [k]) CORINAIR total, based on separate inventory of 12 EEC member States.

Table IV Emissions of SO_2 and NO_x by Mobile and Non-mobile sources in 1990 (in percentage)

	SO_2			NO_x		
	Emissions from non-mobile sources	Emissions from mobile sources		Emissions from non-mobile sources	Emissions from mobile sources	
Austria	94.73684	5.263157	100	30.18867	69.81132	100
Canada	97.23551	2.764486	100	41.65366	58.34633	100
Denmark	95.86776	4.132231	100	64.25702	35.74297	100
Finland	98.01324	1.986754	100	88.04347	11.95652	100
France	90.04893	9.951060	100	24.02476	75.97523	100
Germany	94.30769	5.692307	100	35.31468	64.68531	100
Hungary	98.52216	1.477832	100	53.66795	46.33204	100
Ireland	94.73684	5.263157	100	50	50	100
Netherlands	86.95652	13.04347	100	37.68115	62.31884	100
Poland	97.36842	2.631578	100	68.38709	31.61290	100

Source: Strategies and Policies for Air Pollution Abatement, *op. cit.*

be transported several thousands of kilometres across national boundaries and deposited in other countries.

During their transportation the pollutants may undergo chemical reactions and transformations, such as the transformation of sulphur dioxides into sulphates and nitrogen oxides into nitrates. When there is no precipitation, gaseous and particulate pollutants are removed from the atmosphere through contact with the earth's surface. This process is referred to as dry deposition. This type of deposition is prevalent near the major sources of emissions. Compared to dry depositions, pollutants are removed much more rapidly from the atmosphere by precipitation but can occur at a much greater distance from the emission sources.

The wet deposition of sulphur and nitrogen compounds are easy to measure. The acidity of precipitation is determined by a balance between sulphates, nitrates and possibly chlorides, which tend to make it more acid, and ammonium and soil-derived components, such as calcium, which tend to make it less acid. On the basis of collected data, deposition maps are drawn, which enable us to have a fairly accurate view of the acid depositions throughout Europe and North America.

In order to close the loop and to elaborate a comprehensive view of this phenomenon, an interlinkage between emissions and depositions is being made by mathematical models that connect both sets of data taking into account the meteorological phenomena. The matrices that result from the use of models such as the EMEP Meteorological Synthesizing Centre-East (MSC-E) at the Institute of Applied Geophysics in Moscow and the EMEP Meteorological Synthesizing Centre-West (MSC-W) at the Norwegian Meteorological Institute in Oslo, are shown in Tables V and VI.

It shows how much oxidized sulphur a country receives and from where. One of the conclusions to be drawn from these estimates is that taking the diagonal of the matrix, which represents the part of emission of oxidized sulphur which remains in the emitting country, for most countries it represents only a small amount as compared to the total emission. In other words most emissions will travel across borders. This is also true for oxidized or reduced nitrogen and emphasizes even more the transboundary nature of air pollution and, particularly, the problem of acid rain.

2.3. *Effects*

The effects of acidification can be divided into four major areas: effects on aquatic ecosystems, on terrestrial ecosystems, on health and on materials. All these effects have a socio-economic impact, which is in many cases difficult to estimate.

Acidification of fresh water, be it lakes or rivers, was the first recognized environmental effect of acid rains. Acidification changes the chemistry but also the biology of these waters. Lakes and rivers with soils and bedrock catchment areas which consist of lime-poor weathering-resistant minerals are especially sensitive to acidification. The overall changes in aquatic chemistry have both ecological and economic consequences which are sometimes difficult to assess. The most obvious sign of acidification is the disappearance of fish and the reduced number of plant and animal species.

In such areas also the soils themselves may be subject to acidification processes. The effect of acid compounds on trees are of two types. The direct effects harm needles or leaves as the protective layer of wax is corroded by dry deposition of SO_2, or by acid rains, or there is a damage to the stomata, the minute openings which among other things regulate evaporation of water. The indirect effect occurs when soil has become acidified by acid rains, with a consequent reduction of nutrient and release of substances that are injurious to trees, such as aluminium. Flora and fauna are also damaged by acid rains.

Acid rain also damages materials. It increases the corrosion rates of most materials used in the construction of buildings, industrial equipment, bridges, dams, water supply networks, etc. Severe damage is also caused to cultural objects, such as monuments, sculptures and ornaments.

Polluted air also affects human health in many ways ranging from slight feeling of discomfort to increased morbidity and mortality. Among those especially threatened by this type of pollution are people already suffering from lung diseases, children, elderly persons and those with heart and circulatory diseases.

3. THE CONVENTION ON LONG-RANGE TRANSBOUNDARY AIR POLLUTION

3.1. *Historical Background*

The idea for the Convention on Transboundary Air Pollution developed in the 1960s from the statement by several countries that the phenomenon of acidification and, more particularly, lake acidification could not be solved at the national level. It is interesting to note that during the 1972 Stockholm World Conference on the Human Environment, the Swedish delegation was the first to present a report that directly linked the phenomenon of acid rain to the destruction of Swedish lakes.[10] This link was in many instances noted with scepticism for an obvious reason: some countries, far from the Scandinavian lakes, were thus implicated in the destruction of these natural resources. In the final text adopted by the Conference, no mention was made of the acid rain issue. The only positive aspect included in the Stockholm Declaration relating indirectly to acid rain, which will provide a legal basis for an agreement in the field of transboundary air pollution, is included in article 31 which mentions, *inter alia*, that each state is responsible for the fact that activities undertaken within its territory should not cause any damage beyond its frontiers.[11]

Several studies, such as the OECD's,[12] quickly confirmed the hypothesis, included in the Swedish document, that air pollutants actually travel over several thousands of kilometres before deposition occurs. The fact that the acid rain problem could be solved only at an international level was therefore established. The only thing lacking for negotiations was political will.

The Conference on Security and Co-operation in Europe (CSCE) final document[13] contains elements which gave new impetus to regional co-operation on air pollution within the ECE. For example, it is emphasized that "Participating States will make every effort to co-operate in the area of control of air pollution, desulphurization of fossil fuels and exhaust gases, pollution control of heavy metals, particulate aerosols, nitrogen oxides in particular those emitted by power stations and other industrial plants systems, and methods for observation and control of air pollution and its effect including long-range transport of air pollutants." The Participating States also made the following recommendations on specific measures to be taken: "To develop through international co-operation

Table V SO_2 transport matrix. (Unit 100 tonnes of S)

Receivers	AL	AT	BE	BG	CS	DK	FI	FR	DD	DE	GR	HU	IS	IE	IT	LU	NL	NO	PL	PT	
AL	40	1	0	21	11	0	0	1	11	1	18	12	0	0	26	0	0	0	9	0	AL
AT	0	135	18	4	369	4	1	61	412	110	1	93	0	1	240	1	12	0	166	0	AT
BE	0	0	417	0	15	1	0	138	39	66	0	1	0	2	3	2	34	0	8	0	BE
BG	3	2	2	1212	69	2	1	4	66	8	36	83	0	0	22	0	1	0	103	0	BG
CS	0	27	23	5	3676	9	2	57	1437	136	2	314	0	1	46	1	14	0	721	0	CS
DK	0	0	13	0	22	140	1	17	97	36	0	2	0	2	2	0	12	1	26	0	DK
FI	0	0	10	1	44	20	405	13	138	24	0	7	0	2	1	0	8	5	119	0	FI
FR	0	5	212	1	98	6	1	2334	262	212	0	24	0	11	320	8	60	0	53	15	FR
DD	0	3	47	1	644	25	1	80	6794	280	0	16	0	2	13	2	31	1	168	0	DD
DE	0	18	229	2	392	28	2	438	1393	1705	0	24	0	6	118	11	138	1	128	2	DE
GR	9	1	1	193	29	1	0	3	32	4	473	29	0	0	32	0	1	0	40	0	GR
HU	0	19	5	11	344	2	1	14	196	25	3	1468	0	0	72	0	3	0	254	0	HU
IS	0	0	0	0	0	0	0	1	1	1	0	0	5	1	0	0	1	0	0	0	IS
IE	0	0	5	0	4	1	0	10	9	5	0	0	0	173	0	0	3	0	3	0	IE
IT	2	19	19	22	154	3	1	138	199	60	15	88	0	1	3213	1	9	0	95	3	IT
LU	0	0	3	0	1	0	0	11	2	5	0	0	0	0	1	10	1	0	0	0	LU
NL	0	0	109	0	13	2	0	79	47	142	0	1	0	2	2	1	195	0	11	0	NL
NO	0	0	17	1	32	36	11	26	126	43	0	5	0	10	3	0	14	67	64	0	NO
PL	0	14	51	9	1244	56	10	94	3067	243	2	183	0	4	42	2	36	2	6516	1	PL
PT	0	0	0	0	0	0	0	3	0	0	0	0	0	0	1	0	0	0	0	255	PT
RO	3	9	7	159	334	4	3	15	296	31	24	468	0	0	66	0	5	0	454	0	RO
ES	0	0	10	1	5	1	0	101	17	14	1	4	0	3	41	0	4	0	4	113	ES
SE	0	1	27	2	101	114	52	38	351	77	1	16	0	7	9	0	22	26	239	1	SE
CH	0	4	15	0	30	1	0	96	81	45	0	5	0	0	200	1	6	0	7	1	CH
TR	1	1	1	104	32	1	1	2	51	6	59	29	0	0	13	0	1	0	78	0	TR
SU	1	19	67	120	1047	115	248	108	1949	245	25	544	0	10	97	2	50	11	3183	1	SU
GB	0	0	48	0	17	7	1	95	64	40	0	1	0	64	3	0	29	0	32	2	GB
YU	14	29	7	155	364	5	2	33	310	43	40	478	0	0	385	0	4	0	255	0	YU
REM	0	1	6	5	7	0	0	32	8	7	11	5	0	1	73	0	2	0	4	5	REM
BAS	0	44	114	3	216	283	156	76	895	189	1	32	0	7	14	1	41	10	737	1	BAS
NOS	0	2	222	1	113	105	3	419	416	248	0	7	0	55	11	2	268	27	115	3	NOS
ATL	22	1	96	1	36	34	58	420	162	109	0	3	10	224	17	2	54	31	62	232	ATL
MED	45	18	33	313	286	5	1	284	326	74	527	242	0	3	2340	1	15	0	241	18	MED
BLS	1	2	3	148	109	4	12	4	128	14	21	57	0	0	11	0	2	0	231	0	BLS
SUM	120	333	1776	2495	9825	1016	941	5246	19379	4247	1261	4241	16	593	7437	50	1074	187	14125	657	SUM
	AL	AT	BE	BG	CS	DK	FI	FR	DD	DE	GR	HU	IS	IE	IT	LU	NL	NO	PL	PT	

Emitters

Receivers	RO	ES	SE	CH	TR	SU	GB	YU	REM	BAS	NOS	ATL	MED	BLS	NAT	IND	SUM	
AL	17	2	0	0	0	6	1	37	2	0	0	0	0	0	6	52	276	AL
AT	28	9	2	12	0	21	41	90	1	1	3	1	0	0	7	176	2021	AT
BE	0	5	0	0	0	1	106	1	0	0	12	2	0	0	5	39	898	BE
BG	367	1	1	0	12	140	4	158	2	0	1	0	0	0	12	213	2524	BG
CS	46	7	3	3	0	56	47	72	0	2	5	1	0	0	8	214	6937	CS
DK	0	2	5	0	0	4	98	1	0	5	10	1	0	0	10	49	558	DK
FI	7	1	55	0	0	452	71	3	0	15	5	1	0	0	36	343	1788	FI
FR	4	375	1	18	0	4	510	52	7	1	51	39	0	0	88	729	5503	FR
DO	3	8	3	1	0	14	109	8	0	5	13	2	0	0	11	143	8429	DO
DE	8	42	4	26	0	18	362	18	0	5	41	7	0	0	28	345	5540	DE
GR	88	2	0	0	16	54	2	49	5	0	0	0	0	0	24	209	1297	GR
HU	131	3	1	1	0	52	11	245	1	1	1	0	0	0	5	165	3036	HU
IS	0	1	0	0	0	0	15	0	0	0	0	1	0	0	22	48	97	IS
IE	0	4	0	0	0	1	109	0	0	0	3	9	0	0	23	70	430	IE
IT	35	58	1	21	1	14	51	260	22	1	4	3	0	0	61	554	5128	IT
LU	0	1	0	0	0	0	3	0	0	0	0	0	0	0	0	3	42	LU
NL	0	5	0	0	0	1	162	0	0	0	20	2	0	0	8	45	848	NL
NO	3	4	24	0	0	103	333	5	0	5	19	8	0	0	78	398	1438	NO
PL	60	11	18	3	1	211	164	73	1	17	17	3	0	0	29	483	12663	PL
PT	0	121	0	0	0	0	5	0	3	0	0	23	0	0	15	75	504	PT
RO	2884	4	2	1	13	491	19	416	2	1	2	0	0	0	15	441	6169	RO
ES	2	3013	0	0	0	1	74	11	28	0	5	44	3	0	68	478	4048	ES
SE	9	4	277	0	0	124	225	8	0	24	19	5	0	0	62	408	2249	SE
CH	1	19	0	91	0	1	37	9	1	0	3	2	0	0	5	80	741	CH
TR	109	2	1	0	551	263	5	30	4	0	0	0	0	0	45	697	2089	TR
SU	758	14	103	2	48	18521	347	229	2	58	29	5	0	0	170	4833	32962	SU
GB	0	21	2	0	0	12	4784	1	0	1	50	28	0	0	63	224	5590	GB
YU	240	14	1	2	2	80	18	2714	7	1	2	1	0	0	31	519	5757	YU
REM	4	68	0	1	1	3	12	12	604	0	1	3	1	0	53	624	1553	REM
BAS	18	9	179	1	0	417	295	15	1	119	29	4	0	0	131	487	4525	BAS
NOS	2	44	19	1	0	25	3335	4	0	8	295	42	0	0	254	616	6663	NOS
ATL	4	1543	31	1	0	514	1808	5	4	6	68	685	1	0	2127	3203	11574	ATL
MED	263	738	1	7	119	172	100	646	344	1	9	12	2	0	603	2044	9833	MED
BLS	291	1	2	0	134	907	13	51	1	1	1	0	0	0	108	597	2855	BLS
SUM	5381	6156	737	194	898	22683	13276	5222	1041	281	720	936	8	0	4210	19603	156367	SUM
	RO	ES	SE	CH	TR	SU	GB	YU	REM	BAS	NOS	ATL	MED	BLS	NAT	IND	SUM	

Source: EMEP, Calculated Budgets for Airborne Acidifying Components in Europe, 1985, 1987, 1988, 1989 and 1990, August 1991.

Table VI NO_x transport matrix. (Unit 100 tonnes of N)

Receivers	Emitters																				
	AL	AT	BE	BG	CS	DK	FI	FR	DD	DE	GR	HU	IS	IE	IT	LU	NL	NO	PL	PT	
AL	1	1	0	2	3	0	0	1	2	3	16	2	0	0	15	0	0	0	3	0	AL
AT	0	48	12	0	79	5	1	65	58	232	1	14	0	0	106	1	25	1	50	0	AT
BE	0	1	31	0	4	1	0	71	5	62	0	0	0	1	2	1	29	0	3	0	BE
BG	0	3	1	28	18	1	1	3	10	16	41	13	0	0	13	0	2	0	30	0	BG
CS	0	27	16	1	245	9	2	64	149	283	3	38	0	1	31	1	31	2	161	0	CS
DK	0	1	7	0	6	18	1	18	15	57	0	0	0	2	1	0	19	3	7	0	DK
FI	0	1	9	0	15	23	106	18	33	71	0	1	0	2	1	0	19	18	45	0	FI
FR	0	8	87	0	30	6	1	947	43	356	0	5	0	9	147	5	89	2	18	10	FR
DD	0	4	28	0	96	16	2	80	216	429	0	3	0	2	9	2	57	3	41	0	DD
DE	0	17	111	0	95	18	2	359	170	1230	0	5	0	5	53	8	200	4	41	2	DE
GR	1	2	1	12	9	1	0	4	6	10	146	5	0	0	20	0	1	0	13	0	GR
HU	0	20	3	1	71	2	1	16	29	60	5	76	0	0	51	0	6	0	77	0	HU
IS	0	0	1	0	0	0	0	2	0	3	0	0	2	1	0	0	2	1	0	0	IS
IE	0	0	4	0	1	1	0	15	2	15	0	0	0	17	0	0	8	0	1	0	IE
IT	0	23	13	3	43	3	1	158	32	134	20	15	0	1	599	1	19	1	33	2	IT
LU	0	0	2	0	0	0	0	7	0	5	0	0	0	0	0	0	1	0	0	0	LU
NL	0	1	27	0	3	2	0	47	6	93	0	0	0	2	1	0	62	1	3	0	NL
NO	0	1	14	0	10	29	9	39	31	114	0	1	0	10	2	0	32	57	22	0	NO
PL	0	19	33	1	259	43	12	108	358	505	2	30	0	5	31	2	72	8	617	1	PL
PT	0	0	0	0	0	0	0	4	0	1	0	0	0	0	1	0	0	0	0	44	PT
RO	0	12	4	16	82	4	3	15	45	70	31	71	0	0	44	0	10	1	125	0	RO
ES	0	0	7	0	2	1	0	117	4	34	2	1	0	3	34	0	9	0	2	60	ES
SE	0	2	18	0	31	72	38	45	70	170	1	3	0	6	7	1	43	50	83	0	SE
CH	0	4	10	0	9	1	0	91	13	98	0	1	0	0	59	1	12	0	2	1	CH
TR	0	2	1	8	11	1	1	3	11	16	68	6	0	0	10	0	2	0	29	0	TR
SU	0	32	49	14	312	116	188	142	379	630	34	92	0	10	76	2	116	38	956	1	SU
GB	0	1	20	0	6	6	1	73	12	76	0	0	0	32	2	0	39	2	12	1	GB
YU	2	36	5	16	95	5	2	39	49	100	48	74	0	0	226	0	10	1	86	0	YU
REM	0	1	5	1	2	0	0	50	2	19	17	1	0	1	46	0	6	0	1	3	REM
BAS	0	4	29	0	59	86	48	75	138	306	1	5	0	6	10	1	67	22	156	0	BAS
NOS	0	3	63	0	25	34	2	211	54	305	0	1	1	32	5	2	147	24	30	2	NOS
ATL	0	1	60	0	12	32	25	378	34	262	0	1	7	78	11	2	104	68	21	42	ATL
MED	2	21	19	20	60	4	1	247	40	147	264	30	0	3	574	1	26	1	61	8	MED
BLS	0	2	1	6	15	3	2	3	16	23	20	7	0	0	5	0	3	0	47	0	BLS
SUM	8	297	688	131	1710	542	448	3515	2032	5935	721	504	12	229	2191	35	1271	309	2774	181	SUM
	AL	AT	BE	BG	CS	DK	FI	FR	DD	DE	GR	HU	IS	IE	IT	LU	NL	NO	PL	PT	

Emitters

Receivers	RO	ES	SE	CH	TR	SU	GB	YU	REM	BAS	NOS	ATL	MED	BLS	NAT	IND	SUM	
AL	2	1	0	0	0	2	0	7	0	0	0	0	0	0	0	15	80	AL
AT	5	3	5	20	0	7	29	16	0	2	4	1	0	0	0	68	859	AT
BE	0	2	0	0	0	0	58	0	0	0	7	2	0	0	0	17	300	BE
BG	42	0	1	0	4	43	3	23	0	0	0	0	0	0	0	57	356	BG
CS	7	3	7	7	0	18	36	16	0	3	6	1	0	0	0	80	1249	CS
DK	0	1	4	0	0	1	69	0	0	2	7	2	0	0	0	19	260	DK
FI	1	1	76	0	0	88	56	1	0	13	7	2	0	0	0	105	711	FI
FR	1	139	3	24	0	2	328	13	1	1	42	42	0	0	0	273	2634	FR
DD	1	3	6	3	0	6	73	2	0	4	12	2	0	0	0	62	1161	DD
DE	2	16	7	37	0	7	233	4	0	4	33	8	0	0	0	149	2823	DE
GR	14	1	1	0	4	19	1	10	1	0	0	0	0	0	0	64	345	GR
HU	18	1	2	2	0	16	8	42	0	1	1	0	0	0	0	51	563	HU
IS	0	1	0	0	0	0	16	0	0	0	1	1	0	0	0	28	58	IS
IE	0	3	0	0	0	0	50	0	0	0	3	8	0	0	0	29	159	IE
IT	7	30	2	28	0	5	34	49	4	1	4	4	0	0	0	182	1450	IT
LU	0	0	0	0	0	0	2	0	0	0	0	0	0	0	0	1	21	LU
NL	0	2	1	0	0	0	85	0	0	0	11	2	0	0	0	20	370	NL
NO	1	2	33	1	0	10	254	1	0	5	18	9	0	0	0	150	857	NO
PL	9	4	32	8	0	65	118	17	0	13	17	3	0	0	0	171	2564	PL
PT	0	38	0	0	0	0	4	0	1	0	0	19	0	0	0	44	157	PT
RO	156	1	4	1	5	127	12	61	0	1	2	0	0	0	0	113	1018	RO
ES	1	458	1	1	0	0	56	4	6	0	6	43	3	0	0	215	1071	ES
SE	2	2	139	1	0	40	171	2	0	17	20	5	0	0	0	140	1178	SE
CH	0	9	1	37	0	0	25	2	0	0	3	2	0	0	0	32	414	CH
TR	20	1	2	0	73	117	4	8	1	0	0	0	0	0	0	239	633	TR
SU	109	7	180	8	18	2966	270	54	0	52	33	6	0	0	0	1356	8243	SU
GB	0	9	3	1	0	4	664	0	0	1	23	23	0	0	0	89	1102	GB
YU	40	7	3	5	1	25	12	193	1	1	2	1	0	0	0	151	1236	YU
REM	1	34	0	2	1	1	9	4	37	0	2	3	1	0	0	338	588	REM
BAS	3	4	112	2	0	67	176	4	0	27	23	4	0	0	0	143	1575	BAS
NOS	0	16	16	3	0	7	967	1	0	5	68	34	0	0	0	215	2274	NOS
ATL	1	201	40	3	0	39	888	1	0	6	59	256	0	0	0	1371	4005	ATL
MED	34	146	2	12	18	45	57	86	25	1	8	10	1	0	0	562	2536	MED
BLS	26	0	3	0	18	180	6	7	0	1	1	0	0	0	0	116	514	BLS
SUM	500	1149	686	207	142	3906	4775	629	81	163	425	494	6	0	0	6666	43357	SUM
	RO	ES	SE	CH	TR	SU	GB	YU	REM	BAS	NOS	ATL	MED	BLS	NAT	IND	SUM	

Source: EMEP, Calculated Budgets for Airborne Acidifying Components in Europe, 1985, 1987, 1988, 1989 and 1990, August 1991.

an extensive programme for the monitoring and evaluation of the long-range transport of air pollutants starting with sulphur dioxide and the possible extension to other pollutants." In addition, the Participating States advocated: "the inclusion, where appropriate and possible, of the various areas of co-operation into the programme of work of the United Nations Economic Commission for Europe."

The selection of the UN/ECE as a forum for developing such a programme relies essentially on two elements. First, the UN/ECE already had substantial experience in the field of environment, as its first activities in this area dated back to the late 1950s. Secondly, and perhaps more important, the UN/ECE was the unique forum that regroups all the countries of the region, be they emitters or receivers of air pollutants and was therefore in a position to elaborate a legally binding international agreement.

The negotiations that led to the signature in 1979 in Geneva of the Convention of Long-range Transboundary Air Pollution included several rewordings. Difficulties encountered during the elaboration of the Convention were mainly due to the fact that only acidification of lakes with poor buffering capacity was contemplated at that stage. Finally, a high-level meeting within the framework of the ECE took place in Geneva on 13 November 1979 where the Convention on Long-range Transboundary Air Pollution was signed by the 34 countries and the European Economic Community.[14] Thirty-two countries have now ratified the Convention (Table VII),[15] which entered into force in March 1983, 90 days after the 24th ratification.

3.2. *The Convention*

In 1979, a convention such as the Convention on Long-range Transboundary Air Pollution was an innovation. No international agreement in the field of air pollution had ever been signed, except for the following three agreements: (1) the Treaty Banning Nuclear Weapon Tests in the Atmosphere, in Outer Space, and Under Water in 1963 which, however, concentrates on nuclear problems; (2) the Nordic Environmental Protection Convention of 1974 which mentions only briefly the problems of air pollution; and (3) the Convention concerning the Protection of Workers against Occupational Hazards in the Working Environment Due to Air Pollution, Noise and Vibration of 1977, which does not deal with transboundary air pollution.[16]

The Convention lays down a series of principles. The first fundamental principle of the Convention is that the contracting parties, taking due account of the facts and problems involved, are determined to protect man and his environment against air pollution, which is an attempt to limit, as far as possible, gradually reduce and prevent air pollution including long-range transboundary air pollution.

The ensuing articles of the Convention focus on the necessity for the Parties to exchange information, not only on the volume of the emissions but also on the policies and strategies aimed at limiting these emissions and at consulting each other at the request of one of the Parties to the Convention. Furthermore, the Parties to the Convention undertook to elaborate policies aimed at improving air quality and to co-operate on research and development in the field of technologies for reducing emissions of major air pollutants; on improved models for better understanding of the transmission of long-range transboundary air pollutants; on the effects of sulphur compounds and other major air pollutants on

Table VII Status of the Convention and its related Protocols (as of 26 November 1991)

	Convention (a)		EMEP Protocol (b)		Sulphur Protocol (c)		NOx Protocol (d)		VOC Protocol (e)
	Signature	Ratification*	Signature	Ratification*	Signature	Ratification*	Signature	Ratification*	Signature
Austria	13.11.1979	16.12.1982 (R)		04.06.1987 (Ac)	9. 7.1985	04.06.1987 (R)	1.11.1988	15.01.1990 (R)	19.11.1991
Belarus	14.11.1979	13.06.1980 (R)	28.09.1984	04.10.1985 (At)	9. 7.1985	10.09.1986 (At)	1.11.1988	08.06.1989 (At)	
Belgium	13.11.1979	15.07.1982 (R)	25.02.1985	05.08.1987 (R)	9. 7.1985	09.06.1989 (R)	1.11.1988		19.11.1991
Bulgaria	14.11.1979	09.06.1981 (R)	04.04.1985	26.09.1986 (Ap)	9. 7.1985	26.09.1986 (Ap)	1.11.1988	30.03.1989 (R)	19.11.1991
Canada	13.11.1979	15.12.1981 (R)	03.10.1984	04.12.1985 (R)	9. 7.1985	04.12.1985 (R)	1.11.1988	25.01.1991 (R)	19.11.1991
Cyprus		20.11.1991 (Ac)		20.11.1991 (Ac)					
Czech and Slovak Federal Republic	13.11.1979	23.12.1983 (R)		26.11.1986 (Ac)	9. 7.1985	26.11.1986 (Ap)	1.11.1988	17.08.1990 (Ap)	
Denmark	14.11.1979	18.06.1982 (R)	28.09.1984	29.04.1986 (R)	9. 7.1985	29.04.1986 (R)	1.11.1988		19.11.1991
Finland	13.11.1979	15.04.1981 (R)	07.12.1984	24.06.1986 (R)	9. 7.1985	24.06.1986 (R)	1.11.1988	01.02.1990 (R)	19.11.1991
France	13.11.1979	03.11.1981 (Ap)	22.02.1985	30.10.1987 (R)	9. 7.1985	13.03.1986 (Ap)	1.11.1988	20.07.1989 (Ap)	19.11.1991
Germany (5)	13.11.1979	15.07.1982 (R)(2)	26.02.1985	07.10.1986 (R)(2)	9. 7.1985	03.03.1987 (R)(2)	1.11.1988	16.11.1990 (R)	19.11.1991
Greece	14.11.1979	30.08.1983 (R)		24.06.1988 (Ac)			1.11.1988		19.11.1991
Holy See	14.11.1979								
Hungary	13.11.1979	22.09.1980 (R)	27.03.1985	08.05.1985 (Ap)	9. 7.1985	11.09.1986 (R)	3.05.1989	12.11.1991 (Ap)	19.11.1991
Iceland	13.11.1979	05.05.1983 (R)							
Ireland	13.11.1979	15.07.1982 (R)	04.04.1985	26.06.1987 (R)			1.05.1989		
Italy	14.11.1979	15.07.1982 (R)	28.09.1984	12.01.1989 (R)	9. 7.1985	05.02.1990 (R)	1.11.1988		19.11.1991
Liechtenstein	14.11.1979	22.11.1983 (R)		01.05.1985 (Ac)	9. 7.1985	13.02.1986 (R)	1.11.1988		19.11.1991
Luxembourg	13.11.1979	15.07.1982 (R)	21.11.1984	24.08.1987 (R)	9. 7.1985	24.08.1987 (R)	1.11.1988	04.10.1990 (R)	19.11.1991
Netherlands	13.11.1979	15.07.1982 (At)(3)	28.09.1984	22.10.1985 (At)(3)	9. 7.1985	30.04.1986 (At)(3)	1.11.1988	11.10.1989 (At)(3)	19.11.1991
Norway	13.11.1979	13.02.1981 (R)	28.09.1984	12.03.1985 (At)	9. 7.1985	04.11.1986 (R)	1.11.1988	11.10.1989 (R)	19.11.1991
Poland	13.11.1979	19.07.1985 (R)(2)		14.09.1988 (Ac)			1.11.1988		
Portugal	14.11.1979	29.09.1980 (R)		10.01.1989 (Ac)					
Romania	14.11.1979 (1)	27.02.1991 (R)							
San Marino	14.11.1979								
Spain	14.11.1979	15.06.1982 (R)		11.08.1987 (Ac)			1.11.1988	04.12.1990 (R)	19.11.1991
Sweden	13.11.1979	12.02.1981 (R)	28.09.1984	12.08.1985 (R)	9. 7.1985	31.03.1986 (R)	1.11.1988	27.07.1990 (R)	19.11.1991
Switzerland	13.11.1979	06.05.1983 (R)	03.10.1984	26.07.1985 (R)	9. 7.1985	21.09.1987 (R)	1.11.1988	18.09.1990 (R)	19.11.1991
Turkey	13.11.1979	18.04.1983 (R)	03.10.1984	20.12.1985 (R)					
Ukraine	14.11.1979	05.06.1980 (R)	28.09.1984	30.08.1985 (At)	9. 7.1985	02.10.1986 (At)	1.11.1988	24.07.1989 (At)	19.11.1991 (1)
USSR	13.11.1979	22.05.1980 (R)	28.09.1984	21.08.1985 (At)	9. 7.1985	10.09.1986 (At)	1.11.1988	21.06.1989 (At)	
United Kingdom	13.11.1979	15.07.1982 (R)(4)	20.11.1984	12.08.1985 (R)			1.11.1988	15.10.1990 (R)(4)	19.11.1991
United States	13.11.1979	30.11.1981 (At)	28.09.1984	29.10.1984 (At)			1.11.1988 (1)	13.07.1989 (At)	19.11.1991
Yugoslavia	13.11.1979	18.03.1987 (R)		28.10.1987 (Ac)					
European Community	14.11.1979	15.07.1982 (Ap)	28.09.1984	17.07.1986 (Ap)					
Total:	34	33	22	31	20	20	26	19	21

(a) Convention of Long-range Transboundary Air Pollution, adopted 13.11.1979, entry into force 16.3.1983. (b) Protocol to the 1979 Convention on Long-range Transboundary Air Pollution on Long-term Financing of the Co-operative Programme for Monitoring and Evaluation of the Long-range Transmission of Air Pollution in Europe (EMEP), adopted 28 9.1984, entry into force 28.1. 1988. (c) Protocol to the 1979 Convention on Long-range Transboundary Air Pollution on the Reduction of Sulphur Emissions or their Transboundary Fluxes by at least 30 percent, adopted 8.7.1985, entry into force 2.9.1987. (d) Protocol to the 1979 Convention on Long-range Transboundary Air Pollution concerning the Control of Emissions or Nitrogen Oxides on their Transboundary Fluxes, adopted 31.10.1988, entry into force 14.2.1991. (e) Protocol to the 1979 Convention on Long-range Transboundary Air Pollution concerning the Control of Emissions of Volatile Organic Compounds or their Transboundary Fluxes, adopted 18.11.1991.

Notes: Ratification* = R, Ac = Accession, Ap = Approval, At = Acceptance. (1) With declaration upon signature. (2) With declaration upon ratification. (3) For the Kingdom in Europe. (4) Including the Bailiwick of Guernsey, the Isle of Man, Gibraltar, the United Kingdom Sovereign Base Areas of Akrotiri and Dhekhelia on the Island of Cyprus. (5) The former GDR signed the Convention on 13.11.1979, ratified it on 7.6.1982, accorded to the EMEP Protocol on 17.12.1966 with a declaration upon accession and signed the Sulphur Protocol on 9.7.1985 and the NO_x Protocol on 1.11.198.

human health and the environment; on economic, social, and environmental assessment of alternative measures for attaining environmental objectives; and on educational and training programmes related to the environmental aspect of pollution by major pollutants.

The contracting parties also undertook to exchange available information on emission data on major changes in national policies, control technologies for reducing air pollution, the projected cost of emission control, meteorological, physical, chemical and biological data relating to the effects of long-range transboundary air pollution and national, subregional, and regional policies and strategies for the control of sulphur compounds and other major air pollutants.

The background for the first protocol to the Convention is already a constituent part of the Convention's main document. It concerns implementation and further development of a co-operative programme for monitoring and evaluating the long-range transmission of air pollutants in Europe, which will be referred to as EMEP.

An Executive Body to the Convention was also created: "The representatives of the contracting parties shall, within the framework of the Senior Advisors to ECE Governments on Environmental and Water Problems, constitute the Executive Body of the present Convention and shall meet at least annually in that capacity." The secretariat of the Convention is entrusted to the Executive Secretary of the United Nations Economic Commission for Europe.

At that stage the Convention did not include any specific commitments for the Parties to reduce air pollution. The Convention is more like a framework that allows for future developments. Furthermore, the only pollutant linked to acid rains mentioned in the Convention is sulphur dioxide. However, the Convention started a process of consultation which led to the signature and entry into force of various protocols which did have direct impact on emissions of pollutants and thus on acid rains.

3.3. *The Protocols to the Convention*

3.3.1. *Co-operative Programme for Monitoring and Evaluation of Long-range Transmission of Air Pollutants in Europe (EMEP).*[17] The strength of the Convention on Long-range Transboundary Air Pollution as an international instrument to limit air pollution and consequently acid rain is embodied in its various protocols which have been developed since 1979. The first of these protocols concerns the financing of EMEP. Within the framework of the Convention and its successive protocols EMEP has a key role, since it is a programme that enables parties to assess progress made in implementing the various aspects of the Convention.

The main objective of the Co-operative Programme for Monitoring and Evaluation of Long-Range Transmission of Air Pollutants in Europe (EMEP), established in 1977, is to provide Governments with information on the deposition and concentration of air pollutants, as well as on the quantity and significance of long-range transmission of pollutants and fluxes across boundaries.

The Programme is carried out under the auspices of the Executive Body for the Convention. EMEP is supervised by its Steering Body and operates within the work-plan and the budget approved by the Executive Body. The World Meteorological Organization (WMO) also participates in EMEP by making available the necessary meteorological data, providing a forum for the assessment of the transport of air pollutants by appropriate modelling activities and encouraging the participation in EMEP of competent national meteorological services.

The activities of EMEP are divided into two parts: chemical and meteorological. The Chemical Co-ordinating Centre (CCC) at the Norwegian Institute for Air Research (NILU) is responsible for the chemical part of the programme. The main tasks of the CCC are to collect data from national monitoring stations operated by the Parties to the Convention, in accordance with the established reporting procedure; to assess and store such data, and preparation of data reports; to organize laboratory tests, in order to establish and improve the quality of the chemical analyses; to review and recommend sampling and analytical methods for compounds included in EMEP; to recommend and co-ordinate quality assurance procedures; and to elaborate procedures for estimating and reporting emission data.

The meteorological part of EMEP is carried out, with scientific co-ordination by WMO, by the Meteorological Synthesizing Centres (MSCs). The Meteorological Synthesizing Centre-East (MSC-E) is located at the Institute of Applied Geophysics in Moscow (USSR), and the Meteorological Synthesizing Centre-West (MSC-W) is located at the Norwegian Meteorological Institute (DNMI) in Oslo (Norway). The main tasks of the MSCs have been: model calculations of the transboundary fluxes and the deposition of air pollutants; verification and development of models for the transport of air pollutants, taking into account ground measurement data received from CCC; assessment and utilization of scientific results concerning transport and transformation processes for the development of new models or improvement of existing ones. The collection and storage of emission data received from the Parties to the Convention is a joint effort of all three centres.

The measurement activity of EMEP initially focused on sulphur oxides in air and precipitation, but has been gradually expanded. During the fourth phase of EMEP (1987–1989) and during the fifth phase (1990–1992), the measurement activities have included sulphur dioxide, sulphate, several nitrogen compounds and ozone in rural air, and all important ions in precipitation. The measured data provide valuable information on the regional distribution of air pollutants in Europe, and they are also used for comparison with the model calculations undertaken by EMEP's two Meteorological Synthesizing Centres. In order to improve the reliability of the data, quality assurance activities have a high priority in the work of the CCC.

The EMEP sampling network provides the database for concentrations and depositions of air pollutants, in particular the sulphur and nitrogen compounds, needed for the validation and verification of the model estimates by the MSCs. In 1990, the CCC received data from 101 measurement sites. The network has recently been expanded in the United Kingdom (five new stations), Spain (three new stations), USSR (two new stations), Portugal (two new stations, but one has closed) and the Netherlands (one station). Seventy-nine stations reported both precipitation and air quality data, 10 stations reported only precipitation and 12 stations only air data. There are still areas in Europe where the density of stations is insufficient. The location of the stations is shown in Figure 1.

For the further improvement of data collection and assessment, CCC has elaborated a quality assurance plan for EMEP. It is based on a corresponding plan of the North American acid deposition projects. The quality assurance plan is expected to ensure that collected and reported data are complete, precise and

Figure 1 Location of EMEP Stations.
Source: Norwegian Institute for Air Research, Progress Report from the Chemical Co-ordinating Centre for the Period 1 July–31 December 1990.

representative. It is aimed, *inter alia*, at a review of existing sites, better documentation of sites and site operation, a review of the methods of sampling and chemical analysis, standardization of operating and assessment procedures, training and preventive maintenance.

In the context of EMEP, emission inventories are of utmost importance because emission data form the input to the transmission models and, thereby, decisively affect modelling results. The spatial resolution of the emission data needed is given by the model grid size of 150 km x 150 km (Figure 2). Emission data have been provided by Parties to the Convention or have been estimated by the Centres on the basis of statistical data or other information.

The main goal of the work on atmospheric emissions at CCC during the fourth phase of EMEP (1987–1989) was to prepare guidelines to assist the EMEP countries on estimating emissions and on reporting procedures. This was achieved through the following tasks: collection of information on emissions, emission

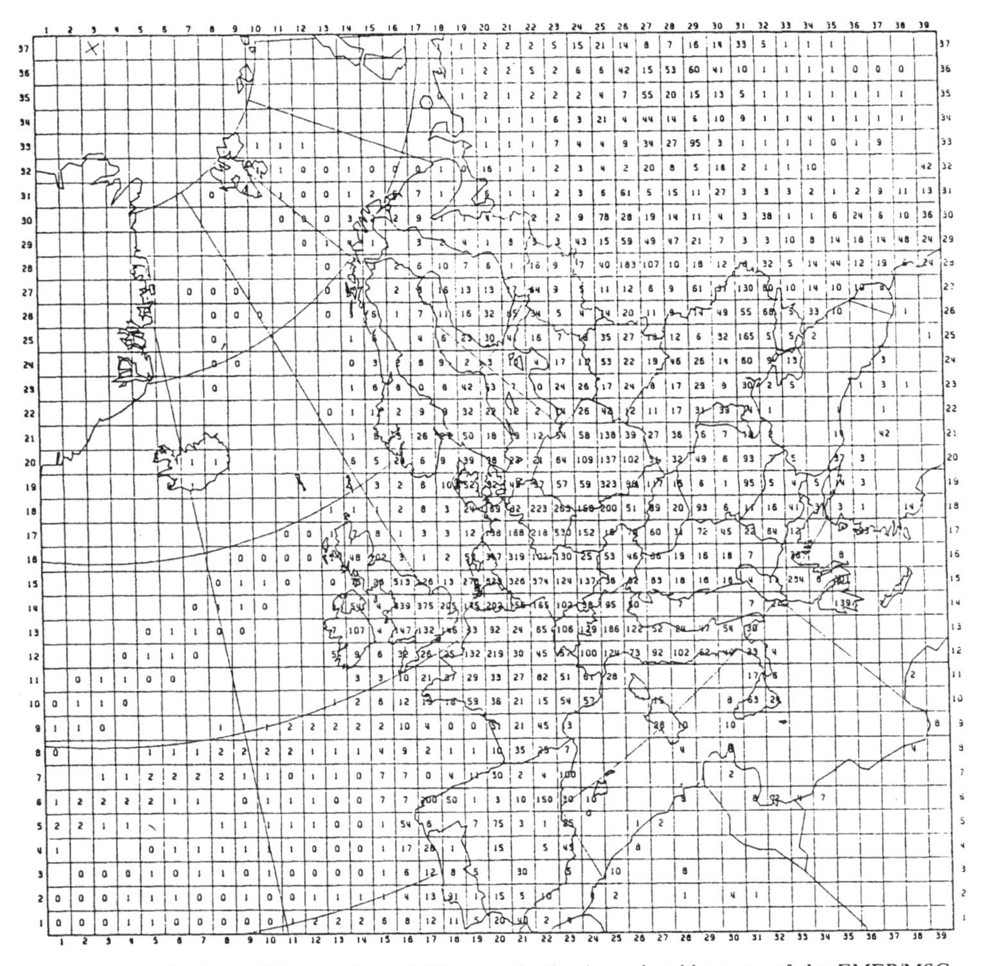

Figure 2 EMEP Map—1989 emissions of Nitrogen Oxides in each grid square of the EMEP/MSC-W grid for Calculations. (Unit 1000 tonnes per annum as NO_2.)
Source: EMEP, Calculated Budgets for Airborne Acidifying Components in Europe, 1985, 1987, 1988, 1989 and 1990, August 1991.

factors, methodologies of their estimation and statistics on the consumption of raw materials and the production of industrial goods in Europe, *inter alia*, through surveys; exchange of information and technical expertise on emission inventories between national experts, as well as between EMEP and other regional programmes, through workshops and several technical meetings, and development of technical guidelines for reporting and estimation of the SO_2, NO_x and VOC emissions in order to harmonize emission inventorying in Europe, taking into account the reporting requirements of the protocols on reduction of SO_2, and NO_x emissions to the atmosphere and the VOC protocol expected to be adopted by 1992.

As far as possible, the emission data employed in the calculations are based on official data submitted by the Parties participating in EMEP. Official national annual total data for sulphur and nitrogen oxide emissions are available for nearly all European countries. When no official information is available, data necessary to carry out the calculations are estimated from open information on the population distribution, and from the locations of power production facilities and relevant industrial activities.

3.3.2. *Protocol to the 1979 Convention on Long-range Transboundary Air Pollution on the reduction of sulphur emissions.*[18] The second protocol signed was, in fact, the first one to contain an explicit notion of emission reduction. This second protocol relates to the reduction of sulphur emissions or their transboundary fluxes by at least 30%. The negotiations for the second protocol lasted two years. It generated a number of discussions, in particular between countries that already had advanced legislation in the field of reducing the sulphur emissions and countries that did not have any, as well as between countries that emit sulphur dioxides and those that receive acid rains or acid depositions. The first group of countries saw this new protocol as an interference in internal affairs and requested more proof, both of the long-range distance aspect of sulphur emissions and the link of these emissions with acid rains. The second group of countries was willing to sign this protocol at once. Finally, in view of the acuteness of the problem and the increased damage to monuments, to water and to forests (in these two latter fields the multilateral conference of June 1984 on the Causes and Prevention of Damages caused to Forests and Water by Air Pollution in Europe, Munich 24–27 June 1984, played a determinant role), the protocol was signed.

With this new protocol Parties undertook to reduce their annual sulphur emissions or their transboundary fluxes by at least 30% as soon as possible and at the latest by 1993, using 1980 levels as a basis for calculating the reductions. The role of the EMEP is reinforced by the fact that EMEP shall in good time before the annual session of the Executive Body provide calculations of sulphur and also transboundary fluxes and disposition of sulphur compounds for each previous year within the geographical scope of EMEP, utilizing appropriate models. On 9 July 1985 the protocol was signed by 20 of the Parties to the Convention and thereafter ratified, approved or accepted in the following months.

Among the most important countries that did not sign the protocol, the United States made it a question of principle not to ratify such an international instrument. They have however come to an agreement on this issue and on a bilateral basis with Canada. The United Kingdom, which is one of major emitters of SO_2 in Europe (Tables I–II), also did not sign the protocol. Most of the other countries that did not sign the protocol did not do so because they were not in a position to substantially reduce their emissions, either because of the structure of their energy production, or because they could not bear the heavy investment necessary to achieve such reductions.

3.3.3. *Protocol concerning control of emissions of nitrogen oxides or their transboundary fluxes.*[19] On 1 November 1988 in Sofia (Bulgaria), 24 countries signed the protocol concerning control of emissions of nitrogen oxides or their transboundary fluxes. Due to the short time since the signature, only 18 Parties have ratified this protocol. The fact that more countries have signed this protocol as compared to the SO_2 protocol is essentially due to the differences in sources of emissions of both pollutants. If in the first case the major energy options are put into question because of the signature of the SO_2 protocol, in the second case as most of emissions of nitrogen oxides come from mobile sources such as transportation, abatement can be more easily achieved by the adjunction, for example, of catalytic converters.

Another major difference between this new protocol and that on SO_2 is that Parties do not undertake to reduce the emissions but to control them so that by 31 December 1994 they do not exceed the level of annual national emissions of nitrogen oxides or transboundary fluxes of such emissions in the calendar year 1987, the year the protocol was negotiated. Another alternative offered to the Party is to choose another year previous to 1987 as a base year and to maintain the level of emissions during the period between 1987 and 1996. This protocol appears therefore to be much more flexible to the Parties.

However, this flexibility is partly offset by a set of measures that Parties will take to reduce the emissions, such as the application of national emissions standards to major new stationary sources or to new mobile sources, the introduction of pollution control measures for major existing stationary sources. Another new element introduced in this protocol is the notion of transfer of technology that Parties undertake to facilitate, through the promotion of commercial exchange of available technology, direct industrial contacts and cooperation, including joint venture planning, exchange of information and experience and provision of technical assistance. Furthermore, Parties undertake to make unleaded fuel sufficiently available in order to facilitate the circulation of vehicles equipped with catalytic converters. Finally, the protocol includes a programme of work that provides an element of dynamism that did not exist in the preceding protocols.

It is to be noted that in article 2 of the protocol, Parties stressed that they are willing to commence negotiations no later than six months after the date of entry into force of this protocol, on further steps to reduce national annual emissions of nitrogen oxide or transboundary fluxes of such emissions. Furthermore, the protocol includes a technical annex which gives information on control technologies for NO_x emissions from stationary sources and from motor vehicles. This protocol thus appears to be much more comprehensive and flexible than the previous one. It provides a complete system of dealing with emission of NO_x which Parties undertake to utilize.

4. THE VARIOUS ACHIEVEMENTS OF THE CONVENTION

4.1. *Preliminary Assessment of its Results*

According to the terms of the 1979 Convention on Long-range Transboundary Air Pollution and subsequent protocols, Parties to the Convention are to submit to the secretariat of the Convention their policies and strategies for dealing with the discharge of air pollutants. Exchange of information concerning this discharge should also be published periodically. The Executive Body of the Convention reviews strategies and policies for air pollution abatement on a regular basis. This document, based on national contributions, reviews national strategies, national policy measures, administrative structures and international activities related to air pollution abatement.

This document also includes the most recent data submitted by Parties to the Convention on ambient air quality standards; fuel quality standards; emission standards; energy consumption and energy makes; total SO_2 emissions and total

NO_x emissions; total NH_3 emissions, total VOC emissions and total CO_2 emissions for 1980 to 2005, and disaggregated air pollutant emissions.

In terms of total sulphur emission, a substantive reduction in sulphur emissions has been achieved throughout the ECE. Taking into account the estimates done for certain countries, this decrease has amounted to 26% for the period 1979–1988. Of the 27 Parties for which definitive data are available for both 1980 and 1987 and for more recent years, 24 have reduced SO_2 emissions compared with 1980. Twelve Parties have already reduced their emissions by 30% or more. In contrast, emission data on nitrogen oxides submitted by 22 Parties for 1987 or a more recent year show reductions compared to 1980 for only 12 Parties and net increases for 10 Parties.

Estimates of future SO_2 and NO_x emissions typically reflect the targets set by the protocols. Parties thus estimate that the 1995 sulphur emissions will be 30% or more below the 1980 level, with twelve Parties forecasting reductions of 50% or more. Most signatories to the 1988 Sofia protocol and the related Sofia declaration expect NO_x emissions to stabilize or decline by 1995, with seven Parties forecasting reductions of about 30% or more. Current emission data on volatile organic compounds show a net increase from 1982 and 1985 levels for almost all parties reporting. Also 10 Parties forecast reductions by 1995.

The EMEP measurements show that between 1983 and 1987 there is no trend (32 sites) or a downward trend (14 sites) for sulphur dioxide at those sites that could be included in the analysis. A significant downward trend for sulphur dioxide was found in the United Kingdom, France, at some sites in Germany, and southern Scandinavia. In other parts of Europe (most parts of central Europe, the Baltic area and at one Polish site close to the Soviet border and in the northern region), there were no significant trends.

Measurements of sulphate in particles give very much the same picture. Downward trends were found at French sites, at the Belgian site and in southern parts of Germany. In the United Kingdom, however, there was no significant trend at the site on the English-Scottish border and at the site in southwest England there was a significant increase in the concentrations from 1982 to 1988. In southern Scandinavia and even in Iceland, there were significant downward trends in the sulphate concentrations in aerosols. The Polish site had a significant increase in concentrations. Again there were no significant trends in central Europe or in the northern part of Scandinavia or Finland.

The nitrate in precipitation had a decreasing trend at one site only (in Germany). At most other sites no significant trend was found or there were upward trends as in the Baltic area and Yugoslavia.

MSC-W has done a study to detect and quantify reduction in concentration levels as a consequence of the sulphur emission reductions reported by Parties (EMEP MSC-W Report 1/89). The reported emission reduction from 1980 to 1986 was 16.4% for Europe as a whole. The resulting concentration reductions were 19% for SO_2 and 16% for particulate sulphate. Thus, the estimates compare well with the reported emission reductions.

4.2. *Other Impacts of the Convention*

The Convention has generated a number of activities which contribute to its dynamism and its adaptation to changing needs. These activities have generated among Parties to the Convention a momentum which increases the knowledge about the problem of transboundary air pollution, its economic consequences, the available abatement technologies and the strategies to do so.

4.2.1. *Abatement Strategies.* In order to prepare abatement strategies, the Executive Body to the Convention has created a special working group. The terms of reference of the Working Group on Abatement Strategies have evolved as a result of the various demands of the Executive Body. Its latest revised terms of references dates back to the eighth session of the Executive Body (1990). It takes into account the elements included in the 1988 NO_x protocol, in particular as it refers to Article 6 of the NO_x protocol. The Working Group is considering the synthesis of research and monitoring through national research programmes, in the workplan of the Executive Body and through co-operative programmes within the framework of the Convention with a view to developing and applying an approach based on critical loads. It aims at developing a common understanding of critical loads and proposals for abatement strategies, taking into account the best available scientific and technological development, internationally accepted critical loads and overall emission reductions.

The strategies elaborated by the Working Group take into account a number of elements such as emission inventories and measurements estimation and modelling of air quality and deposition; assessment of ecological substantiation; scientific information relevant to the determination of critical loads for sensitive receptors; harmonization of procedure methods for the definition of critical loads; mapping of sensitive receptors and their critical load values; assessment of emission reductions in technologies; economic analysis of potential abatement strategies; co-operation on exchange of technologies and existing and proposed national target load values and standards.

The group's work comprises nitrogen oxides and tropospheric ozone and sulphur dioxide and the relevant supporting data, keeping in mind the requirements of the NO_x protocol. This group will present to the ninth session of the Executive Body in November 1991 an outline of a protocol, for further reduction of sulphur emissions after 1993 on the basis of critical loads, best available technology, energy savings and/or other considerations including market-based economic instruments. It will also propose further work for the second step of the NO_x protocol.

4.2.2. *Research on Effects.* As the Convention aims at protecting man and his environment against air pollution, the Parties to the Convention decided to initiate research and exchange of information on the effects of long-range transboundary air pollution. To this end, the Executive Body of the Convention decided to create a Working Group on Effect. On the basis of available data scientific guidance is provided for all questions concerning effects for consideration by the Executive Body. In recent years, the working group has been considering the effects of air pollution on human health, crops, forestry, materials and aquatic ecosystems. Under its aegis a number of International Co-operative Programmes have been set up related to forests (Germany as lead country); materials (Sweden as lead country); crops (United Kingdom as lead country); integrated ecosystems monitoring (Sweden as lead country); and mapping of critical loads and levels (Germany as lead country).

4.2.3. *Control Technologies.* Pursuant to various articles of the Convention and its protocols the Executive Body of the Convention is constantly monitoring the state of air pollution technology and by doing so provides guidance to the Parties in identifying economically feasible technologies. This information is contained in annexes to the protocols and covers in general technologies for the main emission source categories of given pollutants. The technologies listed are considered to be well established on the basis of operational experience. Because of the experience, but also because of the use of new more environmentally sound technologies, the annexes to the protocols are updated regularly.

Regularly seminars on Emission Control technology for Stationary Sources are organized. The fifth one was held in Nuremberg in June 1991. Such seminars provide a forum for exchanging available information on control technology for reducing air pollution, including investment and operating costs of such technologies. It provides also for an integrated pollution control approach for using best available technology (BAT) which would ensure that pollution control strategies would use the most efficient technologies in terms of preventing air pollution. It should be noted that the notion of BAT goes beyond the notion of abatement technologies as it incorporates in its concept the industrial process as a whole from "cradle to grave", namely from the construction of the plant to the consumer habits and the disposal of the product.

4.2.4. *Economics and Air Pollution.* One of the main concerns of Parties to the Convention and to the various protocols is the economic impact of air pollution but also the economic implications of their decisions. The Group of Economic Experts on Air Pollution has been instrumental in this respect. In recent years the work conducted in the economic sphere under the aegis of the Executive Body has concentrated on several very specific issues. Because emission data and their forecast are instrumental in the development and implementation of new protocols, the group has studied ways and means to harmonize emission projections on the basis of common macro-economic hypotheses, the aim being that alternative scenarios based on similar assumptions should be developed on a comparable basis.

Another key issue for future developments of the Convention is the development of efficient economic instruments for enforcing the decisions which will follow the implementation of these scenarios. This approach includes the review of existing economic instruments for stationary and mobile sources and the applicability of taxes on energy, emission charges and emission trading. The benefit of a cleaner environment has also been studied, by evaluating the economic and social activities in specific geographical areas that may have benefited from reducing emission and cleaning up the environmental damage.

Another aspect which has retained the attention of the Group is the integrated assessment of costs and benefits through modelling exercises. The models have been used in order to perform cost effectiveness and cost and benefits analyses. Finally, an analysis of the international distribution of emission reductions which the adoption of the critical loads approach will require has been performed, in order to clarify the equity implications of abatement policies, which include non-uniform percentage reductions, devoting particular attention to the ways Parties might allocate among themselves the resources necessary to achieve such reductions.

5. THE PROTOCOLS OF THE SECOND TYPE

5.1. *The New Protocols*

Three new protocols are under consideration by the Parties to the Convention on Long-range Transboundary Air Pollution. However, these considerations are at different stages of finalization.

5.1.1. *The Control of Emissions of Volatile Organic Compounds or their Transboundary Fluxes.* At its sixth session in 1988, the Executive Body to the Convention identified the need to control emissions of volatile organic compounds (VOCs) with their transboundary fluxes, as well as to control the incidence of photochemical oxidants, and the need for Parties that have already reduced their emissions to maintain and review the emission standards for VOCs. Since then Parties to the Convention, within the framework of the Working Group on VOCs, have negotiated the elaboration of a VOC protocol. This protocol was signed by 21 ECE countries in Geneva on 19 November 1991. The aim of this new protocol is to reduce by 1999 by at least 30%, using 1988 as a basis, the annual emissions of VOCs.

One of the differences from the previous protocols is the introduction of the level of emission of VOCs, namely if the country has an annual emission of VOCs in 1987 and 1988 below 500 000 tons and 20 kg per inhabitant or five tonnes per square kilometre, it will, as soon as possible, and as a first step, take effective measures to ensure at least that by 1999 at the latest its annual emissions of VOCs do not exceed the 1987 and 1988 levels. While in the NO_x protocol the deadline for implementing measures was not specified, in the protocol under consideration, firm deadlines are mentioned for Parties to take appropriate measures, such as to apply appropriate national and international emission standards; use best available technologies which are economically feasible; apply national and international measures to products that contain solvents; or promote the use of products that are low in, or do not contain, VOCs; apply appropriate national or international emissions standards; and encourage public participation in emission control programmes through public information. This latter item is a new item as compared to the former protocols. As in the case of the NO_x protocol, the Parties shall as a second step begin negotiations no later than 6 months after the date the protocol enters into force on further steps to reduce annual emissions of VOCs.

Another innovation introduced in this protocol is the annual review of the protocol that takes into account the best available scientific substantiation and technological development. A set of annexes complement the protocol. They include the designated tropospheric ozone management areas (TOMAS); the control measures for VOC emissions from stationary sources, the control measures for VOC emissions from on-road motor vehicles; the classification of VOCs based on their POCP (photochemical ozone creation potential). These technical annexes are an integral part of the protocol.

5.1.2. *Further Action on Existing Protocols.* As stipulated in the NO_x protocol, the Parties to the protocol as a second step commenced negotiations, six months after the date of entry into force of the protocol, on action to be taken to reduce NO_x emissions. The NO_x protocol entered into force on 14 February, 1991 and a meeting in August 1991 proposed a workplan for the preparation of the second step of the NO_x protocol to the Convention in order to reduce national annual emissions of nitrogen. In order to reduce emissions of nitrogen oxides and their transboundary fluxes critical load maps will be developed. Information provided by ongoing and planned investigations will be compiled and assessed. The concept of acidifying potential will be further elaborated. Information on the role of nitrogen species in eutrophication and tropospheric ozone formation will be substantiated.

A second sulphur protocol is also in its very preliminary phase. This second protocol will include such elements as critical loads, target loads and national emission reductions, as well as further obligations to prevent an increase in national or sub-national sulphur emissions.

As can be seen from the above, the new protocols to the Convention on Long-range Transboundary Air Pollution are more sophisticated. They include more and more elements related to the scientific specification of the emissions concerned, but also possible emission reduction measures to be taken by Parties to protocols. They also include a great deal of research in new areas. The work going on under the auspices of the Executive Body to the Convention is therefore innovative, not only in its legal aspect but also in the scientific, technological and economic aspects related to emissions of transboundary air pollutants.

5.2. *Future Possibilities*

5.2.1. *The Critical Load Approach.* Several problems have been coming up in the implementation of the recent protocols of the Convention which have led to intense discussions among experts. One of these problems is the flat rate approach, namely the fact that in the existing protocols the same abatement rate is applied to all parties, irrespective of their existing level of emission. An alternative to the flat rate reduction approach is the critical load approach. This approach is based on the actual effect of air pollutants, i.e. it is the effects of deposition of polluting substances which should be measured and which should guide parties in their negotiations. The goal is to reduce emissions of polluting substances in a cost-effective manner to levels where ultimately critical loads are not exceeded wherever possible. This approach requires first that a mapping of critical loads be established, and second that depositions be traced back to the emitter countries. Within this framework EMEP is an indispensable tool.

Critical loads can be defined as quantitative estimates of an exposure to one or more pollutants below which significant harmful effects on specified sensitive elements of the environment do not occur, according to present knowledge. They are not always possible to achieve due to economic, technological or other constraints. An intermediate stage is the achievement of what has been called target loads. Target loads take into account not only the environmental sensitivity but also technical, social, economic and political considerations by individual countries. One key concept considered in this context is the application of the best available technology (BAT).

The critical loads approach requires comprehensive information. On the one hand it is necessary to have a clear understanding of emission sources and quantities, which requires an inventory of current emissions and projection of future emission rates. On the other hand, maps of critical loads and target loads are also required. In order to establish the link between these two aspects long range transport models are required. Estimates of potential for and cost of emission reductions, including structural changes and conservation of energy and natural resources, are needed. This process should take place within the framework of an integrated assessment of cost and benefits of emission reduction.

5.2.2. *The Base Year.* Another problem encountered when negotiating the various protocols on VOCs, SO_2 and NO_x was the designation of a baseline year. Depending on the year that is selected, the effort which will be required of Parties to the protocol will differ greatly, depending on whether the Parties have already undertaken some reduction programmes before the selected baseline year. Another negative aspect of these protocols is that they do not take into account the effect of reduction of pollutants of other types; e.g. the reduction of 30% of SO_2 emissions through changes in energy policies will definitely have an impact on NO_x emissions, which means that an integrated approach to emission reduction will eventually be needed. This integrated approach will require that the various protocols on SO_2, NO_x, VOCs and any other pollutants be interlinked, in particular as far as abatement policies and strategies are concerned, as well as technologies.

References

1. Effects and Control of Transboundary Air Pollution, *Air Pollution Studies* No. 4 (Economic Commission for Europe, United Nations, New York, 1987) Chapter 1.
2. Multilateral Conference on the Environment, Munich 24–27 June 1984, organized by the Government of Germany in co-operation with the Executive Secretary of the Economic Commission for Europe, Statement by the delegation of UNEP under agenda item 8.
3. It is interesting to note that in document A/CONF.48/8 of the Stockholm Conference, on the Identification of Pollutants of International Concern, no mention is made of acid rain.
4. Member states of the UN/ECE are: Albania, Austria, Belarus, Belgium, Bulgaria, Canada, Cyprus, Czech and Slovak Federal Republic, Denmark, Estonia, Finland, France, Germany, Greece, Hungary, Iceland, Ireland, Israel, Italy, Latvia, Liechtenstein, Lithuania, Luxembourg, Malta, Netherlands, Norway, Poland, Portugal, Romania, Spain, Sweden, Switzerland, Turkey, Ukraine, Union of Soviet Socialist Republics, United Kingdom, United States and Yugoslavia.
5. Convention on Long-range Transboundary Air Pollution (Geneva, 1979, E/ECE/1010).
6. Formerly the German Democratic Republic. When mentioned, the Federal Republic of Germany is Germany without the former German Democratic Republic.
7. Strategies and Policies for Air Pollution Abatement, 1990 Review (United Nations, New York, 1991) Table V.
8. Strategies and Policies for Air Pollution Abatement, 1990 Review (United Nations, New York, 1991) Table VI.
9. Strategies and Policies for Air Pollution Abatement, 1990 Review (United Nations, New York, 1991) Table X.
10. Sweden's Case Study for the United Nations Conference on Human Environment, 1972: Air Pollution Across Borders.
11. Stockholm Declaration (1972).
12. Programme on Long-range Transport of Air Pollutants—Measurements and Findings.
13. Final Act of the Conference on Security and Co-operation in Europe, Helsinki, 1975, Part 5 Environment.
14. Convention on Long-range Transboundary Air Pollution (Geneva, 1979, E/ECE/1020).
15. Strategies and Policies, *op. cit.*, Table XI.
16. Selected Multilateral Treaties in the Field of the Environment, *UNEP Reference Series 3* (Nairobi, 1983).
17. Protocol to the 1979 Convention on Long-range Transboundary Air Pollution on Long-term

Financing of the Co-operative Programme for Monitoring and Evaluation of the Long-range Transmission of Air Pollutants in Europe (EMEP) (Geneva, 1984, ECE/EB.AIR/11).

18. Protocol to the 1979 Convention on Long-range Transboundary Air Pollution on the Reduction of Sulphur Emissions or their Transboundary Fluxes by at least 30 per cent (United Nations, Helsinki, 1985) ECE/EB.AIR/12.
19. Protocol to the 1979 Convention on Long-range Transboundary Air Pollution concerning the Control of Emissions of Nitrogen Oxides or their Transboundary Fluxes (United Nations, Sofia, 1988) ECE/EB.AIR/21.

7. A REVIEW OF SPATIAL AND TEMPORAL TRENDS OF PRECIPITATION COMPOSITION FOR NORTH AMERICA FOR THE PERIOD 1979–1987

DOUGLAS L. SISTERSON, VAN C. BOWERSOX
and ANTHONY R. OLSEN

This chapter discusses wet deposition in the United States and Canada and provides spatial and temporal (change with time, which includes trend and seasonal variations) analyses. The main purpose is to identify the locations of maximum deposition of pollutants and to determine if the spatial and temporal patterns have changed substantially during the period 1979–1987. Eight major wet deposition networks allow evaluation on a regional scale of seasonal and annual deposition and of precipitation-weighted concentrations of the major inorganic ions found in precipitation.

1. INTRODUCTION

The main purpose of this study is to identify the locations of maximum deposition of pollutants in the United States and Canada and to determine if the spatial and temporal patterns have changed substantially for the period 1979–1987. Several major wet deposition networks that have existed for at least five years allow evaluation of seasonal and annual deposition (precipitation amount times analyte concentration) and precipitation-weighted mean concentrations of the major inorganic ions found in precipitation on a regional scale. Criteria are developed for the selection and analysis of a single concentration and wet deposition data base for spatial and temporal analyses of the major inorganic ions in precipitation over the continental United States and Canada. This study summarizes part of the National Atmospheric Precipitation Assessment Program State of Science Report Number 6 (*Monitoring and Deposition: methods and results*[1]). All figures and tables in the present chapter have been reproduced from the State of Science Report Number 6 with the authors' permission.

2. MAJOR NETWORKS IN NORTH AMERICA

Large networks are defined here as networks that have sufficient density, have operated for five years or more, have a peer-reviewed quality assurance (QA) plan, and are in the Atmospheric Deposition System (ADS) data base.[2] Particularly important is the question of QA. For a network to be considered in this study, a documented and peer- or agency-reviewed QA plan and program must be in place. A common set of procedures must be specified, and a common analytical laboratory must be used. Intercomparison information between co-located sites from different networks must also be available to assess differences between network operations and methods. Networks that fulfill these guidelines are NADP/NTN, MAP3S, UAPSP, APIOS-D and -C, CAPMoN, FADMP, and REPQ; they are all reported in the ADS data base.

The Multi-State Atmospheric Pollution and Power Production Study (MAP3S) initiated the MAP3S precipitation chemistry network in 1976 in the northeastern United States to research the scavenging of pollutants by precipitation. This was the first wet-only, event-sampling network. The National Atmospheric Deposition Program (NADP) established a weekly sampling, nationwide network to monitor trends in precipitation chemistry in the United States in order to conduct research on atmospheric deposition and its effects on ecological, biological, agricultural, and aquatic systems, in cooperation with federal, state, and private research agencies. In 1982, the NADP assumed responsibility for coordinating the operation of the National Trends Network (NTN) of the federally supported National Acid Precipitation Assessment Program (NAPAP). The merged networks now have the designation NADP/NTN. The daily sampling network designated as the Utility Acid Precipitation Study Program (UAPSP) began in 1978 as a research network initiated by the Electric Power Research Institute (EPRI) and was expanded in 1982 to give broader regional coverage. The daily Canadian federal network, CAPMoN, was formed by using two previous networks as a base, the Canadian Network for Sampling Acid Precipitation (1977–1985) and the Air and Precipitation Monitoring Network (1978–1985). In 1980, the Province of Ontario established two networks as part of the Acidic Precipitation in Ontario Study (APIOS). The cumulative network, APIOS-C, with a four-week sampling period is designed to determine the long-term deposition pattern in Ontario. The daily network, APIOS-D, is designed to define the sector of origin of ion species as well as the frequency and intensity of acidic deposition episodes. The Florida Acid Deposition Monitoring Program (FADMP) began weekly sampling in 1981 with a network that collected weekly samples to provide regionally representative precipitation chemistry, with an overall goal to assess and develop an information base on the magnitude, variability, sources, effects, and control options of acidic deposition in Florida. The Reseau d'Echantillonnage des Precipitations du Quebec (REPQ) network is sponsored by the Quebec Ministry of the Environment. Weekly samples are collected to monitor the spatial distribution of and temporal changes in acidic deposition in Quebec and to gather the necessary data to verify and calibrate models of the transport of atmospheric pollutants. These networks are described more fully elsewhere.[1]

Our spatial analysis consists of three-year (1985–1987) composite annual and seasonal plots for individual analytes (H^+, SO_4^{2-}, NO_3^-, NH_4^+, and Ca^{2+}). These plots are based on all eight networks and depict the locations of wet deposition chemistry maxima for North America. The spatial patterns of the analytes are discussed. Trend analysis for wet deposition chemistry for the period 1979–1987 (based on a subset of six of the eight networks) for all analytes and the significance of the trend are also discussed.

A summary of the number of sites operated by each network from 1979 to 1987 and included in the ADS data base is given in Table I. For this table, a site is considered to have operated during a year if one or more samples were collected at some time during the year. Hence, sites are included that operated only for a portion of the year because they were either started or terminated during the year.

Table I Number of Sites by ADS Data Base Network from 1979 to 1987

Network	1979	1980	1981	1982	1983	1984	1985	1986	1987
NADP/NTN	39	82	97	110	142	177	195	203	204
MAP3S	8	8	9	9	9	9	9	9	9
UAPSP	18	10	21	22	25	27	26	25	25
CAPMoN	4	6	6	8	17	18	24	24	25
APIOS-D	0	8	16	16	16	17	15	17	15
APIOS-C	0	30	35	36	38	38	37	38	37
Total	69	144	184	201	247	286	306	316	315

3. CRITERIA FOR DATA BASE MERGER

The ADS 1985 Data Summary Report[3] discusses five quantitative data completeness measures defined by the Unified Deposition Database Committee (UDDC).[4] The five data completeness measures are based on the portion of the selected time period in which the occurrence and amount of precipitation are known, the portion of the collected precipitation volume that is associated with valid deposition samples, the percent of time and percent of collected samples that are associated with valid samples, and collection efficiency. Although the UDDC criteria are reasonable in the sense that a site meeting a network's protocol would be expected to exceed them, the criteria are strict. The primary requirements for data base merger in our study are that precipitation monitoring cover at least 90% of the annual and seasonal periods and that sample chemistry data be available for at least 60% of the annual and 50% of the seasonal precipitation measured.[5]

Figure 1, as an example, shows the location of the network sites operating for a full year in 1987 and the subset of those sites that met the annual UDDC for SO_4^{2-}. Of the 154 sites that met the UDDC criteria for 1987, approximately one-fourth are in the western half of the United States, one-fourth are in Canada, and the remaining one-half are in the eastern United States. In 1987, 54% of the sites operating for a full year met the UDDC criteria. The number of invalid samples occurring during the winter is the most frequent reason for a site to fail the UDDC criteria.

Sites are selected for temporal pattern analyses on the basis of relaxed criteria that are a compromise between ensuring that each summary is based on very complete data (meeting the UDDC criteria) and increasing the number of sites available for temporal pattern analysis. The criteria are that (1) precipitation monitoring covers at least 90% of the annual period and that (2) valid sample chemistry data represent at least 60% of the annual precipitation measured. Although thresholds are the same as the UDDC criteria, the relaxed criteria do not apply to the quarterly UDDC requirements or to the other three UDDC data completeness measures. The number of sites meeting the relaxed criteria for SO_4^{2-}, as an example, is shown in Table II. Over 94% of the sites operating for a full year met the relaxed criteria each year.

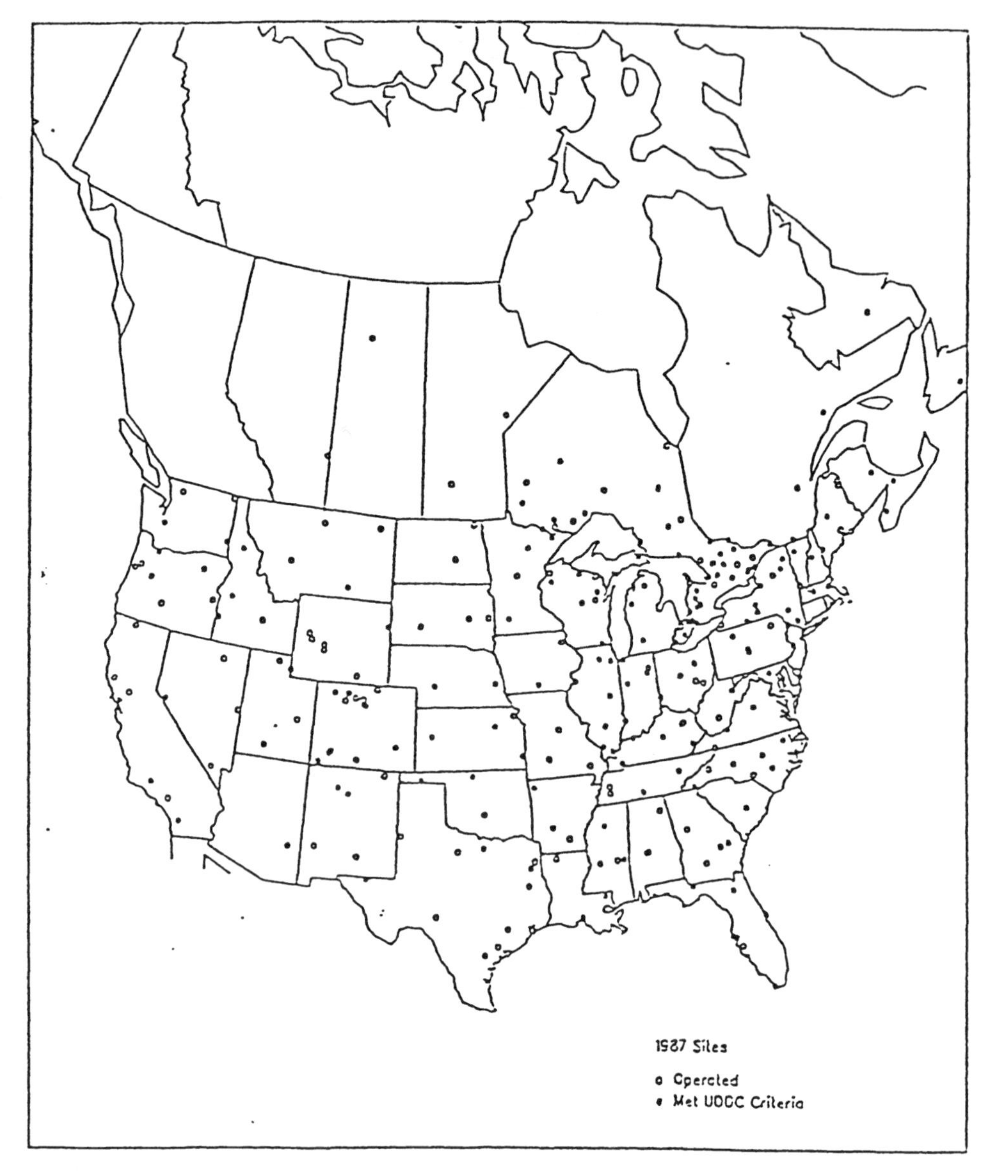

Figure 1 Geographic location of sites operating a full year in 1987 and the subset of sites meeting the annual UDDC criteria for sulfate.

4. METHOD OF PRESENTATION

Seasonal and annual spatial precipitation-weighted concentration (PWC) and deposition data are presented by using a gray-scale scheme. An objective analysis technique (kriging) was used to identify seven ranges of values corresponding to a fixed percentile ranking for North America as a whole. The percentiles correspond to those of the box plots used in trend analysis.

Table II Number of Sites by Network that Meet the Relaxed Annual Summary Data Completeness Criteria for Sulfate for Temporal Analysis

Network	1979	1980	1981	1982	1983	1984	1985	1986	1987
NADP/NTN	16	26	53	60	71	85	101	115	101
MAP3S	5	7	8	9	8	9	9	9	9
UAPSP	16	6	6	21	20	18	19	21	0
CAPMoN	0	0	0	0	4	11	11	17	13
APIOS-D	0	0	0	0	10	10	11	11	10
APIOS-C	0	0	10	14	22	21	20	22	21
Total	37	39	77	104	135	154	171	195	154

4.1. *Spatial Analysis Methodology*

Estimation of the spatial surface displayed in the maps was completed by using simple kriging, with the BLUEPACK kriging software and algorithms[6] applied to local neighborhoods. Other descriptions of the application of kriging to precipitation chemistry data are found elsewhere.[7–10] Simple kriging gave spatial surface estimates based on application of the best linear unbiased inferential rules from generalized least squares. The covariance structure for generalized least squares analysis was derived from the intersite correlations, which were assumed to be a function of distance. The semivariogram was used to develop the covariance structure. Given the semivariogram, a local neighborhood of monitoring sites is used near the grid node being estimated. Specifically, eight sites were used with the restriction that each octant from the grid node contributed a site if one was available within a maximum search radius.

A semivariogram model was estimated for each spatial surface. Intersite distances were grouped into distance categories, and one-half the average of the squared differences between all pairs of observations within the distance category was computed. Parameter values were estimated to fit one of five common semivariogram models: power, spherical, cubic, exponential, or gaussian. When the semivariogram does not tend to zero for measurements taken at co-located or arbitrarily close points, a discontinuity exists at the origin. This discontinuity is called a nugget effect. All semivariograms applied included a nugget effect.

An estimate of the spatial surface was obtained for a systematic triangular grid of points extending across the United States and eastern Canada. Each grid point defines a surrounding hexagonal area; collectively the hexagonal areas cover the surface. Each hexagon has an area of approximately 2670 km^2 and side length of approximately 32 km. A total of 3192 hexagons was used to cover the 48 continental United States. The hexagon was selected because it is the simplest figure that is nearly circular and covers the plane. The estimated surface is displayed by a gray-scale shading of the mosaic hexagon areas, as shown in Figure 2.

Selected percentiles of concentrations or depositions (10th, 25th, 50th, 75th, 90th, and 95th percentiles) defined the category boundaries. The minimum and maximum concentrations (depositions) defined the lowest and highest category boundaries. This approach allows the structure of the estimated surface to define the category and is similar to the use of box plots to represent a distribution. Our ability to estimate the spatial pattern in the western United States is restricted by the greater distances between qualifying monitoring sites (spatial

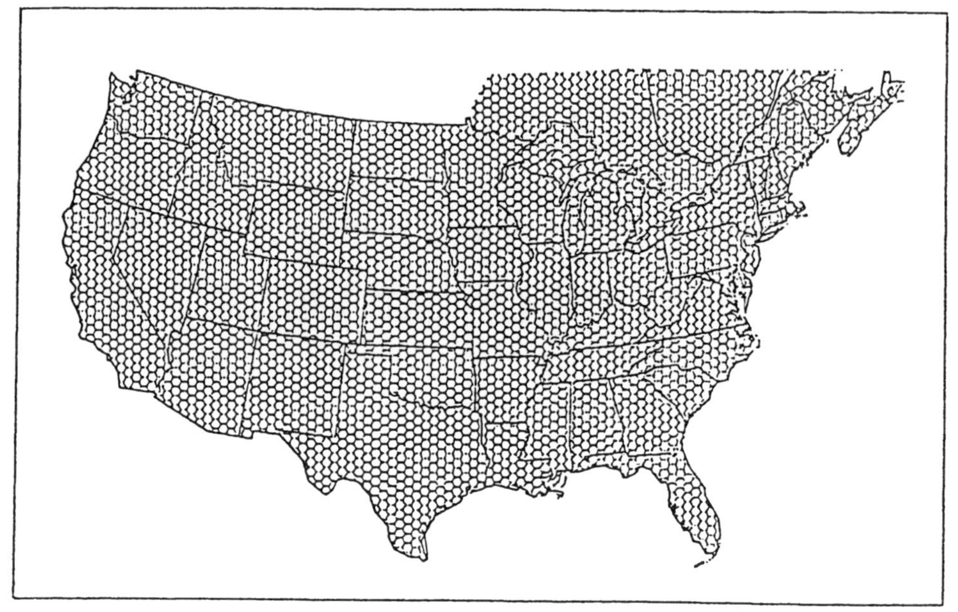

Figure 2 Graphic display of hexagonal areas used for estimation by kriging of the spatial surface.

density is less), by more extreme terrain features, and by more subregional spatial pattern features.

4.2. *Trend Analysis Methodology*

The intent of the temporal pattern methodology was to determine the year-to-year characteristics of mean PWCs and the deposition temporal patterns from 1979 to 1987. The characteristics apply only to the period studied and do not imply either the presence or absence of long-term trend. Data for ten years may not be sufficient to address the issue of long-term trend.

The methodology included two basic, but fundamentally different, approaches. One method used graphic displays of annual data (box plots), and the other used a statistical trend analysis [Kendall seasonal tau (KST) test and Sen's median slope estimation procedure] of monthly data. Box plots allowed visual inspection of annual temporal patterns during the ten-year period. The statistical trend analysis provided estimates of a general linear temporal pattern change over the ten-year period.

Box plots[11] graphically display percentiles of annual data from a set of trend sites for 1979–1987. The box plots simultaneously display the 5th (or 10th), 25th, 50th, 75th, and 95th (or 90th) percentiles of an individual analyte's data for a set of sites. The plots display all site values more extreme than the 5th or 95th percentiles. In some box plots, very extreme data are truncated and plotted at the plot boundary. The box plots, which display the distribution of annual concentrations or depositions at all sites, are notched to display a 95% confidence interval for the median. The upper part of the box plot depicts high-deposition site values, and the lower part of the box plots depicts low-deposition site values. Figure 3 defines the box plot display.

The KST test was developed for trend detection in the presence of constant-

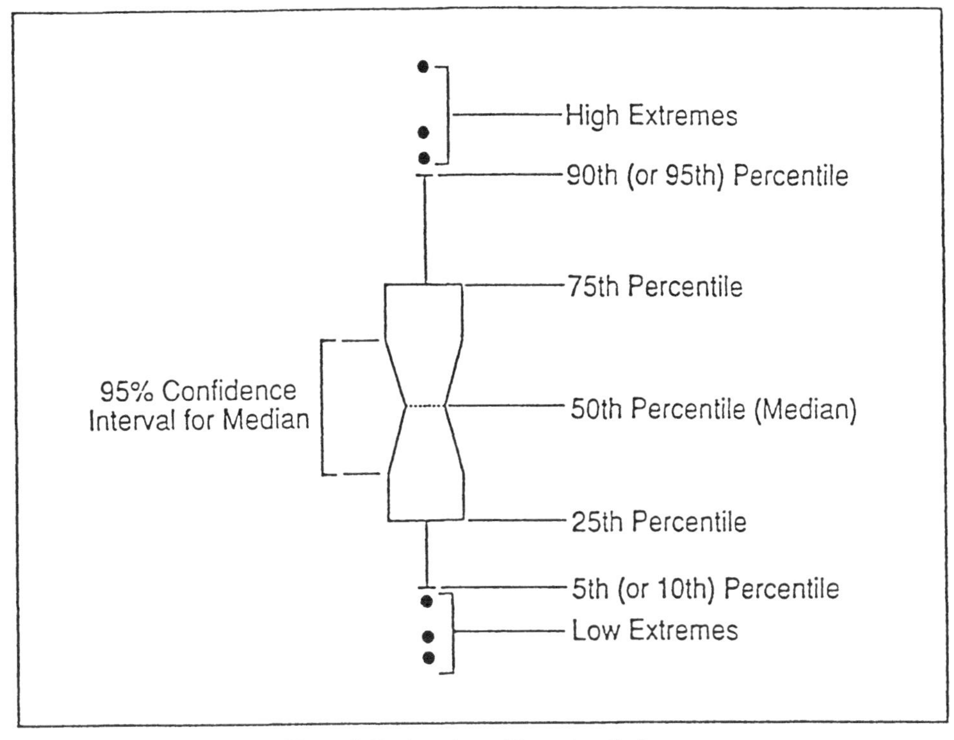

Figure 3 Explanation of box plot display.

length cycles or seasonal effects.[12] The test accommodates missing, tied, or limit-of-detection data values. Furthermore, the validity of the test does not require the data to be normally distributed. The test consists of computing the Mann-Kendall test statistic for year-to-year change separately for each season and combining these seasonal test statistics into a single trend test. In this report, a season refers to the 12 monthly mean PWC values or the 12 monthly mean deposition values potentially available from each year. An estimate of the magnitude of the change is given by Sen's median slope.[13] The median slope of all possible seasonal pairs of values gives an estimate of the annual change.

5. ABILITY OF A NETWORK TO DETECT CHANGE

One of the important uses for spatial analysis is to detect a change. The ability of a network to detect change in annual spatial patterns of analyte deposition and/or concentration depends upon several factors. One factor is the successful operation of all individual sites. A network size (i.e., the number of sites) may be designed for optimal spatial coverage, assuming that each site will operate 100% of the time, will collect all samples, and will have no invalid samples. These are not reasonable assumptions. Quality assurance allows screening of the data (i.e., decisions about which data will be kept and which will be eliminated). The episodic nature of precipitation and the resulting effect on annual deposition are important to QA. The literature agrees quite well that about 20% of the precipitation events account for at least 50% of the annual SO_4^{2-}, NO_3^-, H^+ and NH_4^+ deposition.[14–18] This means that completeness and representativeness factors determining the use of the data should be derived from criteria involving the precipitation amount collected rather than the number of precipitation events sampled.

Another factor that affects the ability of a network to detect a change in annual spatial patterns of precipitation composition is the density of the network. The literature contains very little work that attempts to quantify the uncertainty of regional-scale patterns of wet deposition that is due to the spatial design of a network or networks and the spatial analysis methodology used. Limited information[19] based the density of sites for the NTN network (150 sites distributed throughout the continental United States), suggests that the average uncertainty for annual SO_4^{2-} deposition is about 24% in the Mideast states, 25% in the Southeast, 28% in the West, and 36% in the Northeast. Uncertainty values for annual H^+ and NO_3^- deposition are similar, 25–26% for the Mideast, 26% for the Southeast, 25–29% for the West, and 40% for the Northeast. Concentration uncertainty values are similar to deposition uncertainty values. These values should not be taken as absolute indicators of the ability of the NTN to detect annual spatial patterns of precipitation composition. Equipment and sampling difficulties (unusually large for the Northeast) probably caused uncertainty values to be high. On the other hand, the analysis assumed a flat surface, which may cause uncertainty values to be low because topography affects precipitation amount and composition. Nevertheless, the study offers a sense of the ability of a network to detect changes in wet deposition patterns, even though the uncertainty of the results is high.

The evaluation of the ability of a network to detect change is critical to interpreting any change in spatial patterns of wet deposition, whatever the cause of that change. Any monitoring network should either (1) have a density sufficiently high so that uncertainties associated with spatial analyses are smaller than the expected changes in wet deposition patterns or (2) sample for many years to resolve the causes of changes in wet deposition.

6. RELATIONSHIPS OF ANALYTES

Although spatial and temporal patterns of individual analytes are important, the overall pattern of the integrated chemical composition of precipitation and its relationship to emissions are sometimes ignored. Soluble particulates and gases form ions in solution. Once in solution in ionic form, the original source compounds cannot be analytically identified. The relationships of the various analytes, therefore, must be determined by statistical methods and by application of aerosol composition and scavenging theory. This theory is based on the premise that chemical relationships of ions in precipitation should be largely determined by those in the ambient atmosphere. In general, the literature[20–25] agrees that the various inorganic analytes in regionally representative precipitation can be categorized in terms of probable sources. The primary sources of Ca^{2+}, Mg^{2+}, and K^+ are attributed to wind-blown soil and dust from the tilling of soil (agriculture) and from unpaved roadways. The primary source of Na^+ and Cl^- is sea salt. The primary source of H^+, SO_4^{2-}, and NO_3^- is anthropogenic emissions. The primary source of NH_4^+ is natural soil processes.

Local sources of Ca^{2+}, Mg^{2+}, and K^+ include wintertime road salting (with

$CaCl_2$) and cement plant and other industrial emissions. Wintertime road salting with NaCl contributes Na^+ and Cl^-; swamps, bogs, and marshes contribute SO_4^{2-}; wood burning in fireplaces and stoves contributes NO_3^-, cattle feed lots, areas of large human population, and sludge ponds all contribute NH_4^+. These examples of local emissions (and the list is far from complete) tend to be site specific and can dominate the primary emission sources discussed in the previous paragraph.

Although general groups of analytes that have been identified can be related to compounds in the atmosphere, an anion can be associated with more than one cation and vice versa. Precipitation composition is not stable, and ion concentrations can change significantly with time depending upon sample stability,[26,27] sample frequency[28–33] and collector type.[34–37] North America is made up of many geographically different subregions with different physical and chemical characteristics. These factors make quantative relationships between ions in regional precipitation highly uncertain without knowledge of the compounds (and their relative proportions) that are present in the atmosphere.

Studies that attempt to relate ions to compounds in precipitation indicate that the dominant relationships between analytes (for aerosols and precipitation) for the eastern half (non-coastal sites) of the United States are the following:

$$SO_4^{2-} - NH_4^+ = \text{acidic } SO_4^{2-} \quad (1)$$

$$NO_3^- - (Ca^{2+} + Mg^{2+} + K^+) = \text{acidic } NO_3^- \quad (2)$$

$$Na^+ = Cl^- \quad (3)$$

$$H^+ = \text{acidic } SO_4^{2-} + \text{acidic } NO_3^- \quad (4)$$

The acidic SO_4^{2-} and acidic NO_3^- terms represent the acidic fraction of sulfate (mostly H_2SO_4 from the liquid phase oxidation of SO_2 rather than acidic aerosol) and nitrate (HNO_3) that contribute H^+ to precipitation. The wet deposition of acidic SO_4^{2-} is largely responsible for the acidity of precipitation, particularly in the eastern half of North American.[22] This result may be misleading if it is taken to imply that the contribution of the wet deposition of NO_3^- to precipitation is unimportant. For example, HNO_3 found in the atmosphere is quickly neutralized as it comes into contact with large particles, much as H_2SO_4 is quickly neutralized by NH_3 in the atmosphere. However, many of the neutralizing particles in precipitation that might otherwise be available for the possible neutralization of H_2SO_4 (formed by the liquid phase oxidation of SO_2) are "tied up" with the in-air neutralization of HNO_3. Weak, organic acids also contribute to precipitation acidity, but that contribution is estimated to average less than 15% in North America, where anthropogenic emissions dominate precipitation acidity. Relationships between ions and acidity for the western half of the United States are not as well defined. Most of the uncertainty is with the apportionment of NH_4^+ between SO_4^{2-} and NO_3^-, the apportionment of Ca^{2+} to NO_3^- and SO_4^{2-}, and the presence of catalysts and oxidants.

Precipitation composition is clearly dominated by regional and local land use patterns and proximity to anthropogenic emissions. Although each sampling site has a different apportionment of the various analytes in precipitation because of its location, acidity is highest in precipitation where emissions of precursor anthropogenic gases are highest *and* the concentrations neutralizing soil components are lowest.

7. SPATIAL ANALYSIS FOR THE PERIOD 1985–1987 ANNUAL PATTERNS

Spatial analyses are given for annual PWC and annual wet deposition for H^+ (from laboratory pH), SO_4^{2-}, NO_3^-, NH_4^+, and Ca^{2+}. Fixed percentile ranking spatial analyses by our analysis method are given for annual and seasonal PWC and wet deposition of H^+ (from laboratory pH), SO_4^{2-}, NO_3^-, NH_4^+, and Ca^{2+}. Composite SO_4^{2-}, NO_3^-, and H^+ maps (Figures 4–6 respectively) for the period 1980–1984[38] may be compared to our maps to show changes in the areal coverage of the various analytes between the two time periods. The fixed values correspond to the contour intervals used by Barchet[38] for the 1980–1984 spatial composite maps for the NAPAP interim assessment analyses.

The 1985–1987 annual spatial patterns for pH by fixed percentile ranking and by fixed concentration values (Figure 7) have a range of 4.15 to 5.25 in the eastern United States and a range of 4.63 to 5.62 in the western United States. Well over 50% of the East has pH values that are less than the lowest pH in the West. In the East, the lowest pH values occur in an ellipsoid area across Ohio and Pennsylvania and in an additional area in western New York. The highest pH values in the East occur in Minnesota, Wisconsin, and western Ontario. In the West, the lowest estimated pH values occur in two separate regions: (1) eastern Kansas and (2) Texas and southeastern Arizona. The highest pH values in the West also occur in two separate regions: (1) western Kansas and eastern Colorado and (2) northern California and Nevada.

The estimated 1985–1987 annual spatial patterns for H^+ deposition in the United States by fixed percentile ranking and by fixed deposition values (Figure 8) have a range of 0.045 kg/ha to 0.760 kg/ha in the East and a range of 0.010 kg/ha to 0.276 kg/ha in the West. Over 50% of the East has H^+ deposition greater than the maximum value in the West. In the East, the highest deposition occurs in Pennsylvania, New Jersey, and central New York. The lowest deposition in the East occurs in Minnesota, Wisconsin, and western Ontario. In the West, the highest deposition occurs in eastern Kansas, Oklahoma, and Texas. The lowest deposition in the West occurs in three regions: eastern Colorado; eastern Wyoming; and northern California, northern Nevada and southern Idaho.

The estimated 1985–1987 annual spatial patterns for SO_4^{2-} concentration in the United States by fixed percentile ranking and by fixed concentration values (Figure 9) have a range of 0.74 mg/L to 4.22 mg/L in the East and a range of 0.24 mg/L to 1.55 mg/L in the West, with over 50% of the concentrations in the East being greater than the highest concentration in the West. In the East, the highest concentrations occur in southern Ontario, Ohio and western Pennsylvania. The lowest concentrations in the East occur in three regions: western Ontario, Nova Scotia and Maine, and Florida. In the West, the highest concentrations occur in three regions: eastern Kansas and Nebraska, southern Texas, and southern Arizona. The lowest concentrations in the West occur in northern California, Oregon, eastern Washington, and northern Idaho.

The estimated 1985–1987 annual spatial patterns for SO_4^{2-} deposition in the United States by fixed percentile ranking and by fixed deposition values (Figure 10) have a range of 5.5 kg/ha to 42.7 kg/ha in the East and a range of 0.8 kg/ha to 17.8 kg/ha in the West, with the median in the East (17.3 kg/ha) four to five times the median in the West 4.0 (kg/ha). In the East, the highest deposition occurs in southeastern Michigan, southern Ontario, eastern Pennsylvania, and

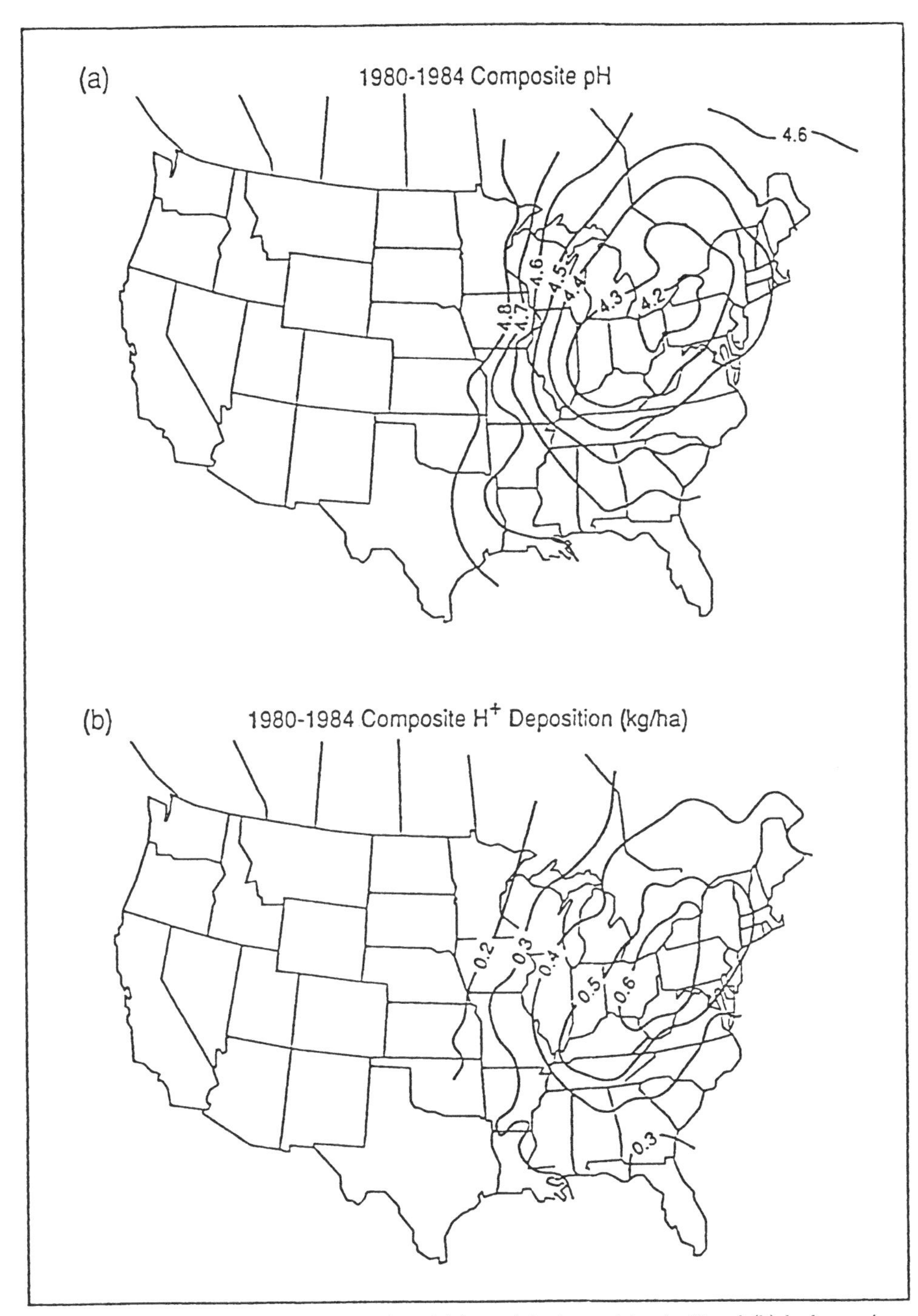

Figure 4 The 1980–1984 annual composite of (a) precipitation-weighted pH and (b) hydrogen ion deposition for eastern North America.

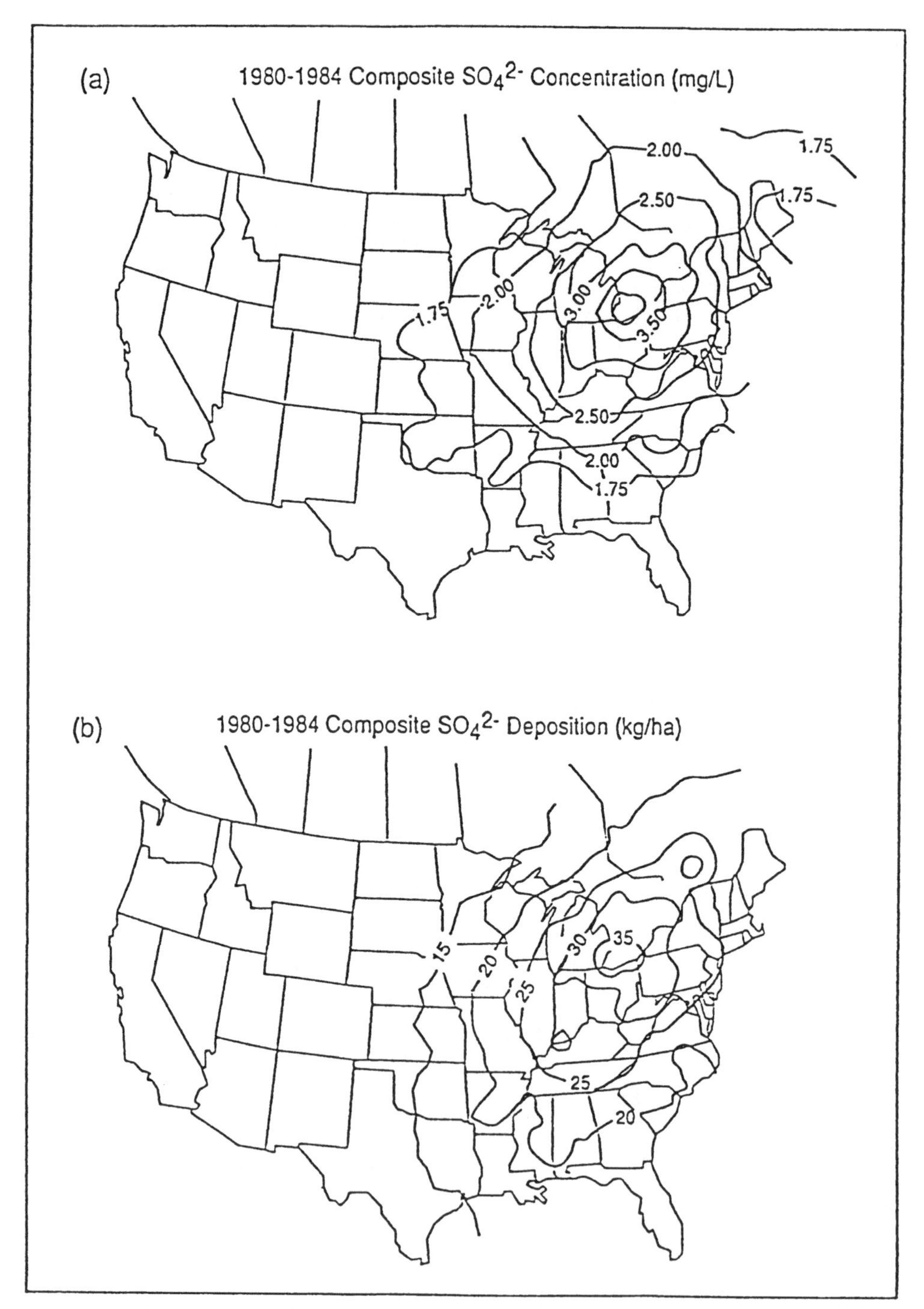

Figure 5 The 1980–1984 annual composite of (a) precipitation-weighted sulfate concentration and (b) sulfate deposition for eastern North America.

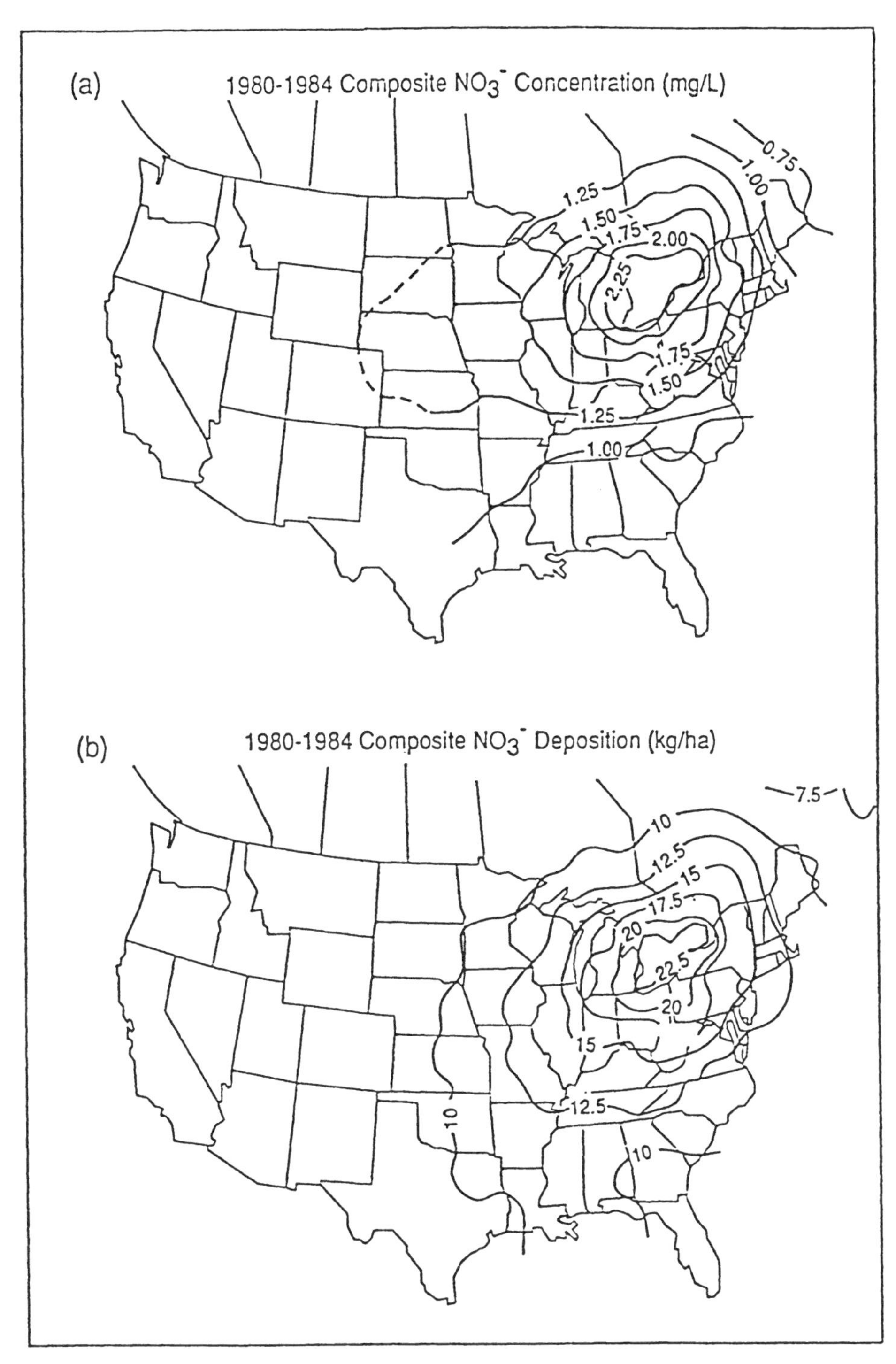

Figure 6 The 1980–1984 annual composite of (a) precipitation-weighted nitrate concentration and (b) nitrate deposition for eastern North America.

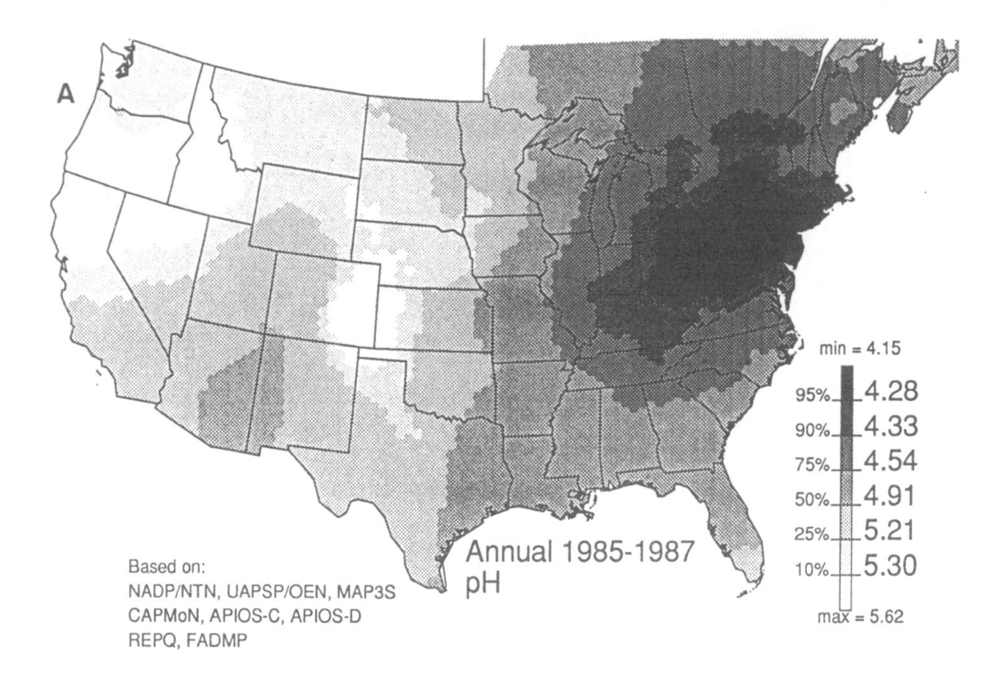

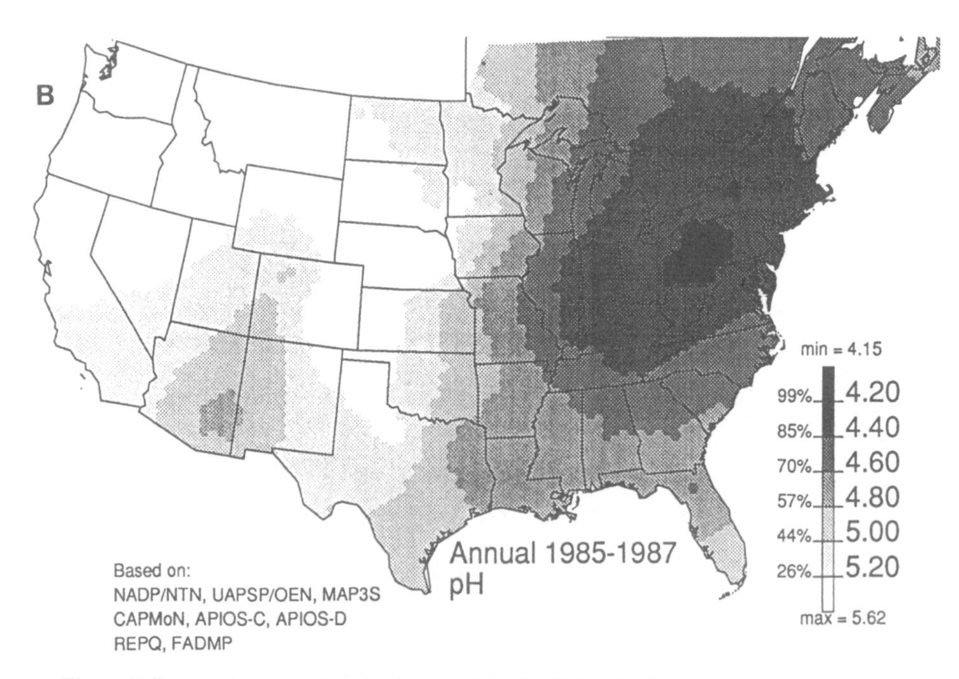

Figure 7 Gray-scale composite of volume-weighted pH for North America in 1985–1987 for (A) fixed percentile ranking and (B) fixed concentration values.

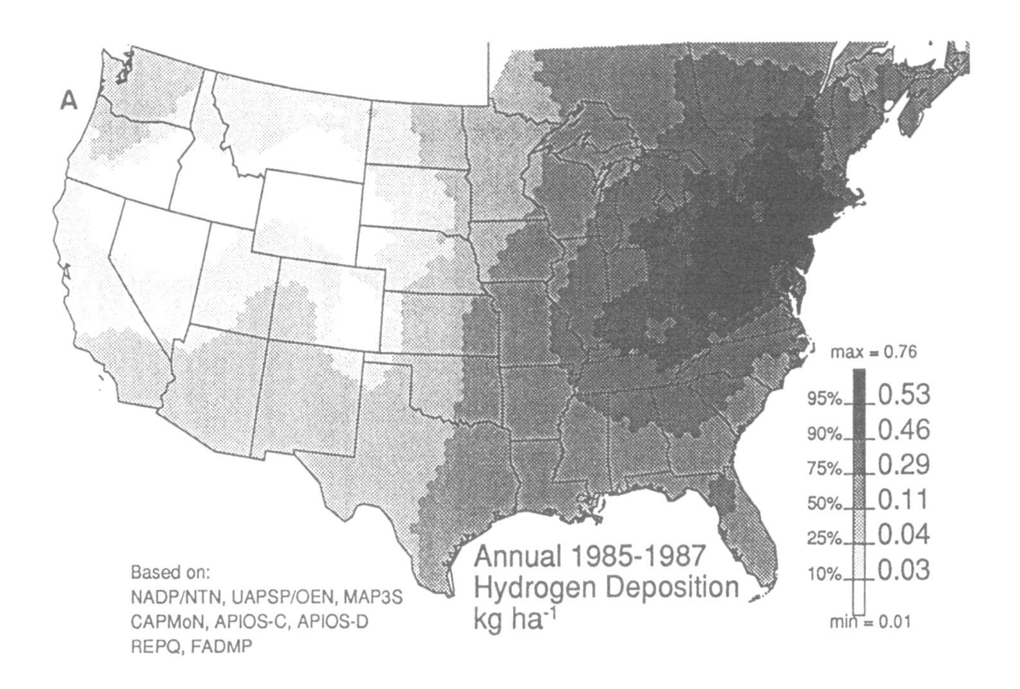

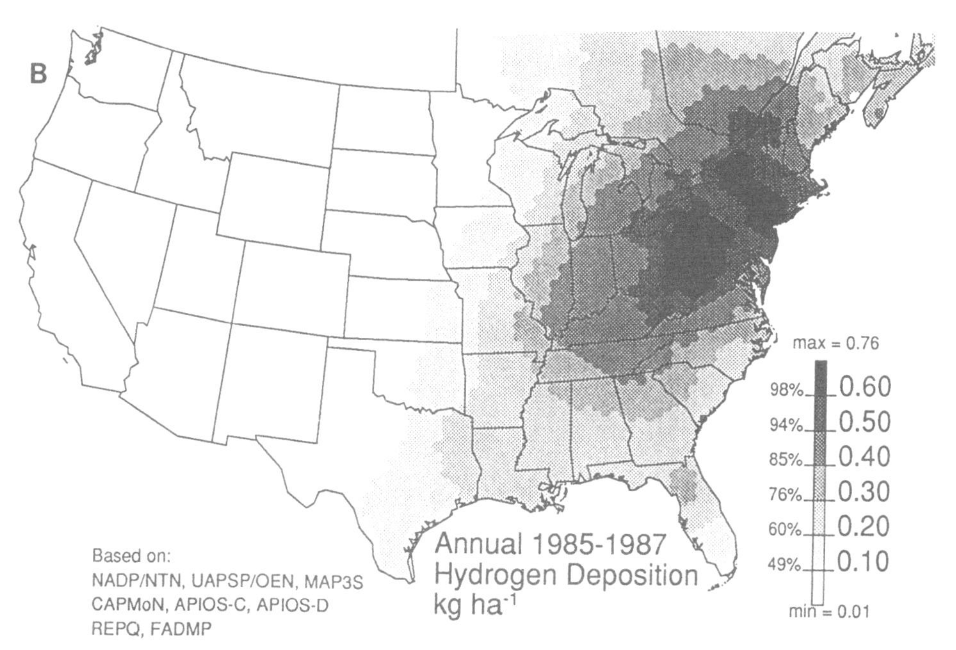

Figure 8 Gray-scale composite of hydrogen ion wet deposition for North America in 1985–1987 for (A) fixed percentile ranking and (B) fixed deposition values.

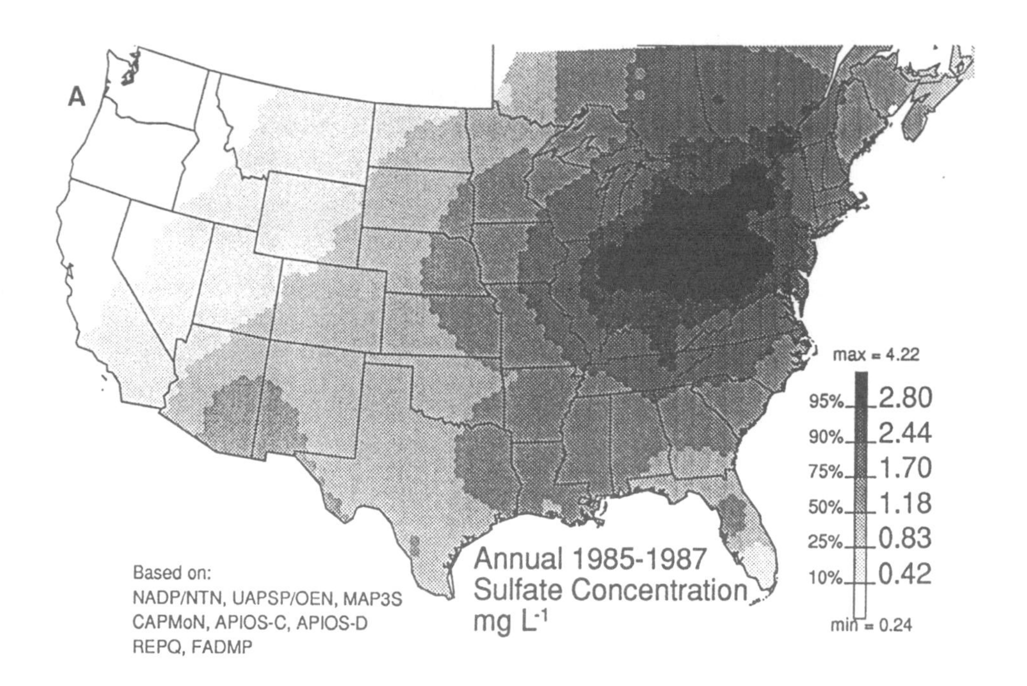

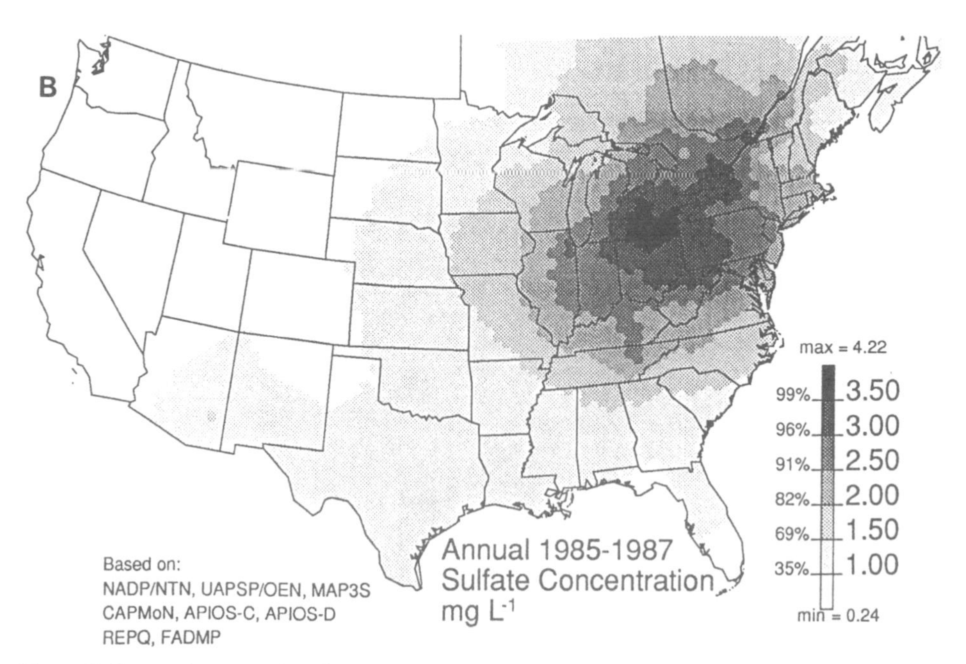

Figure 9 Gray-scale composite of volume-weighted sulfate ion concentration for North America in 1985–1987 for (A) fixed percentile ranking and (B) fixed concentration values.

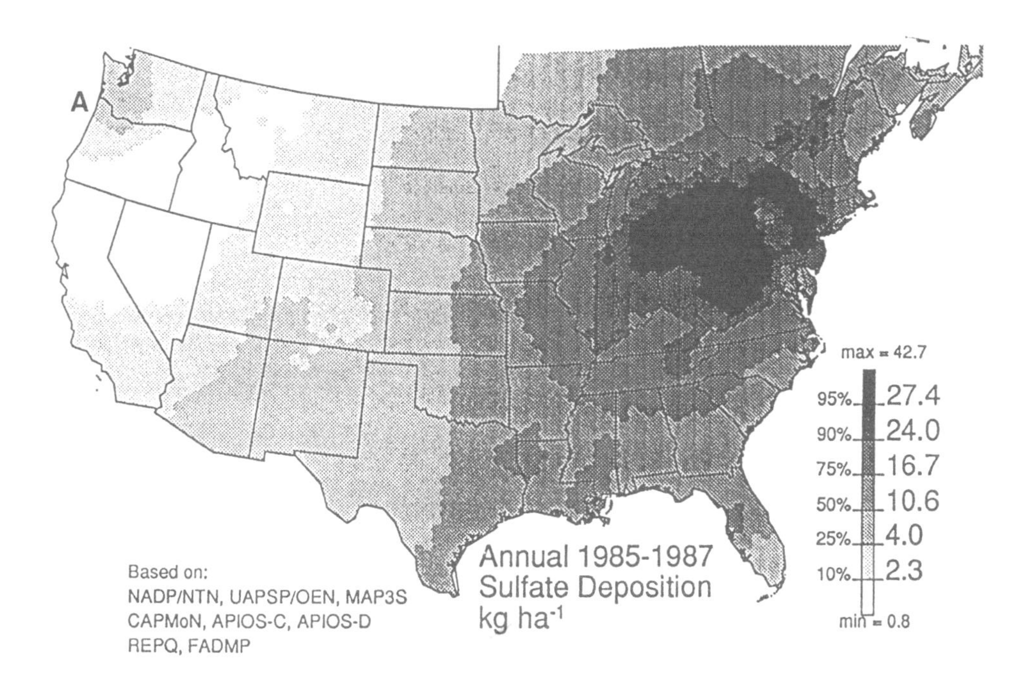

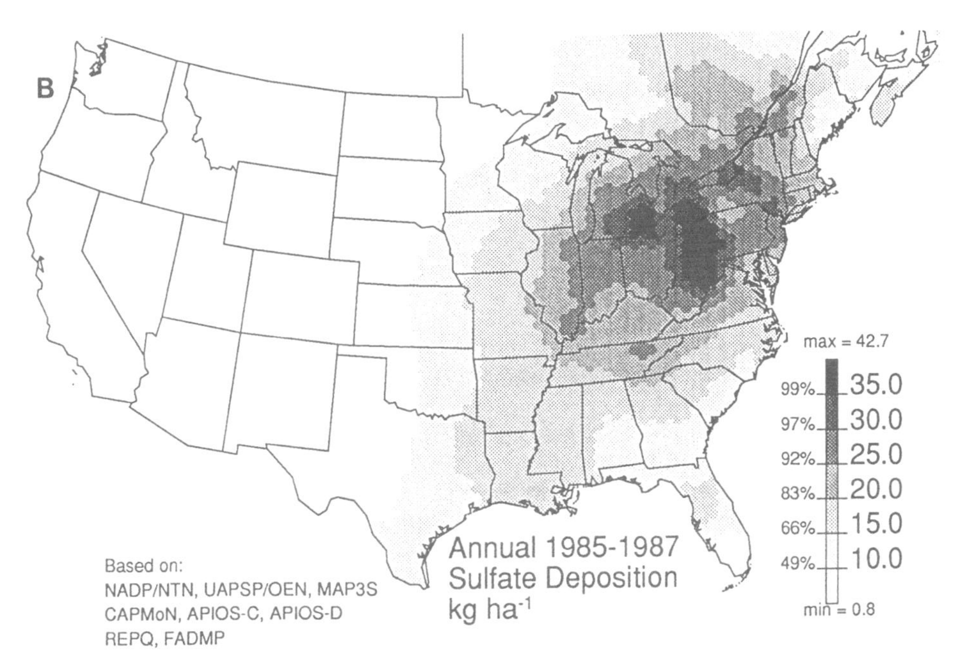

Figure 10 Gray-scale composite of sulfate ion wet deposition for North America in 1985–1987 for (A) fixed percentile ranking and (B) fixed deposition values.

northern West Virginia. The lowest deposition in the East occurs in western Ontario, Minnesota, and western Wisconsin. In the West, the highest deposition occurs in eastern Kansas, Oklahoma, and Texas. The lowest deposition in the West occurs in western Idaho, eastern Oregon, northern California, and northern Nevada.

The estimated 1985–1987 annual spatial patterns for NO_3^- concentration in the United States by fixed percentile ranking and by fixed concentration values (Figure 11) have a range of 0.51 mg/L to 2.84 mg/L in the East and a range of 0.15 mg/L to 1.28 mg/L in the West. Over 95% of the NO_3^- concentrations in the West are less than the median concentration in the East (1.14 mg/L). In the East, the highest concentrations occur in southern Ontario, northern Ohio, western Michigan, and western New York. Seventeen sites in southern Ontario have NO_3^- concentrations higher than the highest New York values. The lowest concentrations in the East occur in two regions: New Brunswick, Nova Scotia, and Maine; and southern Georgia, southern Alabama, and Florida. In the West, the highest concentrations occur in eastern South Dakota, eastern Nebraska, and northern Kansas. The lowest concentrations in the West occur in northern California, Oregon, Washington, and northern Idaho.

The estimated 1985–1987 annual spatial patterns for NO_3^- deposition in the United States by fixed percentile ranking and by fixed deposition values (Figure 12) have a range of 5.5 kg/ha to 29.0 kg/ha in the East and a range of 0.9 kg/ha to 11.6 kg/ha in the West, with almost all of the deposition in the West less than the median deposition in the East (10.9 kg/ha). In the East, the highest deposition occurs in central New York, western Pennsylvania, southeastern Michigan, and southern Ontario. The lowest deposition in the East occurs in three regions: western Ontario; northern Quebec and Maine; and Alabama, Georgia, South Carolina, and Florida. In the West, the highest deposition occurs in eastern Nebraska, Kansas, and Oklahoma. The lowest deposition in the West occurs in eastern Oregon, western Idaho, northern California, and northern Nevada.

The estimated 1985–1987 annual spatial patterns for NH_4^+ concentration in the United States by fixed percentile ranking and by fixed concentration values (Figure 13) have a range of 0.08 mg/L to 0.61 mg/L in the East and a range of 0.03 mg/L to 0.60 mg/L in the West, with the concentrations in the West being only slightly less than those in the East. In the East, the highest concentrations occur in two regions: (1) southern Ontario and eastern Michigan and (2) western Minnesota. The lowest concentrations in the East occur in two regions: (1) Maine and (2) the southeastern United States. In the West, the highest concentrations occur in eastern North Dakota, eastern South Dakota, and eastern Nebraska. The lowest concentrations in the West occur in Washington, western Oregon, and northern California.

The estimated 1985–1987 annual spatial patterns for NH_4^+ deposition in the United States by fixed percentile ranking and by fixed deposition values (Figure 14) have a range of 0.9 kg/ha to 5.4 kg/ha in the East and a range of 0.1 kg/ha to 4.2 kg/ha in the West. The median deposition in the East (2.5 kg/ha is over four times the median deposition in the West (0.6 kg/ha). The highest deposition in the East occurs in southern Ontario and southern Michigan. The highest deposition in the West occurs in eastern South Dakota, eastern Nebraska, and northeastern Kansas.

The estimated 1985–1987 annual spatial patterns for Ca^{2+} concentration in the United States by fixed percentile ranking and by fixed concentration values (Fig-

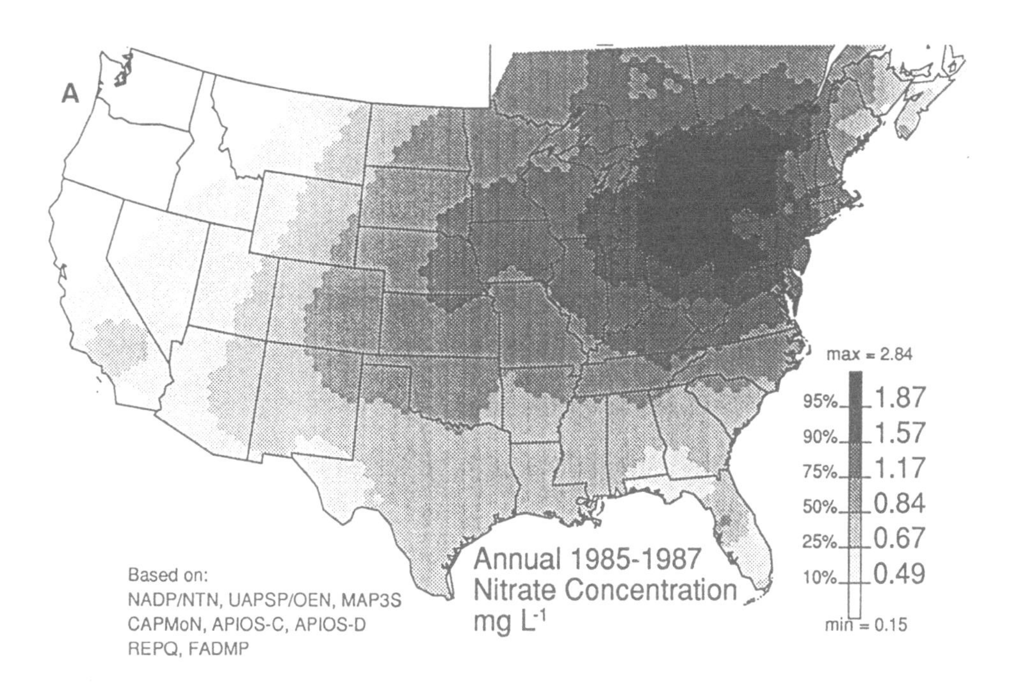

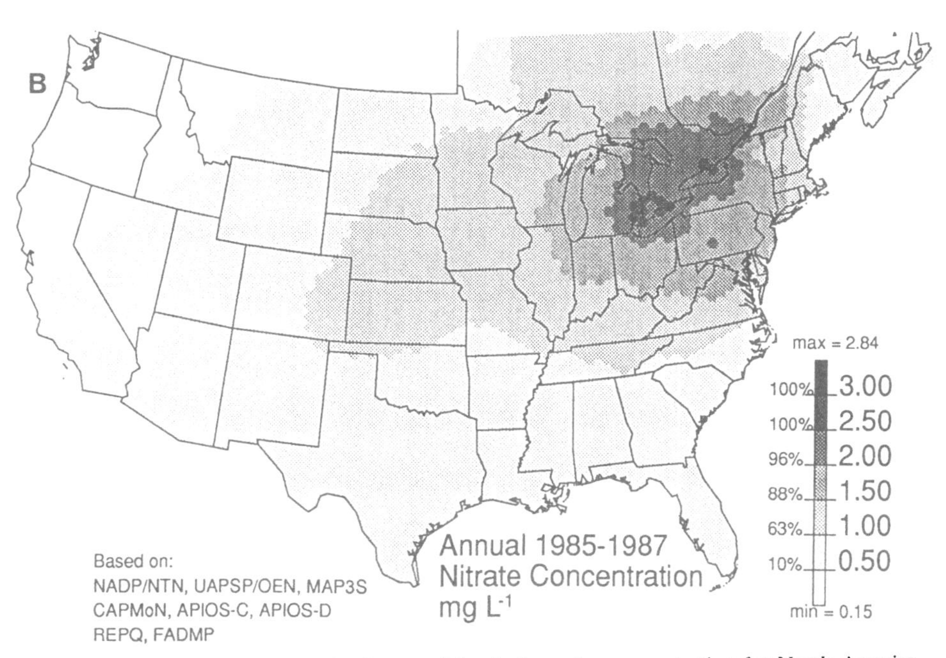

Figure 11 Gray-scale composite of volume-weighted nitrate ion concentration for North America in 1985–1987 for (A) fixed percentile ranking and (B) fixed concentration values.

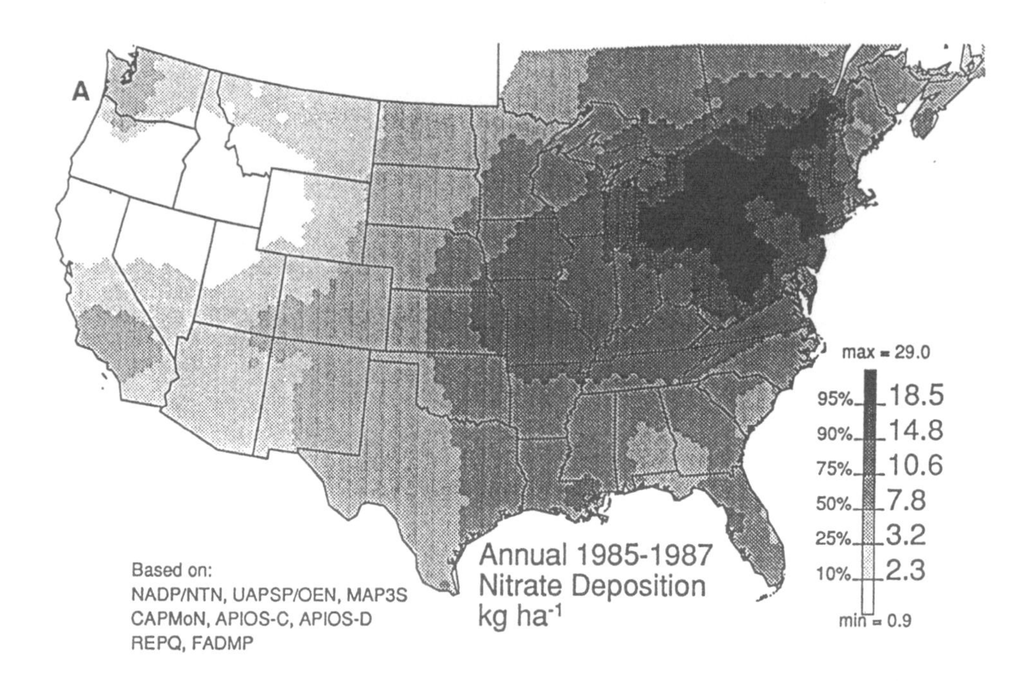

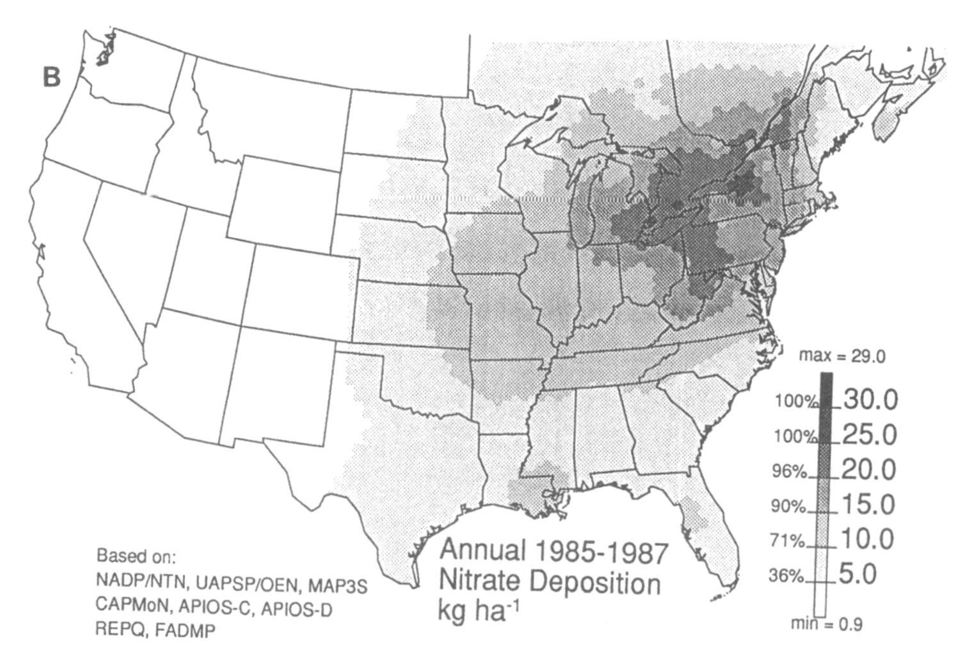

Figure 12 Gray-scale composite of nitrate ion wet deposition for North America in 1985–1987 for (A) fixed percentile ranking and (B) fixed deposition values.

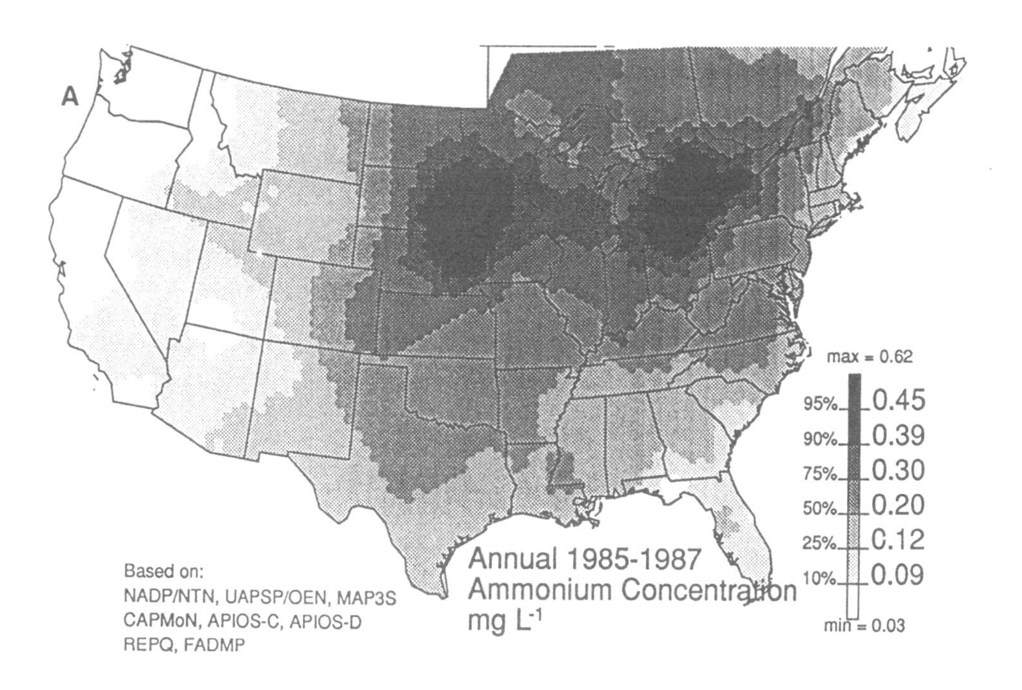

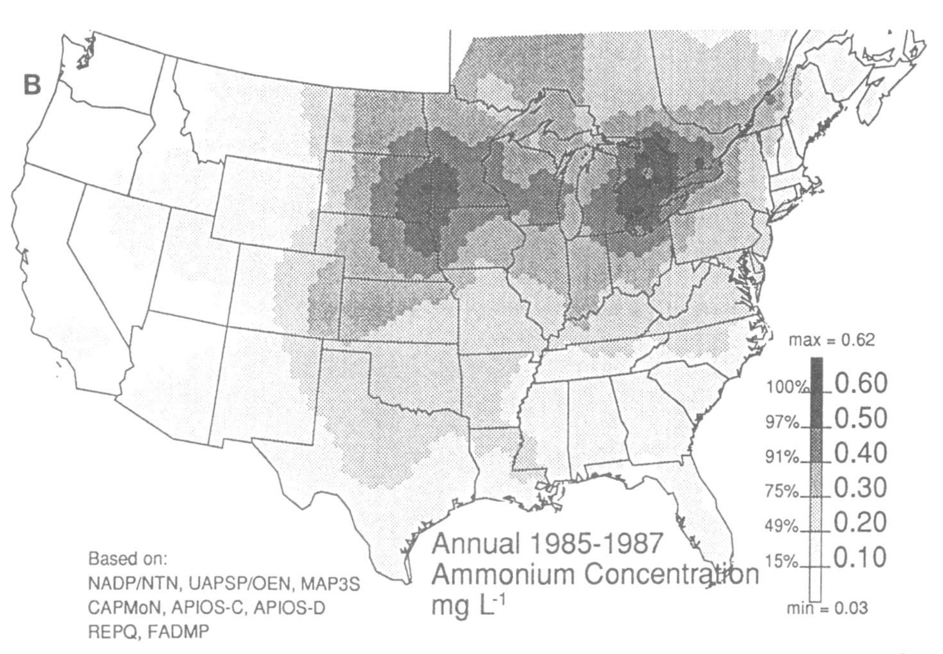

Figure 13 Gray-scale composite of volume-weighted ammonium ion concentration for North America in 1985–1987 for (A) fixed percentile ranking and (B) fixed concentration values.

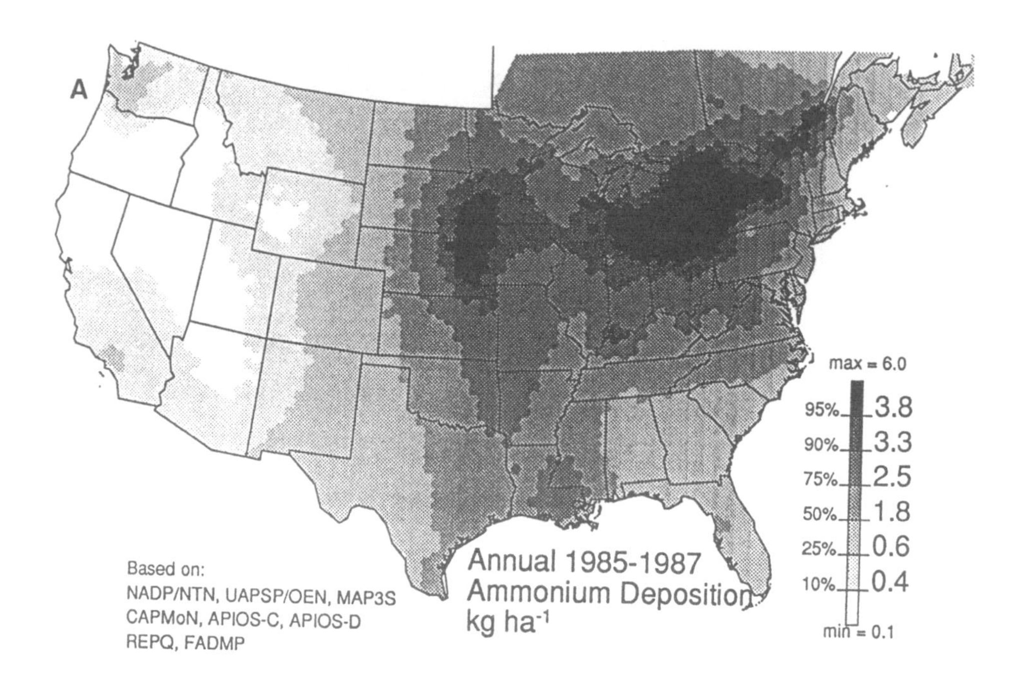

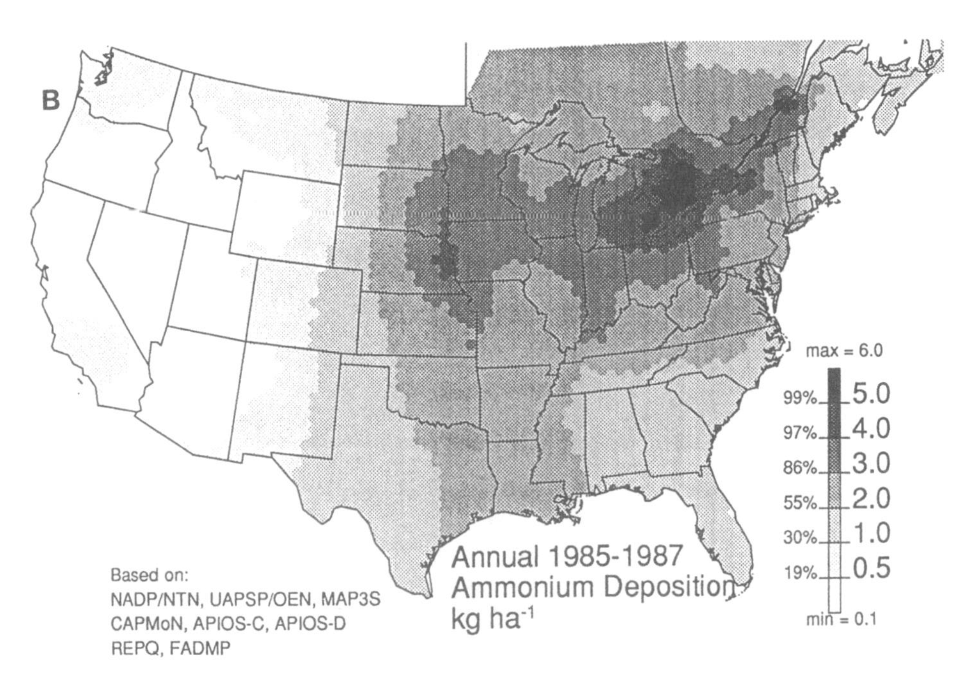

Figure 14 Gray-scale composite of ammonium ion wet deposition for North America in 1985–1987 for (A) fixed percentile ranking and (B) fixed deposition values.

ure 15) have a range of 0.04 mg/L to 0.42 mg/L in the East and a range of 0.04 mg/L to 0.34 mg/L in the West, with the West having a higher median (0.18 mg/L) than the East (0.13 mg/L). In the East, the highest concentrations occur in two regions: southern Ontario and Iowa. The lowest concentrations in the East occur along the eastern seaboard and in the southeastern United States. In the West, the highest concentrations occur in western Kansas, southeastern Colorado, and southern Nebraska. The lowest concentrations in the West occur in northern California and western Oregon and Washington.

The estimated 1985–1987 annual spatial patterns for Ca^{2+} deposition in the United States by fixed percentile ranking and by fixed deposition values (Figure 16) have a range of 0.4 kg/ha to 3.9 kg/ha in the East and a range of 0.3 kg/ha to 2.7 kg/ha in the West, with the median in the East (1.3 kg/ha) larger than the median in the West (0.8 kg/ha). In the East, the highest deposition occurs in two regions: southern Ontaro and central Iowa. The lowest deposition in the East occurs along the eastern seaboard and in the southeastern United States. In the West, the highest deposition occurs in central Oklahoma, eastern Kansas, and eastern Nebraska. The lowest deposition in the West occurs in northern California, Oregon, and western Nevada.

The estimated 1985–1987 spatial patterns for the SO_4^{2-}/NO_3^- concentration ratio in the United States by fixed percentile ranking and by fixed ratios are shown in Figure 17. (Deposition ratio patterns by fixed percentile ranking and by fixed ratios are identical to those for concentration and therefore are not shown here.) The dependence of the patterns on meteorological and climatological factors and emissions is beyond the scope of this chapter.

7.1. *Seasonal Pattern Differences*

Meteorological seasons are used in this analysis, but only summer (June, July, and August) and winter (December, January, and February) spatial patterns are used as the basis for discussion. Only seasonal differences will be discussed here; a more complete description may be found elsewhere.[1]

In both the East and West, the summer and winter pH spatial patterns are distinctly different. In the East, the summer pH values are lower (the median in the summer is 4.45, while the winter median is 4.64), and the center of the lowest pH value has moved east from northeastern Ohio to eastern Pennsylvania. In addition, the pH values in western Ontario decrease in the summer. In the West, the range of pH values is about the same in winter and summer, but the highest pH values in the summer are in the Great Plains, which have much lower pH values in the winter. The highest pH values in the winter occur in northern California and northern Nevada, which have much lower pH values in the summer.

The spatial patterns and the magnitudes of the summer and winter H^+ deposition are generally the same in the West. The one major difference in the West is that deposition increases along the Pacific coast in the winter and increases in Wyoming and Montana in the summer. In the East both the spatial pattern and the magnitude change between winter and summer. The median summer deposition (0.123 kg/ha) is twice the median winter deposition (0.052 kg/ha). The maximum estimate of summer deposition (0.371 kg/ha) is almost three times the maximum winter deposition (0.133 kg/ha).

The estimated spatial patterns of pH and H^+ deposition are similar for the annual and summer summaries. However, the estimated winter spatial patterns

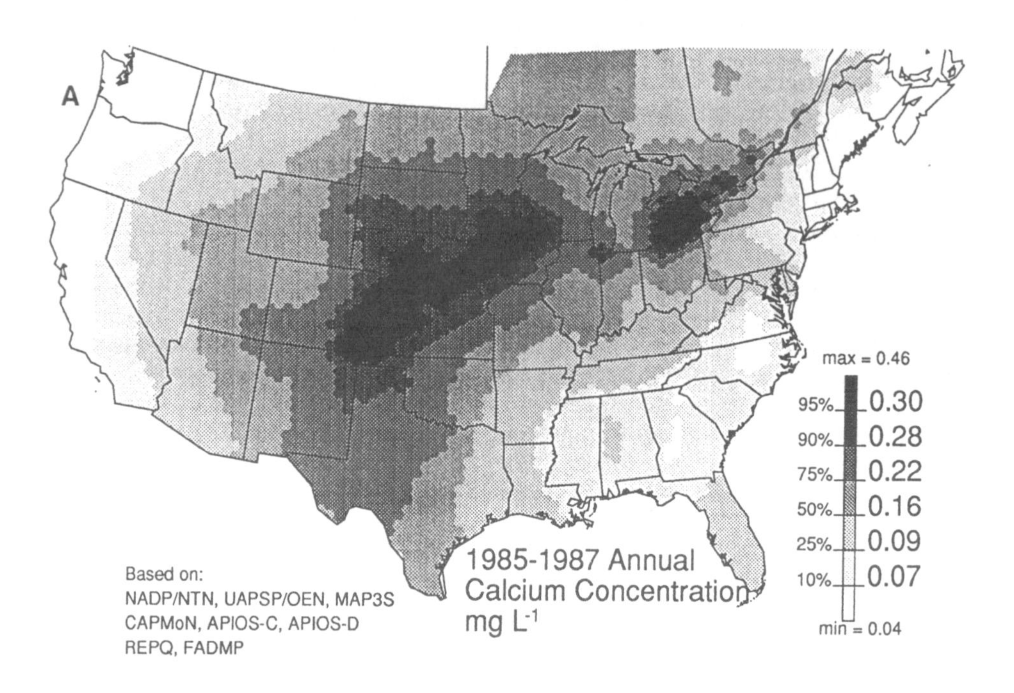

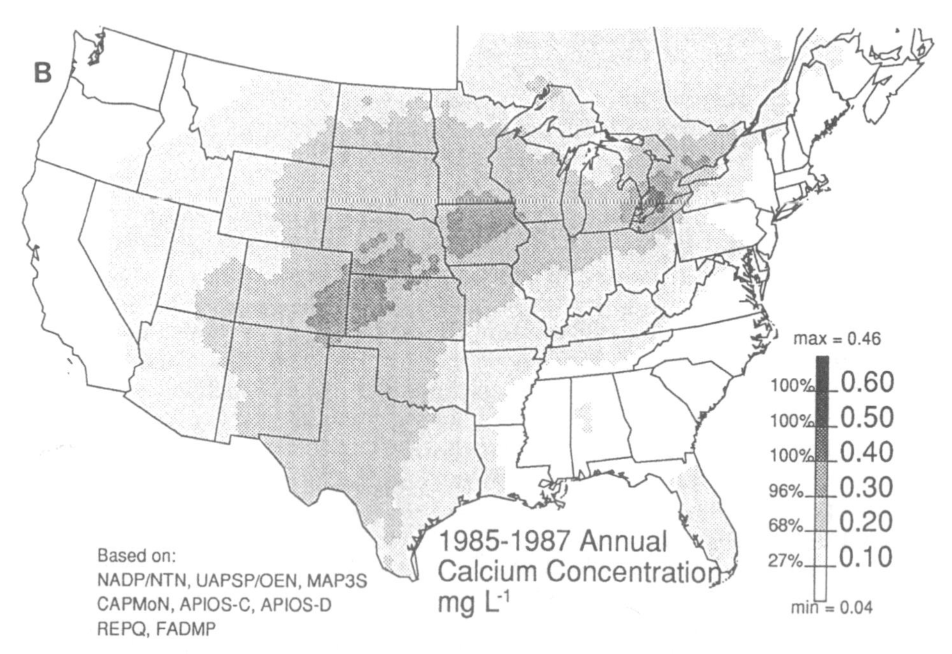

Figure 15 Gray-scale composite of volume-weighted calcium ion concentration for North America in 1985–1987 for (A) fixed percentile ranking and (B) fixed concentration values.

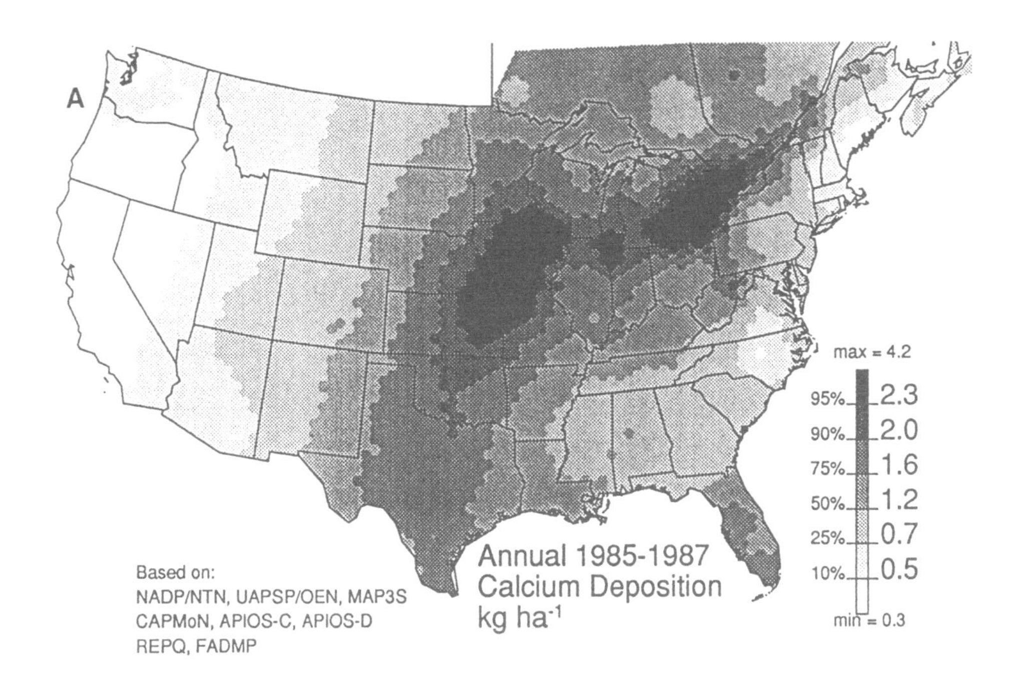

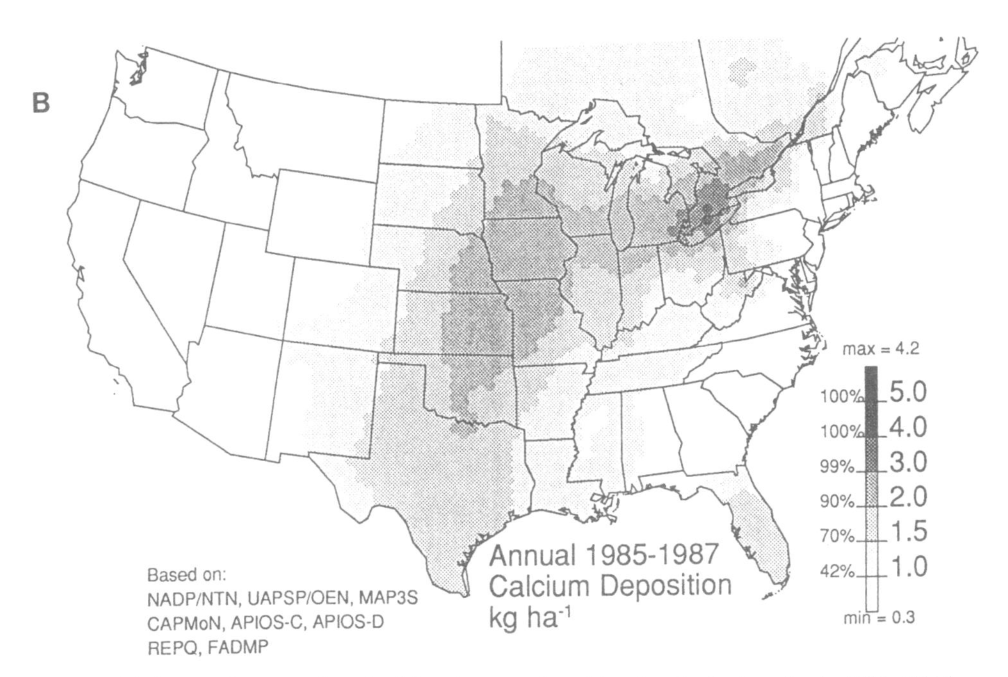

Figure 16 Gray-scale composite of calcium ion wet deposition for North America in 1985–1987 for (A) fixed percentile ranking and (B) fixed deposition values.

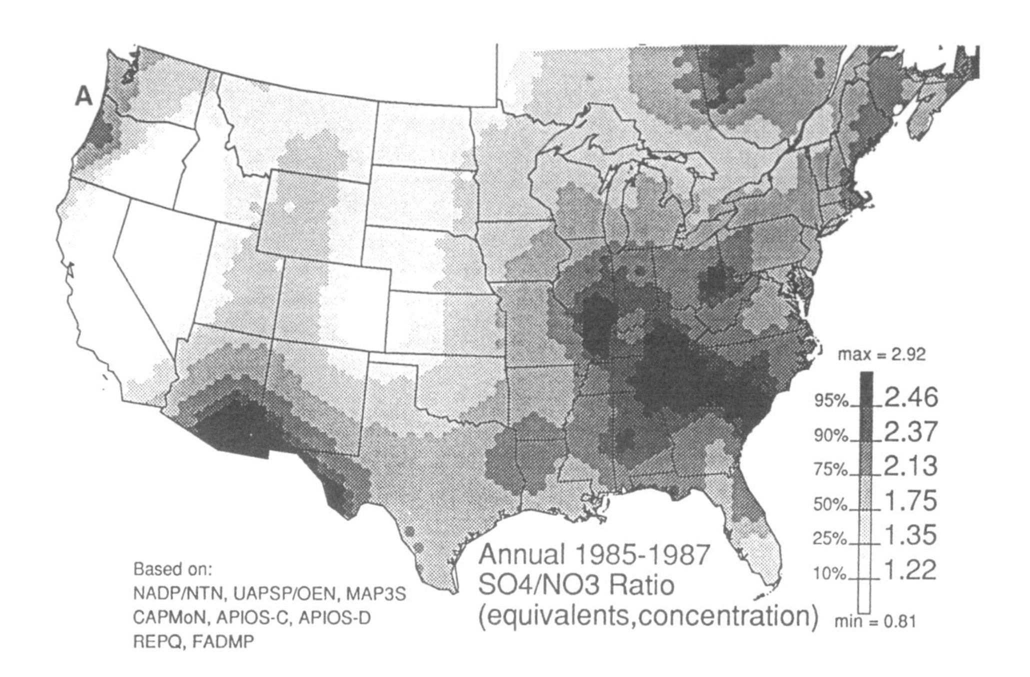

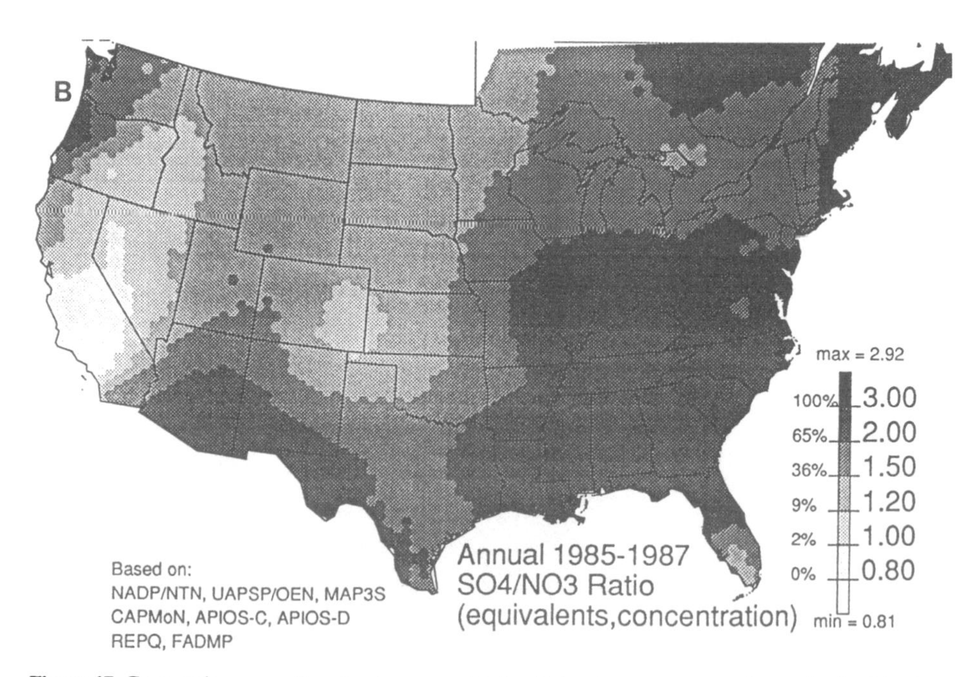

Figure 17 Gray-scale composite of volume-weighted sulfate:nitrate concentration ratio for North America in 1985–1987 for (A) fixed percentile ranking and (B) fixed ratio values.

of pH and H^+ deposition are different in the eastern United States. In the winter, the pH is highest in Ohio and eastern Pennsylvania, but H^+ deposition is highest in New Jersey along the Atlantic coast and in the south.

In the East, the magnitude of the summer SO_4^{2-} concentrations is nearly twice that of the winter SO_4^{2-} concentrations. In the summer the maximum concentrations are centered over Pennsylvania, while in the winter the maximum concentrations are centered to the west, over Ohio. In addition, in the East, the lower summer concentrations are in the South, Minnesota, and western Ontario, while the lower winter concentrations are along the Atlantic. In the West, the magnitude of SO_4^{2-} concentrations increases slightly in the winter, but there is a large increase in the number of low concentrations. (In the summer, only 5% of the concentrations are less than 0.70 mg/L, while in the winter, close to 50% of the concentrations are less than 0.70 mg/L.) In the West, the spatial pattern is relatively smooth in the winter, but the summer spatial pattern shows a number of small regional highs and lows.

In the East, the median summer SO_4^{2-} deposition (6.6 kg/ha) is just slightly greater than the maximum winter deposition. In the summer, the maxima occur in Pennsylvania and southern Ontario, with the minima in the South, Minnesota, and western Ontario. However, in the winter the maxima occur in the South, with the greatest changes in Tennessee and Alabama. In the west, the summer deposition is also greater than the winter deposition, with the median summer deposition (1.4 kg/ha) being twice the median winter deposition (0.6 kg/ha). In the summer, the deposition is at a minimum in the Pacific Northwest, while in the winter, the Pacific coast deposition increases and the minima occur as far inland as eastern Montana.

In both the East and West, there is no significant difference between the magnitudes of the larger percentiles in the summer and winter NO_3^- concentrations. However, the lower percentiles increase in magnitude in the summer for both parts of the United States. In the East, the NO_3^- concentrations increase from the south to the north in the winter, with the highest concentrations in southern Ontario. In the summer, the maximum concentrations move eastward to Pennsylvania, and some of the lowest concentrations occur in the North. In the West, the largest concentrations in the winter occur in the eastern Dakotas, and the concentrations decrease farther west. In the summer, the spatial pattern is dominated by a number of small regional highs and lows.

In both the East and the West, the summer NO_3^- deposition is approximately twice the winter deposition. In the East, the summer maxima occur in Pennsylvania, while in the winter, the maxima move north. In the West, the minimum summer deposition occurs in the Pacific Northwest. However, in the winter, the minimum NO_3^- deposition moves inland as far east as eastern Montana. The high summer deposition in eastern Kansas and Nebraska is not present in the winter.

The summer NH_4^+ concentration is greater than the winter concentration in both the East and the West. The spatial pattern in the East changes somewhat between winter and summer, as does that in the West. The winter spatial pattern is relatively smooth, with high concentrations in Kansas and Nebraska and a decrease in concentrations toward the Pacific Northwest. The summer spatial pattern is not smooth, being dominated by numerous small regional highs and lows.

Ammonium deposition in the summer is more than twice the deposition in winter. In the East, the summer spatial pattern has an elliptical high region centered over the southern tip of Ontario. In the winter, the pattern is less defined and is shifted slightly to the northeast. In addition, a local high deposition region that appears in Louisiana during the winter is less dominant during the summer. In the West, the winter and summer deposition patterns are strikingly different. During the winter, a region of high deposition occurs in east Texas and Oklahoma. During the summer, that region has approximately the same deposition, but the region of highest deposition occurs in the eastern portions of South Dakota, Nebraska and north central Kansas. The region of lowest deposition shifts inland from the northern Pacific coast to the central Rockies during the winter.

In both the East and West, the Ca^{2+} concentration in the summer is approximately twice the winter concentration. In the East, the summer and winter spatial patterns are generally similar, but with higher concentrations in Iowa. The spatial pattern in the West is dominated by high concentrations centered in the summer and winter over the southeastern corner of Colorado.

In both the East and West, the Ca^{2+} deposition in the summer is approximately three times that in the winter. In the East, the spatial pattern of the Ca^{2+} deposition is almost opposite in the summer and the winter. In the summer, the minimum deposition is along the Atlantic coast, while in the winter higher depositions occur along the Atlantic coast and the Gulf of Mexico. In the West, the summer minima again occur along the Pacific coast. However, in the winter, the Pacific Northwest has relatively high Ca^{2+} deposition.

The rather significant changes in the seasonal patterns of both the SO_4^{2-}/NO_3^- concentration and deposition ratios depend upon meteorological and climatological factors and emissions and are beyond the scope of this chapter.

8. TREND ANALYSIS FOR THE PERIOD 1979–1987

The temporal pattern analysis used a subset of a possible 39 long-term sites over a nine-year (1979–1987) period and an expanded subset of a possible 148 sites with greater spatial coverage over a six-year (1982–1987) period. The nine-year period represents the longest period available with wet deposition monitoring data and sufficient sites with data of known quality to produce a descriptive summary of annual temporal patterns. The six-year period restricted the number of years in order to increase the number of sites available with data of known quality. Box plot displays for annual precipitation amount, pH, H^+, SO_4^{2-}, NO_3^-, NH_4^+, Ca^{2+}, Cl^-, Na^+, K^+ and Mg^{2+} are used to provide an overview of the types of temporal patterns observed. This chapter, however, discusses only the primary analytes (pH, H^+, SO_4^{2-} , NO_3^-, NH_4^+, and Ca^{2+}) and only summarizes the temporal analysis; details are found elsewhere.[1]

The KST temporal analyses are summarized for the multivariate patterns, first for concentration and then for deposition. The discussion begins with an overview of the estimated 1987 annual concentrations for the ion species observed at the sites. This overview is followed by a comparison across the ion species of Sen's median slope, a trend estimate associated with the modified KST. The trend estimate is then discussed in terms of percent change during the period. The reference point is the estimated 1987 annual concentration for an ion species at a site. The estimated annual concentration is used because some sites have no annual summaries, because samples were missing or invalid during a significant portion of at least one season. Furthermore, the estimates are more stable than the actual values because the estimates account for within-site variability and spatial correlation with nearby sites. The magnitude of the percent change depends on the reference base at the sites with increasing or decreasing temporal patterns for ion species. For example, using 1987 rather than 1982 as the reference point causes percent changes for cations at most sites to be greater because the cation concentrations decreased markedly from 1982 to 1987.

8.1. *Concentration Trends*

Box plots in Figure 18 give the distribution summaries of the estimated 1987 annual PWCs of all the ions for the 1979–1987 and 1982–1987 trend sites. The box plots use the 10th and 90th percentiles for the 1979–1987 subset and the 5th and 95th percentiles for the 1982–1987 subset. Hydrogen and SO_4^{2-} ions have similar relative median concentrations in precipitation, approximately 40 μeq/L. Nitrate and NH_4^+ have median concentrations of 20 μeq/L and approximately 15 μeq/L, respectively. The median concentration of Ca^{2+} is below 10 μeq/L.

The distributions of Sen's median slope trend estimates for each ion species concentration are presented as box plot displays in Figure 19, and the percentiles are shown in Tables III and IV for the 1979–1987 and 1982–1987 trend sets, respectively. For the 1979–1987 subset, only for NH_4^+ are more than 50% of the slope estimates greater than zero (increasing trends). Approximately 90% of the SO_4^{2-} and Ca^{2+} trends are negative. Hydrogen and NO_3^- ions are approximately centered at zero trend. For the 1982–1987 subset, the percentages of negative slope estimates are reduced for all the ions except NH_4^+. The most striking difference is for NO_3^-, with almost 75% of the trends increasing. This shift toward more positive slope estimates is not due entirely to the additional sites in the 1982–1987 subset. Over 50% of the 39 sites in both subsets have positive increases in their slope estimates for all the ions (smaller negative slopes).

The concentration trend estimates can also be expressed in terms of percent change during the period. The reference point for the percent change is the estimated 1987 annual concentration for an ion species at a site. Tables V and VI and Figure 20 summarize the percent changes in concentrations. For the 1979–1987 trend subset, CA^{2+} has the largest median percent change, −6.1%. Extrapolating this change back to 1979 indicates that the median Ca^{2+} concentration is 54.9% larger in 1979 than in 1987. Sulfate (−2.2%) also has a relatively large median percent change per year. Only for H^+ in the 1982–1987 trend subset are the median percent changes approximately the same as in 1979–1987 trend subset. The percent change decreases by more than one-half for SO_4^{2-}, NH_4^+, and Ca^{2+}. For NO_3^-, the median change in the 1979–1987 trend subset is approximately zero, while that in the 1982–1987 trend subset is 1.7%.

8.2. *Deposition Trends*

Box plots in Figure 21 give the distribution summaries of the estimated 1987 annual total depositions of all the ions for the 1979–1987 and 1982–1987 trend sites. Median depositions are approximately 40 meq/m^2 for SO_4^{2-}, 35 meq/m^2 for H^+, 20 meq/m^2 for NO_3^-, and 14 meq/m^2 for NH_4^+. The median depositions for the other ion species are below 10 meq/m^2. The distribution of Ca^{2+} is skewed because a number of sites have relatively high deposition. For H^+, the total deposition has less variation (a smaller interquartile range) than the precipitation-

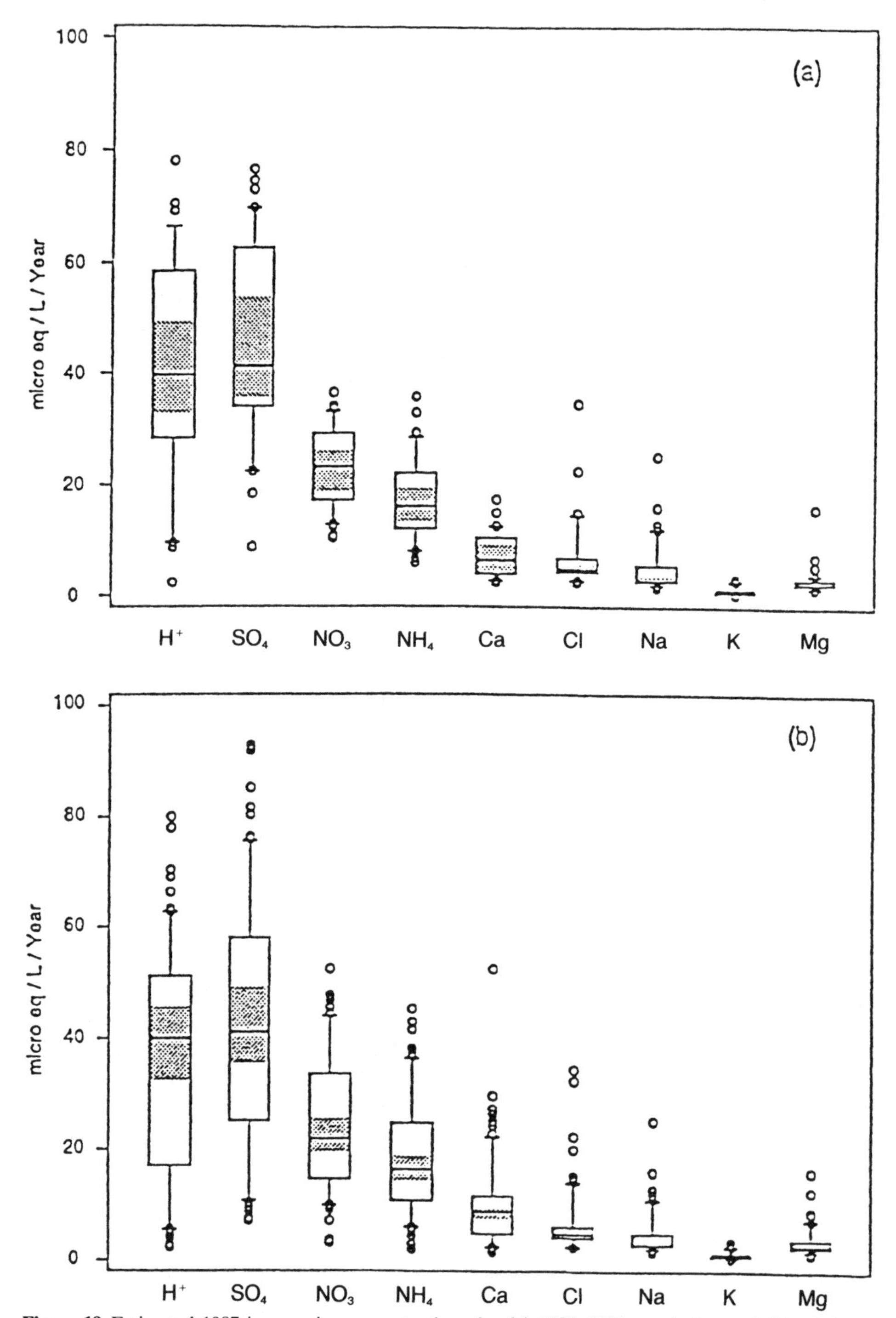

Figure 18 Estimated 1987 ion species concentrations for (a) 1979–1987 trend sites and (b) 1982–1987 trend sites.

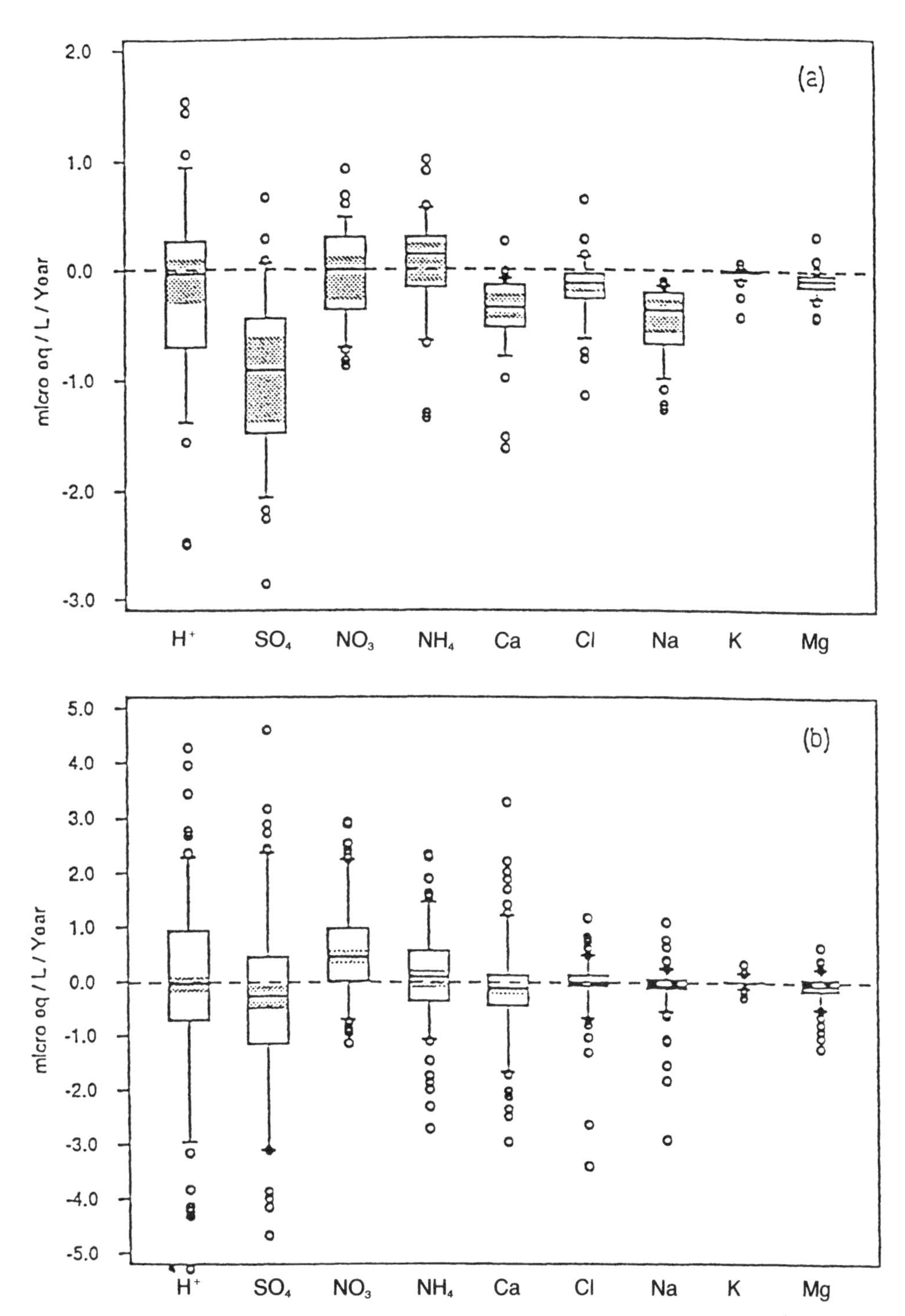

Figure 19 Ion species concentration trend estimates for (a) 1979–1987 trend sites and (b) 1982–1987 trend sites.

Table III Change per Year (μeq/L/yr) of Precipitation-Weighted Annual Mean Concentration for 1979–1987 Trend Sites

Ion	Min	Percentile					Max
		10th	25th	50th	75th	90th	
pH[a]	−0.04	−0.01	−0.01	0.00	0.01	0.01	0.05
hydrogen	−2.51	−1.40	−0.73	−0.05	0.24	0.93	1.53
sulfate	−2.87	−2.07	−1.52	−0.91	−0.44	0.04	0.66
nitrate	−0.87	−0.71	−0.40	−0.01	0.29	0.47	0.92
ammonium	−1.35	−0.65	−0.18	0.12	0.28	0.54	0.98
calcium	−1.63	−0.78	−0.56	−0.35	−0.16	−0.08	0.25
chloride	−1.15	−0.62	−0.28	−0.14	−0.05	0.12	0.64
sodium	−1.28	−0.98	−0.70	−0.38	−0.22	−0.14	−0.10
potassium	−0.43	−0.10	−0.06	−0.03	−0.01	0.01	0.06
magnesium	−0.43	−0.26	−0.18	−0.09	−0.05	−0.01	0.30

[a]pH units.

Table IV Change per Year (μeq/L/yr) of Precipitation-Weighted Annual Mean Concentration for 1982–1987 Trend Sites

Ion	Min	Percentile					Max
		5th	25th	50th	75th	95th	
pH[a]	−0.04	−0.03	−0.02	0.00	0.02	0.04	0.07
hydrogen	−7.41	−3.00	−0.79	−0.04	0.90	2.27	4.26
sulfate	−4.68	−3.12	−1.20	−0.28	0.44	2.33	4.59
nitrate	−1.18	−0.72	−0.05	0.42	0.92	2.23	2.90
ammonium	−2.74	−1.11	−0.42	0.05	0.55	1.40	2.29
calcium	−3.00	−1.69	−0.50	−0.14	0.11	1.18	3.25
chloride	−3.43	−0.71	−0.16	−0.03	0.09	0.45	1.12
sodium	−2.94	−0.58	−0.19	−0.08	0.00	0.21	1.07
potassium	−0.31	−0.13	−0.06	−0.03	−0.00	0.13	0.30
magnesium	−1.25	−0.54	−0.23	−0.10	0.01	0.23	0.62

[a]pH units.

weighted concentration (Figure 18), but the total deposition has larger extremes.

The distribution of Sen's median slope trend estimates for total deposition of each ion species are presented as box plot displays in Figure 22. The percentiles are shown in Tables VII and VIII for the 1979–1987 and the 1982–1987 trend subsets, respectively. For the 1979–1987 subset, fewer than 50% of the slope estimates for every ion are greater than zero (increasing trend). All the Ca^{2+} trends are negative (decreasing), and over 75% of the SO_4^{2-} trends are negative. For only H^+, NO_3^-, and NH_4^+ are less than 75% of the trend estimates decreasing. For the 1982–1987 subset, fewer than 50% of the slope estimates for every ion are again greater than zero. Except for H^+, NO_3^-, and NH_4^+, over 75% of the trend estimates are negative. Unlike the trend estimates for concentration, the percentiles are similar in the two trend subsets, except for Ca^{2+}.

The deposition trend estimates can also be expressed in terms of percent change during the period. The reference point for the percent change is the estimated 1987 annual deposition for an ion species at a site. Tables IX and X and Figure 23 summarize the results for percent change in deposition. For the

Table V Change per Year of Precipitation-Weighted Annual Mean Concentration for 1979–1987 Trend Sites

		Percentile					
Ion	Min	10th	25th	50th	75th	90th	Max
pH[a]	−0.76	−0.28	−0.12	0.04	0.21	0.32	0.78
hydrogen	−5.66	−3.98	−1.65	−0.36	0.82	2.55	5.39
sulfate	−16.87	−6.87	−3.70	−2.22	−0.91	0.10	1.25
nitrate	−6.10	−2.94	−1.81	−0.03	1.34	2.30	4.02
ammonium	−8.19	−3.70	−1.62	0.71	1.91	3.44	5.16
calcium	−25.41	−11.80	−7.40	−6.14	−2.85	−2.17	2.23
chloride	−38.82	−11.67	−6.21	−2.83	−1.00	1.90	5.40
sodium	−80.03	−26.97	−17.29	−9.50	−6.70	−5.27	−0.96
potassium	−66.30	−12.75	−7.69	−3.77	−1.12	0.56	2.12
magnesium	−23.77	−10.24	−6.24	−4.08	−2.39	−1.02	3.03

Table VI Percent Change per Year of Precipitation-Weighted Annual Mean Concentration for 1982–1987 Trend Sites

		Percentile					
Ion	Min	5th	25th	50th	75th	95th	Max
pH	−0.99	−0.68	−0.32	0.03	0.38	0.92	1.32
hydrogen	−14.77	−8.27	−3.34	−0.22	2.54	6.25	9.41
sulfate	−11.99	−9.28	−3.53	−0.98	0.82	4.13	9.47
nitrate	−9.67	−4.39	−0.28	1.79	3.96	8.23	12.74
ammonium	−18.67	−9.37	−3.17	0.28	2.86	7.60	12.76
calcium	−34.56	−17.61	−8.39	−2.62	1.42	7.94	16.77
chloride	−44.75	−13.18	−3.42	−0.74	1.77	9.49	34.66
sodium	−69.88	−14.07	−6.22	−2.86	0.03	6.25	14.34
potassium	−48.83	−20.37	−10.41	−4.64	−0.56	8.05	20.08
magnesium	−40.21	−19.35	−8.60	−3.76	0.42	5.33	16.05

1979–1987 trend subset, Ca^{2+} has the largest median percent change, −5.5%. Extrapolating this change back to 1979 indicates that the median Ca^{2+} deposition in 1979 was about 50% larger than that in 1987. Sulfate (−3.0%) also had relatively large median percent change per year. For all ions except NO_3^-, the median percent changes are larger (more negative) for the 1982–1987 trend subset than for the 1979–1987 subset. This is unlike concentration, for which the median percent change decreases (becomes less negative) for most of the ion species in the 1982–1987 subset.

8.3. *Summary of Trends*

Table XI gives an overview of the KST trend estimate results. The most striking feature of the comparison of concentrations across ion species is that except for NO_3^- and NH_4^+, the distributions are consistent, with over 50% of the estimates indicating a decreasing trend during the period. This is particularly true for the 1979–1987 period; the concentration trend estimates for the 1982–1987 trend sites are more variable. For the 1979–1987 trend sites, the median trend estimate is negative (decreasing) for all ion species except NH_4^+ (Table XI). For the 1982–1987 trend sites, the median trend estimate indicates decreasing concentration

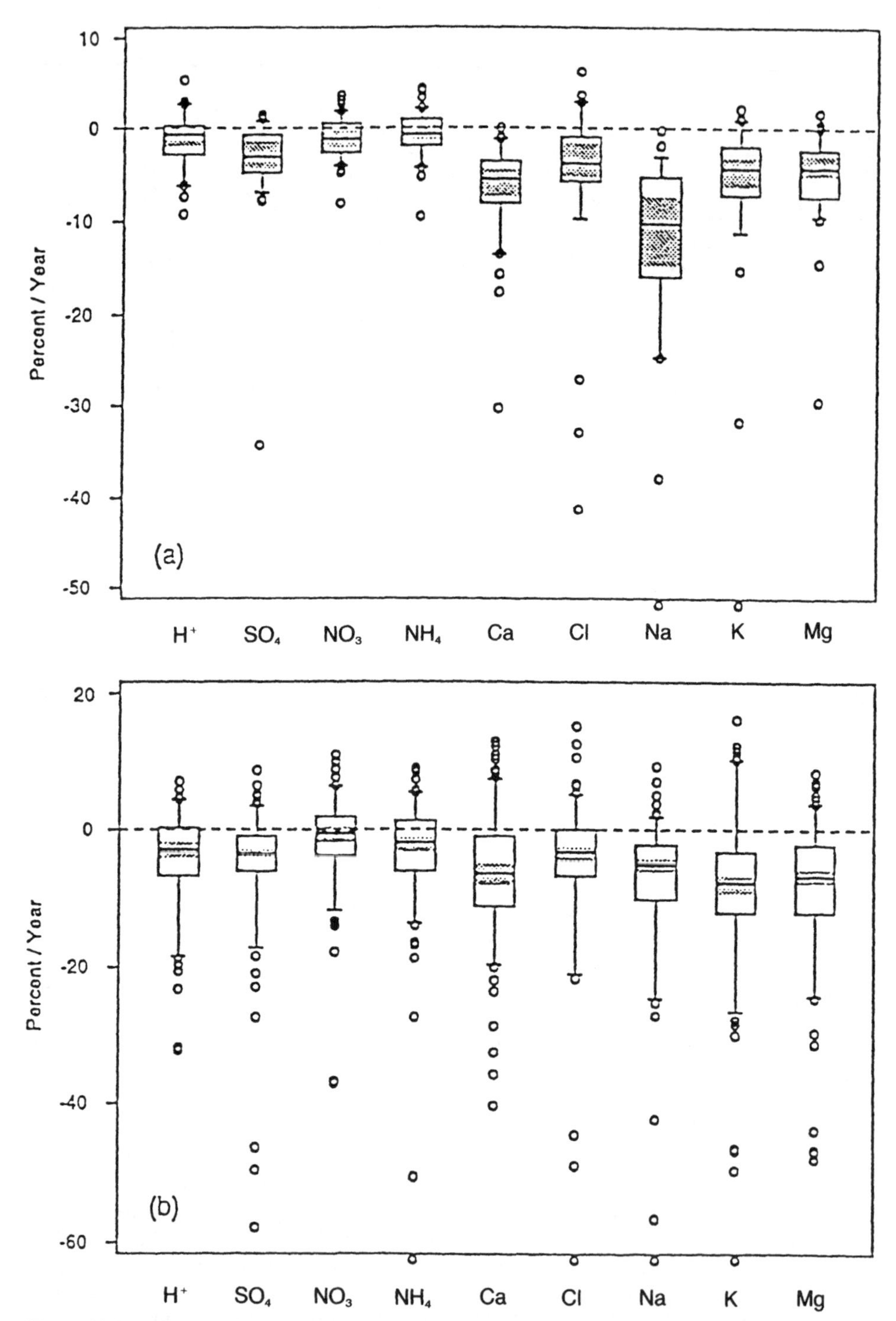

Figure 20 Distribution of ion species concentration trend estimates as percent of 1987 annual concentration for (a) 1979–1987 trend sites and (b) 1982–1987 trend sites.

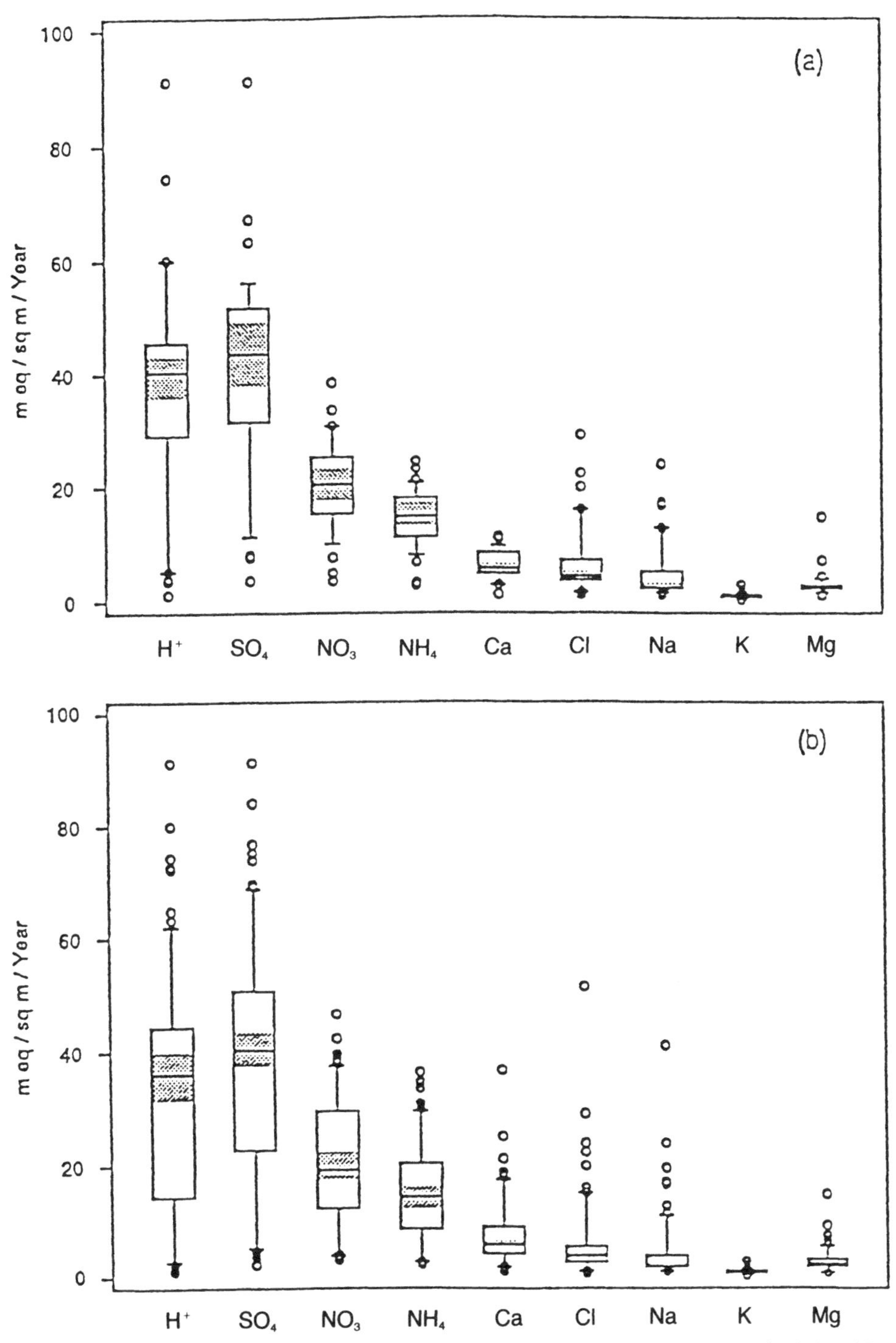

Figure 21 Estimated 1987 ion species deposition for (a) 1979–1987 trend sites and (b) 1982–1987 trend sites.

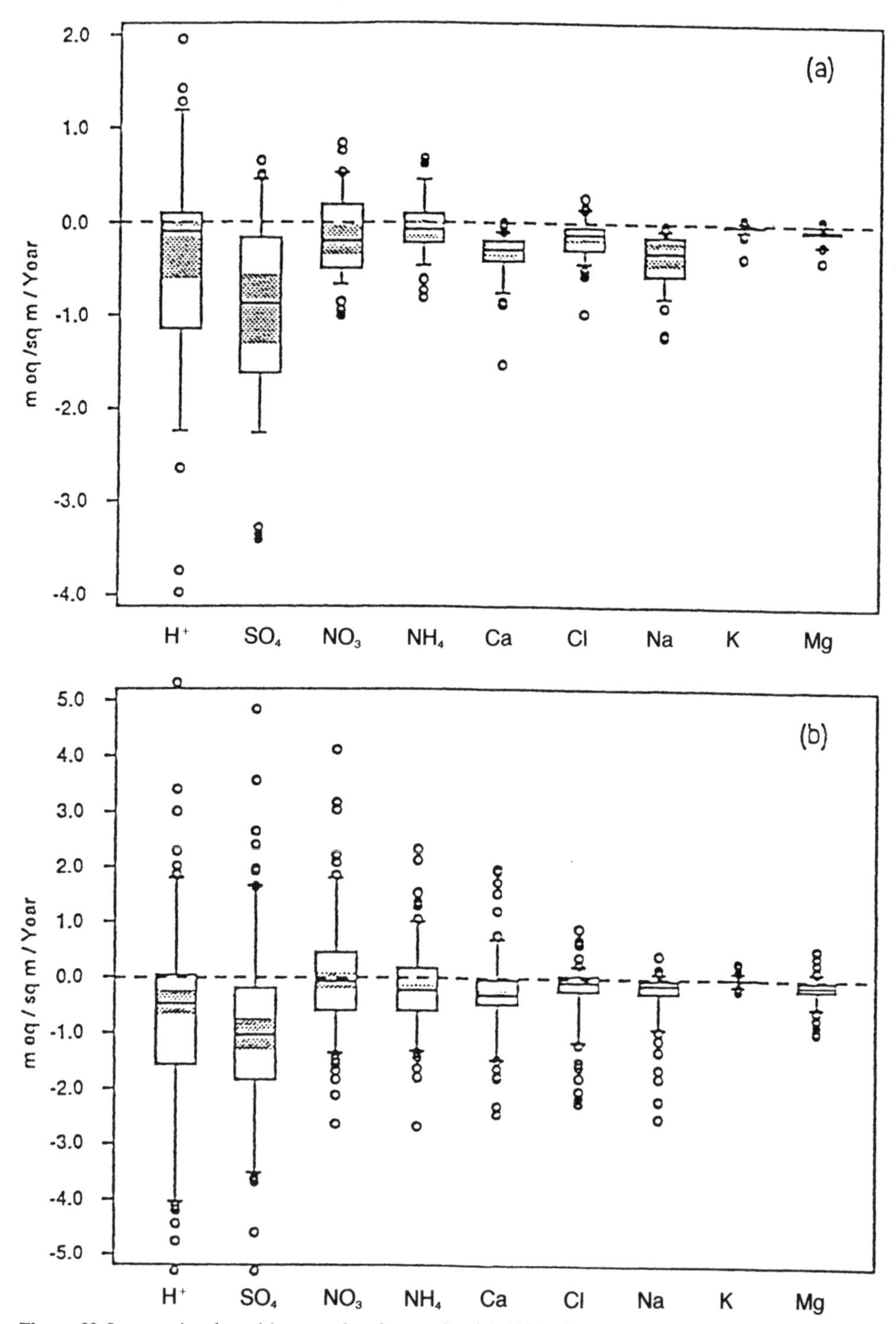

Figure 22 Ion species deposition trend estimates for (a) 1979–1987 trend sites and (b) 1982–1987 trend sites.

Table VII Change per Year (meq/m²/yr) of Precipitation-Weighted Annual Deposition for 1979–1987 Trend Sites

		Percentile					
Ion	Min	10th	25th	50th	75th	90th	Max
precipitation[a]	−5.49	−3.38	−1.75	−0.04	1.10	2.38	4.07
hydrogen	−4.00	−2.23	−1.17	−0.09	0.08	1.21	1.96
sulfate	−3.42	−2.27	−1.64	−0.88	−0.18	0.45	0.64
nitrate	−0.98	−0.65	−0.51	−0.20	0.19	0.52	0.84
ammonium	−0.79	−0.45	−0.24	−0.07	0.10	0.45	0.66
calcium	−1.52	−0.75	−0.43	−0.29	−0.20	−0.11	0.00
chloride	−0.96	−0.44	−0.32	−0.13	−0.06	0.15	0.26
sodium	−1.22	−0.79	−0.59	−0.32	−0.15	−0.07	−0.04
potassium	−0.37	−0.07	−0.04	−0.03	−0.02	0.01	0.03
magnesium	−0.40	−0.22	−0.11	−0.08	−0.05	0.00	0.04

[a]Annual precipitation (cm).

Table VIII Change per Year (meq/m²/yr) of Precipitation-Weighted Annual Deposition for 1982–1987 Trend Sites

		Percentile					
Ion	Min	5th	25th	50th	75th	95th	Max
precipitation[a]	−15.30	−9.28	−4.11	−1.85	−0.28	2.60	5.27
hydrogen	−6.70	−4.04	−1.64	−0.50	0.02	1.76	5.04
sulfate	−5.32	−3.54	−1.89	−1.08	−0.23	1.61	4.82
nitrate	−2.64	−1.40	−0.66	−0.10	0.41	1.79	4.09
ammonium	−2.71	−1.36	−0.67	−0.25	0.14	0.97	2.30
calcium	−2.47	−1.48	−0.52	−0.32	−0.08	0.66	1.95
chloride	−2.25	−1.18	−0.30	−0.11	0.00	0.20	0.87
sodium	−2.54	−0.94	−0.34	−0.13	−0.06	0.04	0.40
potassium	−0.23	−0.13	−0.07	−0.03	−0.02	0.09	0.27
magnesium	−0.98	−0.56	−0.25	−0.13	−0.05	0.40	0.48

[a]Annual precipitation (cm).

trends for all ion species except NO_3^- and NH_4^+. Sulfate concentration has the most extreme median trend estimates, −0.91 and −0.28 μeq/L/y for the 1979–1987 and 1982–1987 trend sites, respectively. The consistent results of the two data sets give credibility to the 1979–1987 trend results, even though substantially fewer sites could be used in the analysis.

The concentration trend estimates can also be expressed in terms of percent change during the period. The reference point for the percent change is the estimated 1987 annual concentration for an ion species at a site. Calcium concentrations for the 1979–1987 trend sites have the most extreme negative percent change, −6.1 percent per year. Distributions for NO_3^-, NH_4^+, and, to a lesser extent, H^+, center on zero percent change. For the 1982–1987 trend sites, the increasing (most negative percent change) rank order of the medians for the distributions is Ca^{2+}, SO_4^{2-}, H^+, NH_4^+, and NO_3^-. Nitrate has a median (1.8%) representing a positive percent change.

The KST evaluates the significance of a trend in the temporal pattern. A significance level of $p = 0.05$ can be used to summarize the test results. For each ion species and the two trend subsets, Table XII summarizes the number

Table IX Percent Change per Year of Precipitation-Weighted Annual Deposition for 1979–1987 Trend Sites

Ion	Min	Percentile					Max
		10th	25th	50th	75th	90th	
precipitation	−7.21	−3.78	−1.94	−0.09	1.25	2.31	3.87
hydrogen	−9.35	−6.29	−3.15	−0.65	0.26	2.73	5.32
sulfate	−34.26	−7.02	−5.11	−2.99	−0.71	0.70	1.44
nitrate	−8.06	−3.95	−2.91	−1.20	0.61	1.96	3.73
ammonium	−9.67	−4.32	−2.05	−0.59	1.09	2.23	4.37
calcium	−30.30	−13.56	−8.26	−5.52	−3.54	−1.23	0.00
chloride	−41.33	−9.90	−6.06	−3.70	−0.96	2.85	6.35
sodium	−96.81	−24.79	−16.38	−10.37	−5.15	−3.14	−0.21
potassium	−75.19	−11.30	−7.36	−4.27	−1.81	0.95	2.26
magnesium	−29.59	−9.61	−7.67	−4.25	−2.40	0.00	1.71

Table X Percent Change per Year of Precipitation-Weighted Annual Deposition for 1982–1987 Trend Sites

Ion	Min	Percentile					Max
		5th	25th	50th	75th	95th	
precipitation	−38.47	−11.01	−5.45	−2.39	−0.54	2.98	5.47
hydrogen	−32.32	−18.70	−6.98	−2.91	0.25	4.46	6.99
sulfate	−58.02	−17.38	−6.55	−3.46	−0.83	3.66	8.74
nitrate	−37.16	−11.98	−4.23	−0.76	1.80	6.38	10.83
ammonium	−90.60	−13.80	−6.43	−1.95	1.13	5.41	8.83
calcium	−40.44	−19.91	−11.47	−6.27	−1.03	7.37	12.89
chloride	−94.61	−21.25	−7.18	−3.34	0.05	5.11	15.08
sodium	−73.73	−24.78	−10.52	−5.22	−2.27	2.08	9.40
potassium	−78.59	−26.59	−12.48	−7.58	−3.11	10.11	15.94
magnesium	−48.10	−24.42	−12.33	−6.70	−2.35	3.97	8.36

and percent of the concentration trend estimates that are significant, either decreasing or increasing. For the 1979–1987 trend sites, only NH_4^+ concentration has a site with a significantly increasing trend. In contrast, all ion species have sites with significantly decreasing trends. However, with 39 sites being tested simultaneously, it is not unusual for a small number of sites to have significant trends even when there is no overall trend (false positive). For 39 simultaneous, independent tests, 5 or more significant results will occur only 4.4% of the time when no trend exists. Thus, the number of significantly decreasing slopes for SO_4^{2-} and Ca^{2+} is greater than expected if no trend exists. The 1982–1987 trend sites show fewer significantly decreasing trends and more significantly increasing trends. For 148 simultaneous, independent tests, 5 or more significant results will occur >86% of the time when no trend exists, 9 or more significant results >20% of the time, and 12 or more significant results only 3.5% of the time. Therefore, the number of significantly increasing slopes for only NO_3^- is greater than expected if no trend exists.

For H^+, SO_4^{2-}, NO_3^-, and NH_4^+, no site that has a significant concentration trend in the 1979–1987 subset has a significant trend in the 1982–1987 subset. For Ca^{2+}, only five sites with significant trends in the 1979–1987 subset have significant trends in the 1982–1987 subset. The reduced number of significant

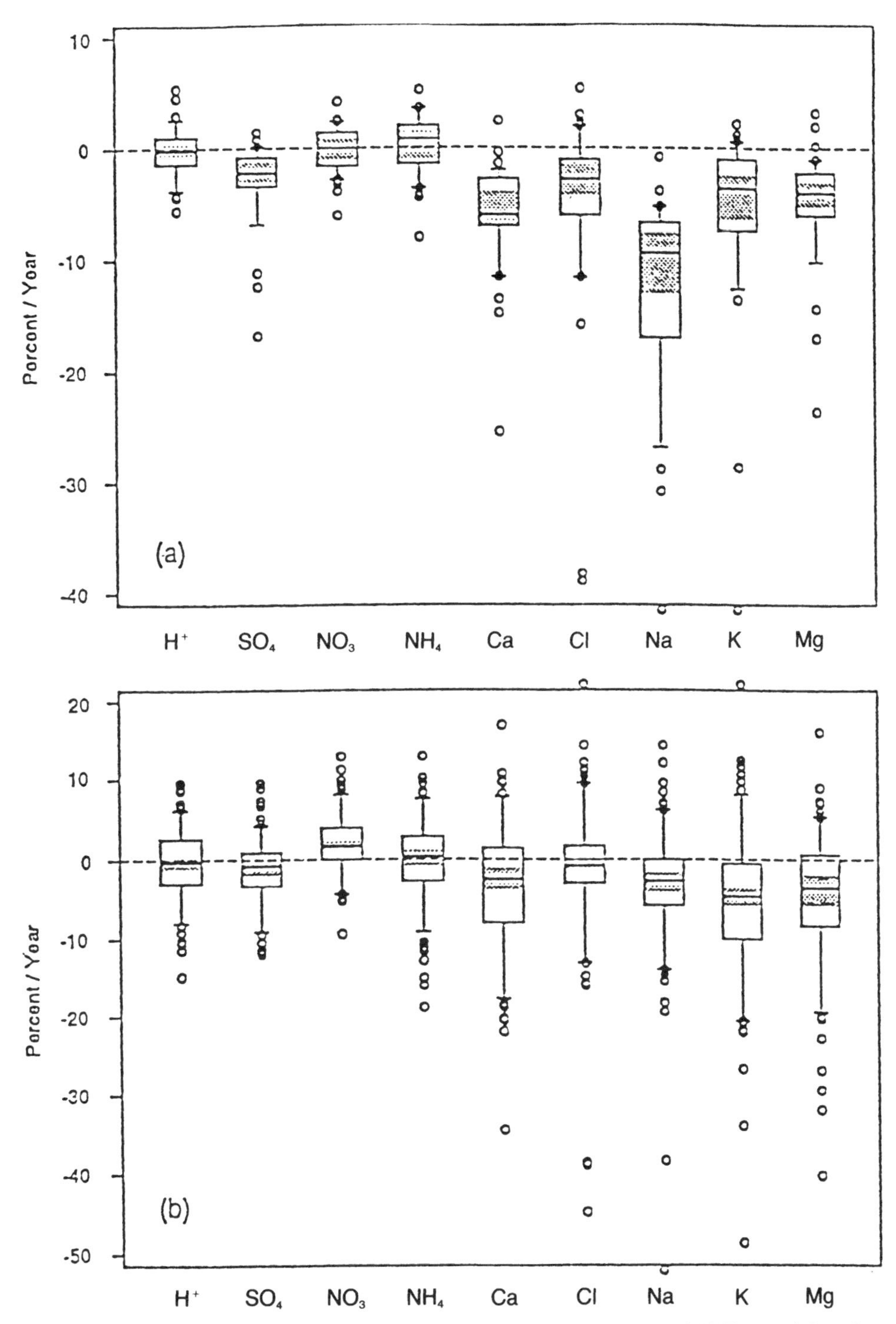

Figure 23 Distribution of ion species deposition trend estimates as percent of 1987 annual deposition for (a) 1979–1987 trend sites and (b) 1982–1987 trend sites.

Table XI Median and Percent Change of Trend Estimates of Ion Concentration and Deposition for 1979–1987 and 1982–1987 Trend Sites

	Concentration (μeq/L)				Deposition (meq/m²)			
	1979–1987		1982–1987		1979–1987		1982–1987	
Parameter	Median Change	% Change	Median Change	% Change	Median Change	% Change	Median Change	% Change
precipitation	—	—	—	—	−0.04	−0.09	−1.85	−2.39
H^+	−0.05	−0.36	−0.04	−0.22	−0.09	−0.65	−0.50	−2.91
SO_4^{2-}	−0.91	−2.22	−0.28	−0.98	−0.88	−2.99	−1.08	−3.46
NO_3^-	−0.01	−0.03	0.42	1.79	−0.20	−1.20	−0.10	−0.76
NH_4^+	0.12	0.71	0.05	0.28	−0.07	−0.59	−0.25	−1.95
Ca^{2+}	−0.35	−6.12	−0.14	−2.62	−0.29	−5.52	−0.32	−6.27

Table XII Number and Percent of Significantly Decreasing and Increasing Trend Estimate for Concentration and Deposition at 1979–1987 and 1982–1987 Trend Sites

	1979–1987 Trend Sites Significance					1982–1987 Trend Sites Significance				
	Decrease		None	Increase		Decrease		None	Increase	
Ion Species	%	N[a]	N	N	%	%	N	N	N	%
				Concentrations						
pH	0	0	36	3	8	0	0	144	4	3
Hydrogen	8	3	36	0	0	3	4	144	0	0
Sulfate	18	7	32	0	0	2	3	144	1	1
Nitrate	5	2	37	0	0	0	0	137	11	7
Ammonium	3	1	37	1	3	2	3	143	2	1
Calcium	33	13	26	0	0	6	9	136	3	2
				Deposition						
Precipitation	5	2	37	0	0	1	1	147	0	0
Hydrogen	8	3	34	2	5	2	3	145	0	0
Sulfate	28	11	28	0	0	5	8	140	0	0
Nitrate	8	3	36	0	0	1	1	145	2	1
Ammonium	5	2	34	3	8	4	6	140	2	1
Calcium	44	17	22	0	0	9	13	134	1	1

[a]N is the number of sites.

trends for the 39 sites in both subsets can be attributed to two factors: first, the power (the ability to detect an existing significant trend) decreases because of the shorter time span in the 1982–1987 subset; and second, more than 50% of the slopes have positive increases (they either becomes less negative or change sign and become positive). As previously noted, the maximum concentrations for many ions occurred in 1980 and 1981, predating the 1982–1987 subset. Thus, the slope estimates and the significance of those slopes in the 1979–1987 subset decreases from the 1980 and 1981 maxima. Of the 1979–1986 trend sites, only one (Bradford Forest, Florida) had a significantly increasing trend in H^+ concentration. In contrast, every ion species had a significantly decreasing trend at some site. Eighty-nine significant decreases occurred in an ion concentration or deposition, representing 33% of the possible 270 tests calculated for the 30 sites and 5 ion species. Calcium had significantly decreasing trend estimates at the greatest number of sites, 17 (57% of sites). Sulfate had significantly decreasing trend estimates at 8 (27%) sites. For the 1982–1987 trend sites, every ion species has a significantly increasing trend estimate at one or more sites. Thirty-five significant increases occurred in an ion concentration or deposition at a site. This number represents only 3% of the 1233 tests calculated for the 137 sites and the 5 ion species. Ammonium and Ca^{2+} have significantly increasing trend estimates at the most sites, 11 and 6, respectively. Significant decreasing trend estimates occur in 126 (10%) of the total (1233) cases. Calcium and SO_4^{2-} have significantly decreasing trends at the most sites, 20 (15% of sites) and 16 (12% of sites), respectively.

Table XII also summarizes the number of sites with significantly decreasing or increasing deposition trend estimates. For the 1979–1987 subset, SO_4^{2-} and Ca^{2+} have more significantly decreasing trends than chance alone would usually explain. For the 1982–1987 subset, Ca^{2+} has more significant decreasing trends than would be expected by chance alone. In addition, SO_4^{2-} has decreasing trends at eight sites and increasing trends at no sites. Although eight or more significant trends can occur 32% of the time when no trend exists, false positive significant trends would be expected to be both negative and positive. Thus the decreasing SO_4^{2-} deposition trend also appears to be significant. Only one site each for SO_4^{2-} and Ca^{2+} has a significant trend in the 1979–1987 subset and also has a significant trend in the 1982–1987 subset.

8.4. *Discussion of Trends*

The question we have attempted to answer in this chapter is the following: On a regional scale, are concentrations and depositions for individual analytes different in 1979 and 1987? (That is, is there a trend?) Because sulfur emissions were reduced in the late 1970s and because emissions influence precipitation composition, one might anticipate a reduction in SO_4^{2-} in precipitation. We have used a linear trend technique for our analyses because a linear method, if valid, allows the trend to be quantified. A great deal of within- and between-year (temporal) variability is in the data, probably because of the varying influences of emissions and meteorological conditions. As a result, no statistically significant trend in regional precipitation composition occurs for any analyte for the period 1979–1987. A few individual sites do show significant trends for some analytes, both decreasing and increasing. However, these sites are not geographic neighbors to imply a subregional trend.

Although no statistical evidence supports a trend, other evidence for a trend cannot be overlooked. For example, SO_4^{2-} concentration at more than 75% of the 1979–1987 trend sites (more than 90% of the 1982–1987 trend sites) exhibits a negative slope (decreasing trend), indicating an average reduction of about 14% over the eight-year period. In other words, most of the sites had lower SO_4^{2-} concentration values in 1987 than in 1979.

The absence of a statistical trend in the wet deposition data does not indicate that deposition and emissions are unrelated. The within- and between-year patterns of SO_4^{2-} in precipitation and SO_2 emissions may be correlated. This is a question addressed elsewhere.[39]

The most dramatic feature of the trend analysis is that although SO_4^{2-} and NO_3^- concentrations and deposition do exhibit an overall decrease between 1979 and 1987, H^+ concentrations and deposition do not. Because the relative relationships of the various analytes must be internally consistent (e.g., ion balances must be good), the large decreases observed for Ca^{2+} concentration and deposition offset the decrease in SO_4^{2-} and NO_3^- concentrations and deposition and result in little or no change in H^+ concentration or deposition. Clearly, Ca^{2+} has decreased in precipitation, but the reason for the decrease is not obvious. Earlier discussions of the relationships between ions in precipitation indicated that Ca^{2+} is related primarily to soils and dust from unpaved roads. In dry years, atmospheric soil loading may be expected to increase,[40] and some of the driest years on record occurred in the late 1980s (M. E. Peden, 1989, *Drought Effects on Precipitation Chemistry*. Unpublished report based on a poster session presented at the Fall Annual Meeting of the American Geophysical Union, December 6–11, 1989, San Francisco, CA. Copies of the report may be obtained from the Illinois State Water Survey, Champaign, IL). Our rationale would suggest that soil-derived components in precipitation should have increased. However, the concentration and deposition of Ca^{2+} decreased between 1979 and 1987, as did those of all other analytes associated with the soil-derived component of precipitation composition (i.e., Mg^{2+} and K^+) and also Na^+ and Cl^-. This observed trend is not unique to any particular network but is observed across all networks. Recent soil conservation practices could have led to a decrease of Ca^{2+} in the atmosphere, as could the paving of rural gravel highways. These changes could diminish Ca^{2+} levels in precipitation in the farm belt, particularly in the Midwest. Our trend analysis for Ca^{2+} revealed no particular regional grouping of sites with lower precipitation Ca^{2+} concentration or deposition. In contrast, the trends in Ca^{2+} concentration and deposition for both the 1979–1987 and 1982–1987 subsets suggest a nationwide decrease, including the Northeast. The decrease in Ca^{2+}, particularly in the Northeast where Ca^{2+} is not associated with soils, suggests a possible link with the diminished fossil fuel emissions. Controls that limit sulfur and nitrogen species in stack emissions may also limit emissions of base cations found in the fuel. Evidence is insufficient to isolate the cause for the apparent decrease in base cations in precipitation; i.e. the observed decrease may be due to a combination of the possible causes discussed here.

9. CONCLUSIONS

The areas of maximum deposition of individual inorganic chemical species in precipitation are located in the northeastern United States and southeastern Canada, and those locations have not changed significantly in the past decade. The actual deposition values and the areal coverages of those categories have changed from year to year depending on the variations in emissions and meteorological conditions, but the locations of the maxima, in general, are quite similar. Strong seasonal variations also occur for all species analyzed.

A number of factors are important in determining the ability of a monitoring network to detect a change in regional-scale precipitation composition. These include density of monitoring sites, complexity of the terrain, meteorological factors, and spatial analysis methodology. Limited information based on the distribution of the 150 NTN sites in the United States suggests that differences in annual mean precipitation composition must be larger than about 25–40% (depending on the subregion) to be reliably detected. The ability of the network to detect changes increases (but not linearly) with increasing numbers of monitoring sites. Combining weekly and event precipitation data and individual network biases does not appear to significantly alter the locations of wet deposition maxima, indicating that uncertainties in the spatial methodology are probably greater than the differences in composition of weekly and event precipitation.

The weight of evidence suggests that all analytes (except H^+) exhibited a decreasing (not significant) trend in deposition and concentration between 1979 and 1987 for the eastern half of the United States. Fewer than 25% of the individual sites showed a statistically significant trend (some increasing, some decreasing); these sites were not located together and therefore did not suggest a subregional trend. However, nearly all of the trend sites showed a decrease in deposition for nearly all analytes. Concentration trend patterns were similar, but the median trend slope estimates were generally smaller. Because both trend data sets include few western sites, all discussion about trend is a reflection of eastern North American sites only. Western sites have been added to regional networks in the past few years, but precipitation is typically much lower in that geographic area than in eastern North America. As a result, wet deposition is more variable, making temporal analysis less certain in western North America.

It is important to note that although SO_4^{2-} and NO_3^- deposition has decreased, acidity has not. The decrease in acid-forming species has been offset by a corresponding decrease in base cations (mostly Ca^{2+}). The reason for the decrease in Ca^{2+} is unclear, although possible explanations include soil conservation practices, the paving of rural gravel roads, and the more efficient but inadvertent removal of base cations from emissions by devices designed to reduce sulfur and nitrogen. The terms acidic SO_4^{2-} and acidic NO_3^- represent the acidic fraction of sulfate (mostly H_2SO_4 from the liquid phase oxidation of SO_2 rather than acidic aerosol) and nitrate (HNO_3) that contribute H^+ to precipitation. The wet deposition of acidic SO_4^{2-} is largely responsible for the acidity of precipitation, particularly in the eastern half of North America. This statement may be misleading if it is taken to imply that the contribution of the wet deposition of NO_3^- to precipitation is unimportant. For example, HNO_3 found in the atmosphere is quickly neutralized as it comes into contact with large particles, much as H_2SO_4 is quickly neutralized by NH_3 in the atmosphere. However, many particles that might otherwise be available in precipitation for the possible neutralization of H_2SO_4 (formed by the liquid phase oxidation of SO_2) are "tied up" with the in-air neutralization of NHO_3. Weak organic acids also contribute to precipitation acidity, but that fraction is estimated to average less than 15% in North America, where anthropogenic emissions dominate precipitation acidity. Relationships between ions and acidity for the western half of the United States are not as well defined. Most of the uncertainty lies with the apportionment of NH_4^+ between SO_4^{2-} and NO_3^-, the apportionment of Ca^{2+} to NO_3^- and SO_4^{2-}, and the apportionment of catalysts and oxidizing species in the ambient atmosphere.

Acknowledgements

The work at Argonne National Laboratory was funded by the U.S. Department of the Interior, U.S. Geological Survey, through interagency agreement WRD9-4800-0321 with the U.S. Department of Energy. The preparation of this chapter has been funded as part of the Atmospheric Chemistry Program, funded by the Office of Health and Environmental Research of the U.S. Department of Energy under contract W-31-109-ENG-38. The work at the Illinois State Water Survey was funded by the U.S. Department of the Interior, U.S. Geological Survey, through cooperative agreement. The work done at Battelle Pacific Northwest Laboratory was supported by the U.S. Environmental

Protection Agency under a related service agreement with the U.S. Department of Energy under contract DE-AC06-76RLO 1830.

References

1. D. L. Sisterson, V. C. Bowersox, A. R. Olsen, T. P. Meyers, R. J. Vong, J. C. Simpson and V. A. Mohnen *NAPAP State of Science/Technology Report Number 6: Deposition Monitoring: methods and results.* (National Acid Deposition Assessment Program, 722 Jackson Place NW, Washington, DC, 1990).
2. C. R. Watson and A. R. Olsen "Acid Deposition System (ADS) for statistical reporting: System design and user's code manual" *Report EPA-600/8-84-023* (U.S. Environmental Protection Agency, Research Triangle Park, NC, 1984).
3. J. K. Sweeney and A. R. Olsen "Acid precipitation in North America: 1985 annual and seasonal data summaries from Acid Deposition System Data Base" *Report EPA 600/4-87-000* (U.S. Environmental Protection Agency, Research Triangle Park, NC, 1987).
4. A. R. Olsen, E. C. Voldner, D. S. Bigelow, W. H. Chan, T. L. Clark, M. A. Lusis, P. K. Misra and R. J. Vet "Unified wet deposition data summaries for North America: Data summary procedures and results for 1980–1986" *Atmos. Environ.* **24**, 661–672 (1990).
5. J. C. Simpson and A. R. Olsen "1987 wet deposition temporal and spatial patterns in North America" *PNL-7208, WC-402* (Battelle, Pacific Northwest Laboratory, Richland, WA, 1990).
6. P. Delfiner, J. P. Delhomme, J. P. Chiles, D. Renard and F. Irigoin *BLUEPACK 3-D* (Centre de Geostatique et de Morphologie, Fontainebleau, France, 1979).
7. S. K. Seilkop, and P. L. Finkelstein "Acid precipitation patterns and trend in eastern North America." *J. Climate Appl. Meteorol.* **26**, 980–994 (1987).
8. A. R. Olsen and A. L. Slavich "Acid precipitation in North America: 1984 annual data summary from Acid Deposition System Data Base." *Report EPA/600/4-86/033* (U.S. Environmental Protection Agency, Research Triangle Park, NC, 1986).
9. P. L. Finkelstein "The spatial analysis of acid precipitation data." *J. Climate Appl. Meteor.* **23**, 52–62 (1984).
10. B. A. Bilonick "Risk qualified maps of hydrogen ion concentration for the New York State area for 1966–1978." *Atmos. Environ.* **17**, 2513–2525 (1983).
11. A. R. Olsen "1986 wet deposition temporal and spatial patterns in North America." *DE-AC06-76RL0 1830* (prepared for the U.S. Environmental Protection Agency under a Related Services Agreement with the U.S. Department of Energy, Washington, D.C., 1988).
12. R. M. Hirsch, J. R. Slack and R. A. Smith "Techniques of trend analysis for monthly water quality data." *Water Resour. Res.* **18**, 107–121 (1982).
13. R. O. Gilbert *Statistical Methods for Environmental Pollution Monitoring* (Van Nostrand Reinhold Company, Inc., New York, NY, 1982).
14. A. Sirois and L. A. Barrie "An estimate of the importance of dry deposition as a pathway of acidic substances from the atmosphere to the biosphere in eastern Canada." *Tellus* **40B**, 59–80 (1988).
15. M. T. Dana and W. G. N. Slinn "Acidic deposition distribution and episode statistics from the MAP3S network data base." *Atmos. Environ.* **22**, 1469–1474 (1987).
16. P. W. Summers, V. C. Bowersox and G. J. Stensland "The geographical distribution and temporal variations of acidic deposition in eastern North America." *Water Air Soil Pollut.* **31**, 523–535 (1986).
17. D. Golomb, J. A. Fay and S. Kumar "Seasonal, episodic, and targeted control of sulfate deposition." *J. Air Pollut. Control Assoc.* **36**, 798–802 (1986).
18. J. Kurtz, A. J. S. Tang, R. W. Kirk and W. H. Chan "Analysis of an acidic deposition episode at Dorset, Ontario." *Atmos. Environ.* **18**, 387–394 (1984).
19. S. K. Seilkop "Evaluation of the National Trend Network's site placement design." *USEPA Contract Number 68-01-6849 (027)* (Program Resources, Inc., Science and Technology Center, Research Triangle Park, NC, 1987).
20. R. P. Hooper and N. E. Peters, "Use of multivariate analysis for determining sources of solutes found in wet deposition in the United States." *Environ. Sci. Technol.* **23**, 1263–1268 (1989).
21. A. S. Lefohn and S. V. Krupa, "The relationship between hydrogen and sulphate ions in precipitation: A numerical analysis of rain and snowfall chemistry." *Environ. Pollut.* **49**, 289–311 (1988).
22. D. L. Sisterson, "A method for evaluation of acidic sulfate and nitrate in precipitation." *Water Air Soil Pollut.* **43**, 61–72 (1988).

23. G. H. Wagner and K. F. Steele, "Rain chemistry at a midcontinent site, U.S.A., 1980–1984," *Atmos. Environ.* **21**, 1353–1362 (1987).
24. J. Crawley and H. Sievering, "Factor analysis of the MPA3S/RAINE precipitation chemistry network: 1976–1980," *Atmos. Environ.* **20**, 1001–1013 (1986).
25. J. W. Munger and S. J. Eisenreich, "Continental-scale variations in precipitation chemistry," *Environ. Sci. Technol.* **17**, 32A–42A (1983).
26. T. B. Ridder, T. A. Buishand, H. F. R. Reijnders and M. J.'t Hart, "Effects of storage on the composition of main components in rainwater samples," *Atmos. Environ.* **19**, 759–762 (1985).
27. M. E. Peden and L. M. Skowron, "Ionic stability of precipitation samples," *Atmos. Environ.* **12**, 2343–2349 (1978).
28. R. J. Vet, A. Sirois, D. S. Jeffries, R. G. Semkin, N. W. Foster, P. Hazlett and C. H. Chan, "Comparison of bulk, wet-only, and wet-plus-dry deposition measurements at the Turkey Lakes Watershed," *Can. J. Fish. Aquat. Sci.* **45**, 26–37 (1988).
29. L. E. Topol, M. On-Lev and A. K. Pollack "Comparison of weekly and daily wet deposition sampling results" **In**: (R. W. Johnson and G. E. Gordon, eds.) *The Chemistry of Acid Rain: Sources and Atmospheric Processes* (ACS Symposium Series No. 349. American Chemical Society, Washington, D.C., 1987).
30. R. G. de Pena, K. C. Walker, L. Lebowitz and J. G. Micka, "Wet deposition monitoring: Effect of sampling period," *Atmos. Environ.* **19**, 151–156 (1985).
31. D. L. Sisterson, B. E. Wurfel, and B. M. Lesht, "Chemical differences between event and weekly precipitation samples in northeastern Illinois," *Atmos. Environ.* **19**, 1453–1469 (1985).
32. G. J. Stensland and V. C. Bowersox, "A comparison of methods of computing precipitation pH averages" **In**: *Proceedings of the 77th Annual APCA Meeting*, Paper 84-19.1 (Air Pollution Control Association, Pittsburgh, PA, 1984).
33. B. C. Madsen, "An evaluation of sampling interval length on the chemical composition of wet-only deposition," *Atmos. Environ.* **16**, 2515–2519 (1982).
34. D. S. Bigelow, D. L. Sisterson and L. J. Schroder, "An interpretation of differences between field and laboratory pH values reported by the National Atmospheric Deposition Program/National Trend Network monitoring program," *Environ. Sci. & Technol.* **23**, 881–887 (1989).
35. R. C. Graham, J. K. Robertson and J. Obal, "An assessment of the variability in performance of wet atmospheric deposition samplers," *U.S. Geological Survey Water-Resources Investigations Report 97-4125* (Copies of this report may be purchased from U.S. Geological Survey Books and Open-File Reports Section, Federal Center, Building 810, Box 24525, Denver, CO 80225, 1987).
36. W. H. Chan, F. Tomassini and B. Loescher, "An evaluation of sorption properties of precipitation constituents on polyethylene surfaces," *Atmos. Environ.* **17**, 1779–1785 (1983).
37. J. N. Galloway and G. E. Likens, "Calibration of collection procedures for the determination of precipitation chemistry," *Water Air Soil Pollut.* **6**, 241–258 (1976).
38. W. R. Barchet, "Acidic deposition and its gaseous precursors, pp. 5-1 through 5-116" **In**: *NAPAP Interim Assessment Volume III: Atmospheric Processes and Deposition* (National Acid Precipitation Assessment Program, Washington, DC, 1987).
39. A. Venkatram, D. McNaughton, P. K. Kamramchandani, J. D. Shannon, D. L. Sisterson and M. Fernau, *NAPAP State of Science/Technology Report Number 8, Relationships between Atmospheric Emissions and Deposition/Air Quality* (National Acid Precipitation Program, 722 Jackson Place NW, Washington, DC, 1990).
40. G. J. Stensland and R. G. Semonin, "Another interpretation of the pH trend in the United States," *Bull. Amer. Meteorol. Soc.* **63**, 1277–1284 (1982).

8. ATMOSPHERIC EMISSIONS FROM ROAD TRANSPORT

CLAIRE HOLMAN

Road transport gives rise to high levels of air pollution in the form of carbon monoxide, volatile organic compounds, nitrogen oxides and particulates. In addition, road transport is an important source of carbon dioxide, the major greenhouse gas. Engine and vehicle design improvements, new fuels, and after-treatment devices such as catalytic converters have been developed that can successfully reduce the level of emissions from individual vehicles. However, demand for both passenger and commercial transport is growing rapidly and with it the number of vehicles on the roads. Today there are 550 million vehicles in use around the world which could rise to one billion in the next 20 to 40 years. Technical solutions alone cannot be relied upon to achieve environmental improvements in the absence of other measures, particularly attempts to curb the growth in car traffic.

1. INTRODUCTION

Transport has a wide range of adverse effects on human health and the environment, and its provision is considered to be one of the most polluting of all human activities. The growth in road vehicle use has meant that efforts to reduce pollution have often been overtaken by the sheer volume of traffic on the roads. For example, in just forty years the world's vehicle population has grown from around 50 million to over 500 million.[1] A parallel growth has been seen in the U.K. In 1989, just over 24 million vehicles were licensed to use the public road network in Great Britain. Of these nearly 80 per cent were cars.[2]

Much of the focus to date has been on the emission of gaseous pollutants from vehicle exhausts. However, the environmental impacts of transport provision are considerably more far reaching. Pollution occurs throughout the entire life cycle of a vehicle, beginning with the extraction of raw materials, through its operational life, to its final disposal. It has been estimated that of all the energy used by a vehicle during its production and use, 20 percent is consumed during manufacture. The production of this energy, primarily from fossil fuels, will contribute emissions of the principal gaseous pollutants. Maintenance and spare parts also have associated environmental impacts.

The production of vehicles involves the use of a number of volatile organic compounds (VOCs) such as solvents in paints and chlorofluorocarbons (CFCs) in high density foam products and air conditioning units and for the cleaning of electronic circuitry. Many of these VOCs are known to have an adverse effect on air quality. Chlorofluorocarbons are involved in the destruction of stratospheric ozone and are powerful greenhouse gases. Most major vehicle manufacturers have pledged to phase out CFCs but are turning instead to hydochlorofluorocarbons (HCFCs), which will have a lesser but still significant impact on global warming and ozone depletion. Given the huge scale of vehicle manufacture, with at least 35 million new cars being produced worldwide in 1989, the combined environmental burden of these effects should not be overlooked.

Further threats to the environment result from the provision of transport infrastructure. Demands for large areas of land for roads, car parks, and service stations destroy not only unique wildlife habitats and areas of outstanding natural beauty, but also parts of towns, historic sites and the more mundane landscapes and recreational areas which add to the quality of life for many people.

Transportation systems also give rise to significant levels of noise and vibration. These too are known to be damaging to the built environment and to contribute to stress in humans. In addition, road travel causes a large number of deaths and serious injuries. In the UK, for example, there are approximately 5,000 deaths and serious injuries per annum and many more thousands of other more minor injuries. Apart from those directly affected by these casualties, many others may suffer loss of amenity through being discouraged from travelling, particularly as pedestrians or cyclists.

This chapter, however, will focus on the adverse environmental effects resulting from atmospheric emissions arising from the operation of road vehicles, primarily cars. Carbon dioxide (CO_2) is the main gaseous emission, but as it is not toxic in the atmosphere it has not in the past been commonly regarded as a major problem. However, its ability to alter the global heat balance—the so called 'greenhouse effect'—makes it a pollutant of at least equal importance as the precursors to acid rain and photochemical smog. There is also a relationship between emissions of carbon dioxide and the traditional atmospheric pollutants. While carbon dioxide emissions are directly related to the amount of fuel burnt, the relationship between the emissions of the other atmospheric pollutants and fuel consumption is considerably more complex. However, in general terms, these also increase with increasing fuel consumption. Thus strategies to reduce fuel consumption will have a beneficial effect on emissions of the more traditional pollutants.

In the atmosphere chemicals of both natural and anthropogenic origin react together to form new compounds in a very complex series of reactions. Thus the borders between emissions of acid rain precursors and those of other pollutants is difficult to define. This chapter considers a much wider range of pollutants than may traditionally be associated with acid rain. In the discussion of pollutants arising from road vehicle operations and their control technologies it includes carbon dioxide and particulate emissions from diesel vehicles. In addition, it looks at the growth of the road vehicle population, the effects of this on emissions and briefly discusses the role of traffic restraint in reducing emissions from the transport sector.

2. THE PRINCIPAL POLLUTANTS

Nitrogen oxides (NO_x), volatile organic compounds (VOCs); carbon monoxide (CO) and carbon dioxide (CO_2) are emitted in ever increasing amounts from road traffic. As Figure 1 shows all emissions, except those of sulphur dioxide, have increased by at last 38 per cent since 1979 in the UK. Table I shows that road transport is a major source of CO, NO_x and VOCs. Similar proportions of total emissions have been attributed to road vehicles in other countries. In OECD countries, for example, road transport contributes, on average, 75 per

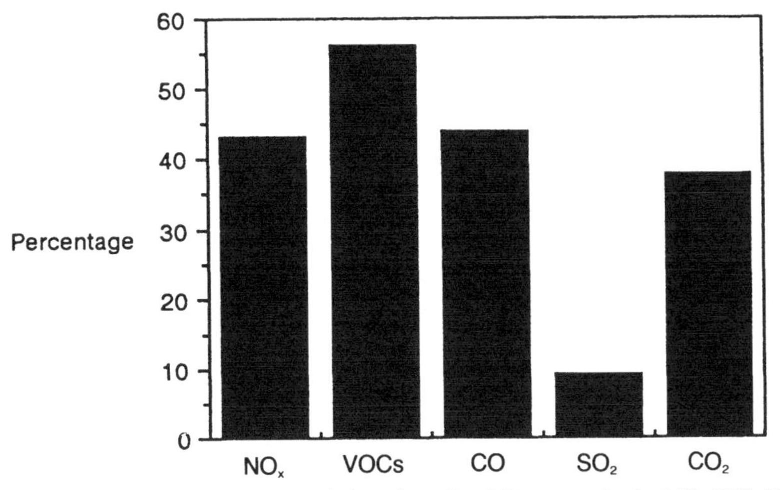

Figure 1 Percentage Increase in Emissions from Road Transport in the UK, 1979–1989
Source: Digest of Environmental Protection and Water Statistics, 1991[10]

Table 1 Emissions of the principal gaseous pollutants from road transport in the UK

Pollutant	1989 Emissions Thousand tonnes	Percentage of total UK emissions
NO_x	1,298	48
VOCs	762	37
CO	5,751	88
SO_2	60	2
CO_2	106	19

Source: Digest of Environmental Protection and Water Statistics, 1991.[10]

cent of total CO emissions; 48 per cent of NO_x; and 40 percent of VOCs.[3] Road vehicles are not a major source of sulphur dioxide.

2.1. *Nitrogen oxides (NO_x)*

Nitric oxide (NO) and nitrogen dioxide (NO_2), jointly termed NO_x, are one of the main precursors of acid deposition. In addition NO_x is involved in the complex photochemical reactions which give rise to ozone and other photochemical oxidants. These oxidants play a major role in the transformation of primary pollutants, particularly the oxidation of sulphur and nitrogen compounds to acidic species.

In motor vehicles NO_x is formed in the engine as a result of the combination of atmospheric oxygen with either atmospheric nitrogen or nitrogen in the fuel at high combustion temperatures. Thus emissions depend on the combustion conditions as well as the fuel properties. For this reason estimates of emissions of NO_x are less certain than those of some other pollutants. Most of the NO_x is emitted in the form of nitric oxide, and this is rapidly converted in the air to nitrogen dioxide. Of the 1,300 million tonnes of NO_x emitted from road transport in the UK in 1989 about one half came from heavy goods vehicles and one half from cars.

Emissions of NO_x from vehicles depend on the type of engine used. Thus a diesel engined car will emit about 80 per cent less NO_x than a comparable petrol engine car built to current emission regulations (i.e. UNECE Regulation 15.04), under most driving conditions. However, a car fitted with a controlled three way catalyst will emit about 95 per cent less NO_x compared with a current car.[4]

NO_x emissions from petrol engined cars show a general rising trend with speed, reflecting the higher load factor on the vehicle as speed increases. Emission levels are typically 2.5 g/km at low speed (around 20 kph) rising to about 4.0 g/km at 112 kph. Diesel cars generally show little change in emissions with speeds above about 40 kph. In addition, emissions from all types of engines are higher when started from cold. Diesel cars emit about 50 per cent less NO_x with a cold engine compared to a petrol engine without a catalyst.

Total emissions of NO_x in the UK have not shown any clear trend over the past twenty years. However, there has been a major change in the relative contributions from different sources. Emissions from motor vehicles first exceeded those from power stations in 1980, and are now the single most important source.

2.2. *Volatile Organic Compounds (VOCs)*

The unburnt and partly burnt fuel emitted from motor vehicles are volatile organic compounds. There are two main sources: the exhaust and evaporative losses from the fuel system. It has been estimated that petrol-fuelled vehicles account for 90 per cent of total traffic emissions of VOCs. In the European Community about 30 per cent of the VOC emissions are from evaporation and 70 per cent from exhausts. The proportion of evaporative emissions can increase significantly during the summer months. In modern vehicles crankcase emissions have largely been eliminated by the use of positive crankcase ventilation values which allow the crankcase gases to be recycled to the air intake.

Two stroke engines, widely used in Eastern and Central Europe, emit particularly high levels of VOCs. Cars with four stroke engines, built prior to any VOC emissions controls, emit four times as much VOCs as current Western European cars (built to emission regulation 15.04), while two stroke engines emit nine times more.

Certain classes of VOCs play an important role in the formation of acidic species through their involvement in the photochemical production of ozone and other oxidants. Some VOCs can cause unpleasant effects such as drowsiness, eye irritation and coughing and others, for example benzene, are carcinogenic.

Volatile organic compounds come from both natural and anthropogenic sources and it has been estimated that approximately one third of all European VOC emissions, mainly terpenes, come from natural sources.[5] Road traffic is the single most important anthropogenic source of VOCs in many countries.

Speciation of the VOCs emitted from petrol cars suggest that ethylene, toluene, xylene (meta and para), benzene, and propylene are among the main species emitted. These belong to groups of VOCs with the greatest potential to form ozone. The VOCs found in car exhausts have two different origins. Some, such as toluene, arise mainly from the unburnt fuel, while others such as ethylene, are largely combustion derived. Emissions of VOCs present in the fuel decrease with vehicle speed while those formed during the combustion process tend to increase with speed.[6]

Another important VOC emitted into the atmosphere as a result of petrol vehicle use is benzene, which occurs naturally in crude oil. Its major sources are vehicle exhausts and evaporative losses during the handling, distribution and storage of petrol. In the European Community there is a mandatory limit of 5 per cent benzene by volume in petrol. Within this limit there are quite wide variations depending mainly on the type of crude oil used and on the particular refining processes involved. Unleaded petrol, particularly super unleaded, is likely to contain a higher concentration of benzene than leaded petrol.

Benzene is a known carcinogen and several epidemiological studies on benzene-exposed workers have revealed a statistically significant association between acute leukaemia and occupational exposure to benzene. There is no evidence that exposure to benzene at petrol stations and in urban air has caused any additional leukaemias. The World Health Organization has concluded, however, that there is no known safe level.[7]

Aldehydes are another group of VOCs that are emitted from vehicles as a result of incomplete combustion of fuels. Aldehydes are characterized as colourless gases with pungent odours that contribute to the odour nuisance from vehicles. In addition, aldehydes can affect human health. For example, one of the more common aldehydes, formaldehyde, can cause irritation of the eyes, nose and throat, together with lachrymation, sneezing, coughing, nausea and breathing difficulty. Reportedly children are most sensitive. There is evidence suggesting that formaldehyde is carcinogenic in animals but there is insufficient evidence at present of carcinogenicity in humans.[7] Aldehydes also contribute to the formation of tropospheric ozone.

2.3. *Carbon Monoxide (CO)*

Carbon monoxide is one of the most directly toxic substances emitted from motor vehicles and is another product of incomplete combustion. It adversely affects human health by impairing the oxygen-carrying capacity of the blood, resulting in impaired perception and thinking, slowing of reflexes and drowsiness. It can also increase the occurrence of headaches and affects the central nervous and cardiovascular systems.

Carbon monoxide plays an important role in the chemistry of the atmosphere. As concentrations increase, levels of hydroxyl radicals decrease, resulting in the build-up of other trace gases such as methane. Methane is a powerful greenhouse gas, and emissions of carbon monoxide therefore contribute indirectly to the greenhouse effect.

2.4. *Particulates*

In diesel engines the main incomplete combustion products are particulates. These are mainly tiny particles of carbon onto which potentially toxic chemicals are absorbed. In major cities diesel vehicles have become the major source of particulates taking over from domestics fires. Vehicles may now account for up to 90 per cent of the black smoke in urban areas.[8]

There is a strong correlation between high levels of particulates in air, usually when there are also high levels of sulphur dioxide, and mortality. At lower levels children's lung function has been shown to be affected. Particulates in air can aggravate diseases such as bronchitis and asthma but possibly the most worrying aspect of particulate emissions from diesel vehicles is that they may be carriers for cancer-causing agents, particularly polynuclear aromatic hydrocarbons (PAHs).

Several hundred PAHs have been detected in the air but most is known about benzo(a)pyrene. Some of these compounds are known carcinogens and according to the World Health Organization "Owing to its carcinogenicity, no safe level of PAH can be recommended."[7] Vehicle exhaust, particularly from diesel engines, has been shown to be mutagenic and to cause tumours in some animal studies. The results of recent epidemiological studies have lead international health experts to conclude that there is an increased risk of cancer from occupational exposure to diesel particulates.[9]

One of the most immediately obvious effects of air pollution in cities is the soiling of buildings. Diesels emit black, fine oily particles which have a greater soiling effect than those from other sources. Although the soiling effects of various types of smoke are difficult to assess accurately it has been estimated that diesel smoke has a soiling effect three times that of smoke from coal combustion and seven times that of smoke from petrol vehicles. The soiling of buildings has a considerable economic cost as well as causing public nuisance.

2.5. *Sulphur Dioxide*

Sulphur dioxide is produced when fossil fuels containing sulphur are burnt. Motor vehicles are not a major source of sulphur dioxide at a national or international level, and are estimated to contribute only 1 per cent of all sulphur dioxide emissions in the UK.[10] However, as emissions of sulphur dioxide from vehicles occur near the ground they may cause a local problem near the roadside. Emissions of sulphur dioxide are determined by the amount of sulphur in the fuel. Diesel is the only automotive fuel with a significant sulphur content. A pollution inventory by London Scientific Services suggests that about 8 per cent of the emissions of sulphur dioxide in the Greater London area are from traffic, mainly diesel vehicles.

High concentrations of sulphur dioxide can cause breathing problems, particularly in people with pre-existing respiratory diseases such as asthma and bronchitis. The most serious health effects occur when sulphur dioxide is present with elevated levels of particulates.

Sulphur dioxide also damages a wide range of materials and is most noted for its effects on historic stone buildings and statues. It can also adversely affect plant growth. Sulphur dioxide is a major contributor to acid rain, adversely affecting aquatic life in rivers and lakes over large areas of the world with sensitive ecosystems.

2.6. *Ozone*

While ozone in the stratosphere protects us from the sun's dangerous ultra-violent rays, at ground-level ozone can damage plants and destroy a wide range of materials. Ozone, and other photochemical oxidants, can affect human health causing coughing, impaired lung function, and respiratory (eye, nose and throat) irritation. It can also affect plant growth and ruin the appearance of leaves, causing economic damage for leaf crops and has been implicated in the widespread tree damage seen over the past decade in much of Western Europe. Photochemical oxidants are involved in the oxidation of nitrogen and sulphur compounds to acidic species.

Ground-level ozone is formed when VOCs and NO_x are present together with high temperatures and intense sunlight. In Northern Europe high levels of ozone usually occur when there is a high pressure weather system (anticyclone) in summer. In this type of weather condition high levels of ozone may be observed over much of the region as formation continues downwind of urban areas until all the VOCs have been consumed. Ozone is just one of many potentially harmful chemicals formed in these conditions. It is measured as an indicator of photochemical smog, thus when high levels of ozone are observed there are also likely to be high levels of other photochemically generated pollutants such as peroxyacetyl nitrate (PAN).

Volatile organic compounds are essential for the build up of ozone but some of these compounds have greater ozone formation potential than others. Methane, for example, is considered 'unreactive' although it does have an important role to play in controlling 'background' levels of ozone. It may take several days to form ozone while 'highly reactive' volatile organic compounds may take an hour or less. Therefore, when comparing different fuels it is important to consider the types of chemical compounds present in the exhaust. In the UK aromatic hydrocarbons make the largest contribution to ozone formation.[11] It has been shown that the VOC to NO_x ratio increases downwind of urban areas and the higher the ratio the more efficient the ozone production becomes. In most circumstances control of VOC emissions will decrease ozone formation whereas control of nitrogen oxides emissions may in some cases increase or in other cases decrease ozone concentrations.

Tropospheric ozone is also a greenhouse gas. Background concentrations of ozone are thought to have doubled over the last century, largely as a result of increased nitrogen oxides emissions into the atmosphere from anthropogenic sources. Ozone is an extremely powerful greenhouse approximately 2000 times more effective than carbon dioxide at retaining the Earth's heat.

2.7. *Carbon Dioxide*

Carbon dioxide is the major greenhouse gas. Current estimates suggest that it has contributed 55 per cent of the enhancement of the greenhouse effect over the last decade. Carbon dioxide concentrations in the atmosphere are rising by about 0.5 per cent per annum. This is largely due to the combustion of fossil fuels. Road transport is responsible for about 20 per cent of UK emissions. It has been predicted that without major improvements in vehicle fuel economy carbon dioxide emitted from traffic could double by the end of the century. By contrast it is generally agreed that emissions need to be cut or at the very minimum to be stabilized as soon as possible.

The amount of carbon dioxide emitted is directly related to the amount of carbon in the fuel and the amount of fuel burnt. There is currently no add-on technology to reduce emissions of carbon dioxide from vehicle exhausts. The only way to reduce emissions is to reduce the amount of fuel used. In OECD countries energy use for transport is growing faster than total energy requirements and faster than consumption in other end use sectors. Road transport accounts for over 80 per cent of the energy used by the transport sector as a whole.[12]

3. CONTROL TECHNOLOGIES

3.1. *Petrol Vehicles*

3.1.1. *Closed Loop Three Way Catalysts*

The best available current technology

to reduce exhaust emissions is the closed loop three way catalytic converter. This technology is already in widespread use in North America, Japan and parts of Europe, and has been required on all new cars in the European Community since 31st December 1992. Eastern and Central Europe have yet to use this technology, which requires the widespread availability of unleaded fuel. In the US further emission reductions have recently been mandated which will require advanced closed-loop three way catalytic converters. These systems will become available by the mid-1990s.

Catalytic converters (also known as catalysts and catalysers) remove pollutants by a chemical reaction on the surface of the catalyst. Catalysts do not act as filters. The modern three-way catalyst is a honeycomb structure with a surface area equivalent to the size of two football pitches (250,000 square feet). This is enclosed in a stainless steel box, about 30 cm long by 25 cm wide. Usually platinum and palladium is used to oxidize unburnt hydrocarbons and carbon monoxide into water vapour and carbon dioxide, while rhodium is used to reduce nitrogen oxides to nitrogen and oxygen.

Catalysts are 'poisoned' if the engine runs on leaded petrol. Even one tank of leaded fuel will reduce the efficiency of a catalyst, but returning to unleaded petrol will allow the catalyst to regain much of its efficiency. The use of narrow fillers and nozzles makes accidental misfueling virtually impossible and the widespread availability and cheaper price of unleaded petrol make deliberate misfuelling unlikely.

There are essentially two types of three-way catalyst: controlled or uncontrolled. The catalysts are most effective when the air to fuel ratio is 14.7:1 (known as the stoichiometric ratio). In a controlled system, the oxygen in the exhaust is measured using a lambda sensor to ensure that the engine is running stoichiometrically. If not, the electronic fuel management system in the engine adjusts the fuel injection accordingly. In a controlled system, the engine is running under optimum conditions for efficient catalyst operation. In an uncontrolled system there is no lambda sensor and the catalyst is likely to be operating in less than ideal conditions for much of the time.

A controlled three-way catalyst can reduce emissions of nitrogen oxides, VOCs and carbon monoxide by 95 per cent, 90 per cent and 80 per cent respectively, as measured under the test driving cycle. An uncontrolled catalyst will reduce emissions by between 50 and 70 per cent for all three pollutants. This is insufficient to meet current emissions standards in North America, Japan, and parts of Europe, and the forthcoming regulations in the European Community.

Emission controls can significantly affect the relative abundance of different VOC species in the exhaust. Catalytic converters are most effective at oxidising the more reactive species, such as aldehydes, alkenes, aromatics, and the longer-chain alkanes. Thus, exhaust VOCs from vehicles equipped with catalytic converters tend to have relatively high proportions of the less reactive (and less toxic) species such as methane, ethane, propane and acetylene.

Controlled three-way catalysts may not reduce emissions effectively when they are below 250°C. This has important implications for air quality in urban areas, where most driving with a cold catalyst takes place. For example, tests have shown that emissions of nitrogen oxides and VOCs can be 5 times higher when tested on a cold start urban test cycle compared with a hot start urban test cycle. For carbon monoxide, when tested from a cold start, emissions were ten times higher.[4]

The introduction of catalytic converters on current European production vehicles has lead to a small fuel penalty. However, new models equipped with controlled three-way catalysts are likely to become more fuel efficient than current models. By using a catalyst to control emissions motor vehicle engineers will be able to design for optimum economy (or performance) while leaving the catalyst to deal with the emissions. In addition, the use of electronic fuel management systems on all cars rather than just those at the top end of the market will also improve fuel consumption across the whole fleet.

One of the major disadvantages of the three-way catalyst is that emission control is totally dependent on an add-on device. If it fails, emissions could be worse than if there had been no catalyst fitted. The success of this control technology depends on regular in-service testing of vehicles and their catalysts.

Catalysts can be retrofitted to some existing cars. Usually it is only feasible to fit a simple oxidation catalyst (see below) or an uncontrolled three-way catalyst, and costs vary depending on the make of car.

Catalytic converters can only be regarded as a temporary solution to vehicle pollution since there are only limited supplies of the catalyst metals, the only reserves of which are in South Africa and Russia.

3.1.2. *Oxidation Catalysts* An oxidation catalyst is used to reduce emissions of incomplete combustion products such as VOCs and carbon monoxide by using a catalyst to oxidise them to carbon dioxide and water. Unlike the three-way catalyst, an oxidation catalyst is not restricted to operating at the stoichiometric air to fuel ratio, and therefore can be used with lean-burn petrol and diesel engines. However it does nothing to reduce nitrogen oxide emissions. In some European countries oxidation catalysts are finding a new application on motorcycles. Little effort has been directed towards controlling emissions from two-wheeled vehicles but in Austria and Switzerland regulations on VOC emissions from motorcycles and mopeds means that new two-wheel vehicles require oxidation catalysts.

3.1.3. *Carbon Canisters* Evaporative losses from petrol engines are determined mainly by ambient temperature and the fuel's vapour pressure and composition. Increasing the temperature and/or vapour pressure disproportionately increases the emissions.

There are four major sources of evaporative emissions of hydrocarbons from petrol vehicles:

- fuelling losses during refuelling when the fuel vapour occupying the space above the fuel in the tank is displaced into the air;
- running losses when a vehicle is in use due to fuel evaporation from the fuel system;
- hot soak losses when the vehicle has stopped, but is still warm and fuel evaporation continues; and
- diurnal losses due to the effects of variations in ambient temperature on the fuel and vapour in the fuel tank.

Evaporative emissions can be reduced by using a charcoal (carbon) canister and a purge system to burn the VOCs in a controlled manner in the engine. These on-board fuel vapour collection systems prevent vapours from escaping to the atmosphere by using a sealed filler tank cap. A vapour line runs from the top of the fuel tank filler to an activated charcoal filter (in the canister), which absorbs and stores the fuel vapour. When the engine is running, fresh air is drawn into the canister and mixed with the petrol vapours which have been absorbed onto the charcoal filter and the vapour/air mixture flows to the engine.

These canisters have been used in North America for a number of years, and are being used increasingly in Europe. From 31st December 1992 all new cars in the European Community have been fitted with a small canister (one litre in size).

However, experience from North America suggests that they have not been as successful in controlling emissions as originally thought, and in the summer the canisters may suffer from overload. It is thought that controlling fuel volatility in summer may be necessary until large canisters are used. Small canisters are capable of achieving a 75–80 per cent reduction in evaporative emissions, but this can be as high as 95 per cent with petrol of low volatility.

Canisters of the size required in Europe from the end of 1992 will not be sufficient to control refuelling emissions. These can be recovered by systems installed at petrol stations (called stage II controls) or by large canisters on vehicles. Stage II controls involve fitting dual filling hoses to the petrol pump to draw vapour from the vehicle's fuel tank as it is filled, instead of allowing it to escape to the atmosphere. This is then returned to the filling station's storage tanks. This is a well established technology, while on-board control systems have been proven only using prototypes. There has been a bitter debate between the oil and car industries as to which approach is best. The advantage of installing controls at petrol stations is that emissions will be reduced faster than if controls are only fitted to new cars, which only displace old vehicles at a rate of about 10 per cent per annum.

3.1.4. *In-Service Emissions* Emissions from vehicles in use can be reduced through fuel volatility controls, the use of low-level oxygenated fuel blends (for rich running engines) and retrofitting emission control equipment. In addition inspection and maintenance programmes can identify excessively polluting vehicles and help ensure that the emissions are reduced to acceptable levels. The major problem with this approach is that the checks are infrequent and even the most responsible motorist may drive around for several months unaware of a failure of the engine/emission control system. This limitation can be overcome by use of on-board diagnostic equipment, that can determine component failure.

Evaporative emissions are influenced by the fuel's vapour pressure. In summer, in particular, fuel evaporation gives rise to significant emissions of VOCs. Petrol sold in both Europe and USA in the summer has a Reid Vapour Pressure (RVP) value in the range of 60 to 90 kPa. This is higher than strictly needed for good vehicle performance. Petrol has a relatively high vapour pressure because butane, a cheap and plentiful hydrocarbon, is blended with petrol. Using butane in petrol gives it a higher economic value than selling it for use in other applications.

In a European test programme using vehicles without evaporative emissions controls, a decrease of RVP from 82 to 62 kPa lowered the evaporative emissions by a factor of 2. Limiting RVP to the lowest level specified anywhere in Europe could lower overall evaporative emissions in western Europe by 25 per cent.

There are two main advantages in reducing RVP. Firstly, it immediately controls the emissions from all vehicles in use, and from the gasoline distribution system. Secondly, on cars fitted with evaporative emission controls it allows the small canisters to operate more efficiently without the risk of overloading. Within the European Community discussions are already well under way to establish new fuel volatility limits, which are proposed to come into effect at the beginning of 1993. The RVP is likely to be reduced down to between 35 and 70 kPa in many countries.

A relatively small number of vehicles on the roads contribute a disproportionately large amount to total emissions. In the US less than 10 per cent of on-road vehicles account for half the carbon monoxide emissions,[13] while in the UK, where there are currently less stringent emission requirements, approximately 17 per cent of the vehicles cause half the carbon monoxide emissions.[14] Detection of these 'gross' polluters and a requirement that they be repaired can reduce average vehicle emissions significantly. Inspection and maintenance programmes of this type can also discourage owners from tampering with, or disabling, emission control equipment.

Another method of reducing in-use vehicle emissions is to retrofit cars with emission control devices. Retrofitting of catalysts is limited to vehicles able to use unleaded petrol. However, there are serious difficulties in ensuring consistent performance in use.

3.2. *Diesel Vehicles*

3.2.1. *Emission Control by Engine Modification* The objective of automotive engineers designing new commercial engines is to build low emission engines that will meet the world's strictest emission standards without resorting to exhaust after-treatment. This approach is considered to be potentially more reliable and durable than depending on a particulate trap.

Engine improvements that lead to decreased NO_x emissions mainly involve *reducing* the combustion temperature. However, control of particulate emissions requires *increasing* the combustion temperature. This trade-off between the production of NO_x and particulate matter is a fundamental feature of the engine operation and therefore there may eventually be a limit to the gains which can be achieved by technical improvement of engines. Exhaust after-treatment devices are likely to be necessary to achieve reductions in emissions below these levels. The engine could be designed to achieve greater control of NO_x if the requirement for it to control particulates were relaxed as a result of fitting after-treatment for particulate matter. On the other hand, the engine could be designed to achieve greater control of particulates if aftertreatment of NO_x were available.

Low emission engines rely on the precise control of the combustion process. Much effort has already gone towards engine design which can deliver an adequate charge of air throughout the combustion cycle and which thoroughly mixes the fuel with the air. The greatest advances have been made through redesigning the combustion chamber, improving fuel injection and increasing the charge of air by means of turbocharging and aftercooling.

Further engine modifications required for lower emissions include:

- variable geometry turbocharging;
- four-valve combustion systems;
- reduced oil consumption; and
- advanced computer controls.

3.2.2. *Particulate Traps* To reduce particulate emissions from diesel vehicles, particularly lorries and buses, particulate traps have been developed. Unlike the catalysts discussed above this is a filter system. A 'simple' filter would rapidly become blocked and eventually exert an unacceptably high back pressure on the engine. This would adversely affect engine performance, fuel consumption and emissions. It is therefore necessary to clear the trap of particulates before blocking becomes a problem. This can be done by oxidizing or burning the particulate matter and is known as trap regeneration.

The exhaust gases are not hot enough to burn off the particulate matter directly, except under conditions of full load. Higher temperatures may be reached by restricting the flow of exhaust gases or by installing a supplementary heater powered by diesel fuel or by electricity. A fuel penalty of about 1 per cent is incurred by this. Regeneration is initiated when a certain back pressure is reached. The monitoring and control system required for this makes such traps complex and expensive devices.

The ignition temperature of the particulate matter may be reduced by coating the filter with a metal catalyst, such as copper, or by adding a catalyst to the fuel. The latter reduces and may eliminate altogether the need for regeneration equipment. These catalytic systems are the preferred options.

Many particulate trap systems are under development in both Europe and the USA by independent trap manufacturers and by the engine producers, with trap efficiencies of 70 to 80 per cent commonly observed. Currently their durability is being assessed in a series of field trials.

Particulate traps increase the amount of the sulphur in the diesel fuel that is oxidised into sulphate particulates such that emissions of particulates can increase, although the particulates have a different chemical composition. This can be overcome by reducing the sulphur content of the diesel fuel from the present maximum of 0.3 per cent to about 0.05 per cent. In the US low sulphur diesel fuel has been available since 1993 while in Europe it will become available by 1996.

3.2.3. *Oxidation Catalysis* Oxidation catalysts can be used on diesel vehicles to reduce hydrocarbon and particulate emissions. Carbon monoxide emissions from diesel vehicles are already relatively low and therefore catalysts are not needed to reduce these emissions.

Diesel exhaust temperatures are very low at light loads, due to the large excess of air, and the catalysts may not light up under these conditions.

The catalyst will reduce particulate emissions under operating conditions where a high proportion of the particulate is liquid or gaseous hydrocarbons and therefore this technology may be well suited for the mixed load operation experienced by city buses.

At least one production diesel car is already equipped with an oxidation catalyst and this technology is likely to become increasingly common in the coming years.

3.2.4. *Exhaust Gas Recirculation* Exhaust gas recirculation is used to reduce emissions of NO_x by returning a proportion of the exhaust gas to the combustion chamber and hence displacing some of the fresh air charge from the next cycle. This reduces the peak combustion temperature and hence reduces the level of NO_x produced.

3.2.5. *NO_x Aftertreatment* NO_x can be removed from the exhaust of a diesel engine and such systems have been used successfully for stationary systems. Little interest has been shown to date in its application for mobile sources.

In the Raprenox system (Rapid Removal of NO_x), isocyanic acid gas (HNCO) is mixed with the engine exhaust at temperatures above 400°C. It is claimed that more than 90 per cent of the NO_x is removed. In the SCR system (selective catalytic reduction), ammonia (NH_3) is used to remove the NO_x. Again NO_x reduction in excess of 90 per cent has been claimed. Both these systems require an on-board supply of the reducing chemical, and a reaction chamber as well as electronic controls for the supply of the reducing chemical, the reaction temperature and NO_x monitoring.

3.2.6. *In-Service Emissions* Inspection and maintenance programmes also have an important role to play in ensuring that in-service emissions from diesel fuelled vehicles remain low throughout a vehicle's lifetime. As electronic controls become more widespread in heavy duty engines, diagnostics will become increasingly important to alert the driver of failures in the engine management or emission control system.

3.3. *Fuel Economy*

The only method of reducing carbon dioxide emissions from vehicles is by burning less fuel. By increasing vehicle fuel efficiency the emissions of carbon monoxide and VOCs will also be reduced.

A number of factors influence how much energy is used by vehicles including the use of energy efficient technology, drivers' skill, level of maintenance, road conditions, and traffic levels. Reducing energy demand and hence emissions in the sector as a whole will require a combination of measures including traffic restraint, promotion of public transport, walking and cycling, as well as a rapid introduction of new vehicle technology. However, this section will focus on energy efficient technology, particularly for passenger cars as these vehicles dominate road traffic.

Since the first oil price shock of 1973, the fuel efficiency of road vehicles has improved markedly. Most of the improvements occurred between 1975 and the early 1980s when oil prices were still rising and there was strong consumer interest in saving energy. During this period motor manufacturers were forced by legislation and voluntary agreements to reduce vehicle weight (in USA) and invest in new technologies (in Europe). In the decade from 1978, new car fuel economy improved by about 20 per cent in the UK,[15] and similar improvements were made in other European countries. In the US fuel economy improvements of 36 per cent were reported over this period albeit from a much lower starting point.[16]

The average fuel consumption of new cars in the US today is 8.4 litres/100 km, while in other OECD countries it is slightly lower, for example, in the UK it is 7.4 1/100 km. It is lowest in Italy (6.8 1/100 km) where there is a predominance of small cars, many of which are diesel fuelled. In many countries the sales weighted fuel consumption of new car fleets has actually been rising in the last few years due to increased demand for larger, higher performance cars. Data on the fuel economy of new cars in developing countries is poor, but it is probably somewhat lower than that of OECD countries. In East and Central Europe, new car fuel economy is likely to deteriorate rapidly as their relatively energy efficient two stroke engines are replaced by four stroke engines.

These improvements in new car fuel efficiency have not generally been reflected in on-road fuel consumption. In Europe and Japan the fuel saving since 1978 has been around 10 per cent on a vehicle kilometre basis and, due to declining load factors, less on a person kilometre basis. In the developed countries there is a relatively rapid turnover of the vehicle stock, with about 10 per cent being renewed every year. In the developing nations, on the other hand, vehicles are typically much older, and there is a substantial market for imported second hand cars. Thus in these countries it will take longer for fuel efficient technology to penetrate the market, and the on-road reduction in fuel consumption is likely to be even lower.

Over the last 15 years, since the first oil crisis, fuel efficiency of road vehicles has improved, with some individual car designs now using at least 50% less fuel. However, new technological developments such as turbocharging, which could have been used to produce cars with smaller and hence more fuel efficient engines, have been used instead to increase the vehicle's power. As the relative cost of fuel has decreased there has been an increasing trend in many countries towards larger more powerful vehicles. For example, the specific power output of European cars has increased from 50 horsepower per litre engine capacity (hp/1) ten years ago to the range of 60–75 hp/l today.[17]

A wide range of new technologies have been or are in the process of being developed to improve the fuel economy of road vehicles powered by internal combustion engines and conventional fuels. These include engine improvements, vehicle weight reductions, reductions in drag and rolling resistance, and the introduction of better transmission systems such as continuously variable transmissions.

Table II lists the potential fuel economy benefits of some of the technologies relative to a 1986 petrol engined car. The estimates given are for the benefit from individual technological improvements. A combination of the benefits is not necessarily additive, and some of the developments listed are mutually exclusive.

Energy losses in a car engine can be reduced by making it operate more efficiently under part load. One of the reasons for the inherent fuel consumption superiority of diesel engines is that they can operate better under these condi-

Table II Technologies available for improved car fuel economy

Technology Design Approach	Potential Gain (as % of 1986 gasoline car)
Engine Improvements	
Improvements to current gasoline engines (e.g. precision cooling, reduced engine friction, reduced pumping losses)	up to 6
4 valve per cylinder 4 stroke lean-burn gasoline engines	5 to 15
Direct injection 2 stroke gasoline	0 to 10
Diesel engines (conventional and improved direct injection high speed designs)	20 to 30
Electronic engine management	5 to 20
Vehicle and Transmission Design Improvements	
Transmission Improvements (e.g. automated manual)	10 to 15
Continuously variable transmission	10 to 15
Weight reduction	15 to 20
Aerodynamic improvements	5 to 10
Improved tyres, lubricants, accessories	5 to 10

Source: Martin and Shock, 1989[18]

tions. While indirect injection diesel engines can offer fuel economy improvements of up to around 20 per cent compared with traditional petrol engines, direct injection diesel engines in turn may use 10 to 15 per cent less fuel than indirect injection models. Direct injection diesel engines were used in most of the fuel efficient prototypes developed by motor manufacturers during the 1980s.

Further into the future, fuel economy improvements may be based on the development of 'adiabatic' engines, stratified charged engines, and two stroke engines. The latter bear little resemblance to the familiar two stroke lawn mower engines, and may offer significant fuel economy benefits without the more undesirable attributes of their predecessors.

Other ways of making engines operate more efficiently include using power boosters such as turbochargers to reduce engine size; electronic engine management to ensure that the vehicle is running as efficiently as possible; and reduced engine warm-up time. For petrol engines considerable effort has also gone into improving compression ratios and using lean-burn combustion technology.

Fuel consumption could also be cut by reducing vehicle weights. Further improvements will be achieved by material substitution through, for example, the extensive use of plastic materials for structural and body components and the adoption of aluminium engine blocks.

Further efficiency improvements are also possible using transmissions which allow the engine to operate at full load more often. One such technology already used in some production vehicles is continuously variable transmission (CVT). Modern versions have overcome the reliability problems of the very early types.

Overcoming the vehicle's resistance to aerodynamic drag, especially at high speeds, is yet another way of improving car fuel consumption. At present the best European cars have a coefficient of drag of 0.28; but at least one manufacturer believes that its company's average will be down to around 0.25 by the turn of the century.[19] Given the appropriate economic and political climate, these and other technologies offer considerable scope for fuel economy improvements using the internal combustion engine. Prototypes capable of travelling 100 miles per gallon (2.7 1/100 km) already exist; and some idea of the scope for further progress is suggested by the world record of 6,409 miles on a single gallon of petrol (i.e. a fuel consumption of 0.044 litres per 100 kilometres).[20]

3.4. *Alternative Fuels*

Continuing problems with air quality in the US despite the use of controlled three way catalysts for almost a decade has lead to the promotion of alternative-fuelled vehicles as a solution. In the United States, the recently agreed Clean Air Act mandates the use of alternatively fuelled vehicles in areas where air quality is particularly poor. Alternative fuels *may* also offer a solution to the apparently intractable problem of increasing carbon dioxide emissions.

Vehicles capable of running on a wide range of fuels are currently being developed by vehicle manufacturers around the world. Fuels under consideration include methanol, ethanol, electric vehicles, hydrogen, natural gas, liquid petroleum gas, vegetable oils, and fuel cells. Some of these fuels are already in limited use, while others such as hydrogen and fuel cells are many years away from incorporation into production vehicles. A number of these fuels are most suited to heavy duty applications as they have a low energy density, and large and heavy fuel storage containers are required.

In California, which for many years has lead the world on vehicle emissions standards, the main alternatives for passenger cars currently being promoted are methanol and electric vehicles. For buses and other heavy duty vehicles the preferred option is natural gas. These vehicles are generally further developed than those that use other alternative fuels, and more is known of the adverse effects on human health and the environment. The plentiful supply of natural gas in North America and Europe make the use of methanol, which can be manufactured relatively efficiently from natural gas, and compressed natural gas favoured alternatives.

When considering the environmental benefits in terms of reduced atmospheric emissions it is important to compare the life-cycle emissions, that is those arising during the production, distribution and use of the different fuels, and not just those arising from vehicle operation. Therefore, the fuel feedstock is of vital importance in determining the relative cleanliness of the fuels.

Alcohols, natural gas and hydrogen can be made from coal, natural gas and biomass, while liquid petroleum gas is a by product of natural gas and crude oil exploitation. The fuel feedstocks for electric vehicles and those using hydrogen manufactured by the electrolysis of water, depend on how the electricity is generated but are also likely to be coal, and in forthcoming years, natural gas.

The production of petrol from crude oil generally gives rise to lower emissions of particulates, sulphur oxides, VOCs and NO_x than the production of alternative fuels from biomass and coal.[21] Compared to the production of electricity (assuming the current UK electricity generation mix) for use in electric vehicles and for the production of hydrogen, the production and use of petrol gives rise to lower levels of nitrogen and sulphur oxides but higher emissions of VOCs and carbon monoxide.[22]

In addition, for alcohol fuels, natural gas, and liquid petroleum gas vehicles there are significant emissions from the vehicles in use, which can be compared directly to those from conventionally fuelled vehicles. Nitrogen oxides emissions are higher for all these fuels in both light and heavy duty applications compared with petrol cars with a catalyst and a diesel heavy duty engine respectively. Emissions of carbon monoxide and non-methane hydrocarbons, as conventionally measured, are generally lower than from conventionally fuelled vehicles. However, the combustion of alcohol and natural gas fuels gives rise to high levels of aldehydes and methane respectively.

The widespread use of natural gas vehicles is likely to reduce the occurrence of short lived elevated ozone episodes but its use may increase background levels of ozone, a greenhouse gas. The limited evidence available suggests that ethanol vehicles may increase the occurrence of elevated levels of ozone. Great uncertainty revolves around the role of methanol fuelled vehicles in reducing ozone formation, with some studies suggesting that its use will increase the ozone formation potential and others suggesting the opposite.

Recently, a number of studies have considered the greenhouse effects of emissions from both conventional and alternative transport fuels, including those of methane and nitrous oxide as well as carbon dioxide, to reflect the total direct effect. The results from DeLuchi *et al.*,[23] which include emissions from the production, distribution, and end use of each fuel, show that the source of the fuel is vitally important. For example, hydrogen is either the best means of reducing greenhouse gas emissions or the worst, depending on the fuel feedstock and production process used in its manufacture.

The use of fossil fuel feedstocks will in most cases not significantly reduce the greenhouse gas emissions, compared to the current use of petrol and diesel. The use of natural gas as a feedstock would result in only slight emission reduction relative to petrol. The use of coal as a fuel feedstock either directly or as a fuel for electricity generation can more than double the total direct greenhouse emissions compared to a conventionally fuelled vehicle fleet.

The in-service emissions of carbon dioxide from liquid petroleum gas vehicles may be slightly higher than those from a comparable petrol vehicle, as vehicles currently running on this fuel are less fuel efficient. Theoretically, however, an optimalised liquid petroleum gas vehicle could reduce carbon dioxide emissions. The in use emissions of carbon dioxide from alcohol fuels and natural gas are lower but with the latter the increased leakage of methane into the environment may counteract some of the benefits of carbon dioxide reductions.

The analysis of DeLuchi *et al.* estimates that doubling current US fleet fuel efficiency would reduce emissions proportionately, and this would be preferable to the adoption of all alternative fuels except those derived from electric power generated from renewable sources and possibly biomass. When the agricultural inputs into biomass production are taken into account the relative benefits of this feedstock are considerably reduced. One analysis suggests that when all factors are taken into consideration carbon dioxide emissions from a biomass derived ethanol fuelled fleet would increase by 23 per cent compared to an equivalent petrol one.[24]

The benefits of using most alternative fuels in terms of reduced atmospheric emissions over the complete life cycle of the fuel is far from clear cut. Only the use of electricity generated from renewable non-fossil sources for electric vehicles and hydrogen manufacture offers a major reduction in emissions over the range of pollutants. Currently little electricity is generated in this way, and there is little prospect of renewable energy sources making a major contribution to electricity supply in the coming years.

With the possible exception of alcohol vehicles there are still a number of outstanding technical problems to be solved before there will be widespread customer acceptance of the alternative transportation fuels. Some of these are shown in Table III below.

All internal combustion engine fuels can be explosive. Methanol and liquid hydrogen are more explosive than conventional fuels in a confined area, while in the open air natural gas, liquid hydrogen and liquid petroleum gas are the most explosive. As a result of LPG's high explosiveness and propensity to accumulate near the ground, its use is banned in some countries in certain underground car parks and tunnels and on some bridges.

The weight and volume of fuel containers are important factors in terms of customer acceptability, since weight will affect performance and vehicle handling, while excessive fuel volume limits the available passenger and luggage space. All the alternative fuels are less energy dense than petrol and their containers are heavier than a comparable range petrol tank. For example, full methanol and compressed hydrogen tanks are between 1.5 and 2.5 times the weight of the equivalent petrol tank and compressed natural gas up to 11 times the weight depending on the tank type. However, a liquid hydrogen vehicle would probably weigh only 7 to 35 kg more than a petrol vehicle of equal range as its high thermal efficiency counteracts most of the effects of low energy density.[23]

Similarly, all the alternative fuels are more bulky than conventional fuels, particularly hydrogen and compressed natural gas. Hydrogen fuel tanks occupy up to nine times as much space for an equivalent energy content and compressed natural gas up to five times compared to petrol.

Table III Some technical obstacles to adoption of alternative fuels

Fuel	Problem
Methanol (often mixed with petrol as M85, i.e. 85% methanol and 15% petrol)	Vapour lock in hot weather Difficulty in starting in cold weather Poisonous Corrosive Burns with an invisible flame (when used in the pure form)
Ethanol	Difficulty in starting in cold weather Requires assisted ignition
Compressed Natural Gas	Large fuel storage volume required Very heavy storage system High pressure required for compression
Liquefied Natural Gas	Boil off Large fuel storage space required
Liquefied Petroleum Gas	Insufficient fuel pressure in cold weather
Liquefied Hydrogen	Boil off Large fuel storage space required Dangerous to handle
Hydrides	High pressure required Large fuel storage volume required Needs extremely pure hydrogen Prone to backfiring Slow refuelling
Electric vehicles	Inconvenient and slow refuelling Battery disposal Very heavy/bulky batteries

Source: Adapted from Holman, Mitchell and Fergusson, 1991[22]

Most of the alternative transportation fuels are inherently more thermally efficient than petrol, and this can more than offset the weight disadvantages experienced when the fuel is used in a dedicated engine. For example, methanol vehicles have about 10 per cent higher specific power output than petrol vehicles, while it has been suggested that optimalised natural gas vehicles could have a 0 to 60 mph acceleration time up to 15 per cent better than a comparable petrol vehicle.

Most of the alternative fuels still face technical obstacles and performance limitations for private car use. However, these problems are far less insurmountable for larger vehicles and these new fuels could find limited application in specialist functions, such as local delivery vans or urban buses. There may, for example, be considerable potential for developing some electric vehicle fleets for local authority services or public service vehicles powered by compressed natural gas to improve local problems of air quality.

There is no doubt that in the short term alternative fuels such as methanol, compressed natural gas and electric vehicles, will be introduced in some areas as an emergency measure to alleviate acute urban air pollution problems, as indeed is already the case in Southern California. Programmes such as this are unlikely to overcome the underlying drawbacks set out above, however, and these measures can be regarded only as a stopgap for the time being. In the case of methanol, there are serious questions as to whether its use will alleviate the air quality problems even in the short term.

4. TRAFFIC GROWTH

The major difficulty of reducing emissions from road vehicles is that while technology, such as catalytic converters and fuel switching, has an important role to play, in the long term the improvements are likely to be outstripped by the rapid growth in traffic. This has already been seen in California where catalytic converters have been fitted to cars for over 15 years and yet the national ambient air quality standard for ozone continues to be exceeded on one out of every three days in the Los Angeles area.

Demand for both passenger and commercial transport is growing rapidly. In the United Kingdom, for example, the decade 1980–89 has seen road transport grow by around 40 per cent while transport as a whole has grown by approximately one third. This growth in road vehicle use is reflected in the increasing number of vehicles on the roads. In 1989, there were 24 million vehicles licensed to use the public road network, nearly one million more than in the previous year.

Over the past decade much of the growth in road traffic has been on the motorways and trunk roads (in non-built up areas). In the United Kingdom, traffic flow on all roads has increased by 49 per cent, with that on motorways and trunk roads by 75 per cent and 77 per cent respectively.

This growth in both road traffic and the number of vehicles on the roads is broadly reflected around the world. For example, in the OECD countries, in the decade to 1985, road traffic grew by approximately one third, with the largest growth taking place in Australia.

In 1950, there was a global car population of 53 million. Today there are over 550 million vehicles in use around the world, of which over 400 million are cars. It has been estimated that this could rise to 1 billion in the next 20 to 40 years.[25] In 1990 around 35 million new cars rolled off the world's production lines, increasing the global vehicle fleet, after retirement, by around 4 per cent.[1]

Car ownership is strongly linked to purchasing power and economic growth and a number of developing countries are experiencing a high growth rate in car ownership. In Thailand, for example, new car and truck sales increased by 40 per cent in 1990 compared to the previous year. In the industrialised nations car ownership varies from about 625 cars per thousand population (USA) to around 300 per thousand population. In the UK it was 400 per thousand population in 1989. By way of comparison, the car ownership level in Brazil is 75 per thousand population while in the Soviet Union it stands at 48 per thousand population. Table IV shows car ownership rates for some selected countries.

In addition to the increase in the total number of vehicles each vehicle is now being driven further. In the United Kingdom the average distance travelled by each car has increased by 16 per cent over the last decade. Fuel consumption and emissions from motor vehicles have grown along with the global vehicle population and the number of vehicle kilometres travelled.

The emissions trade-off between technological improvements and traffic growth has been estimated for the United Kingdom using a model of atmospheric emissions developed by Earth Resources Research.[26] Figure 2 shows predicted emissions of nitrogen oxides from cars and for all road vehicles for the years 1988 to 2020. It is assumed that three way catalysts are introduced from the end of 1992 on all new cars and that new heavy duty engines meet more stringent emission requirements than those currently in force from 1993. Figure 3 shows the estimated emissions of carbon dioxide for the same time period.

2a: Low Forecast Traffic Growth

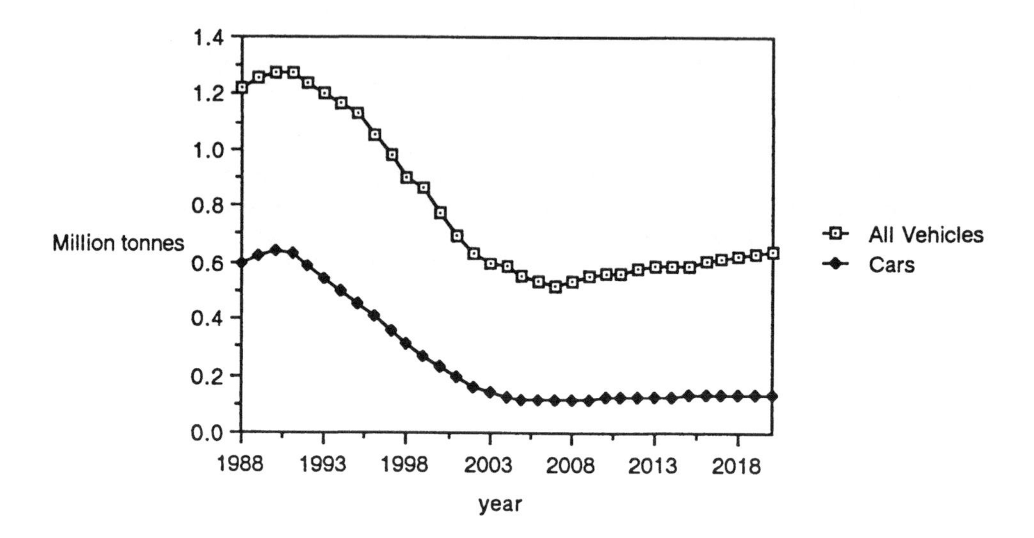

2b: High Forecast Traffic Growth

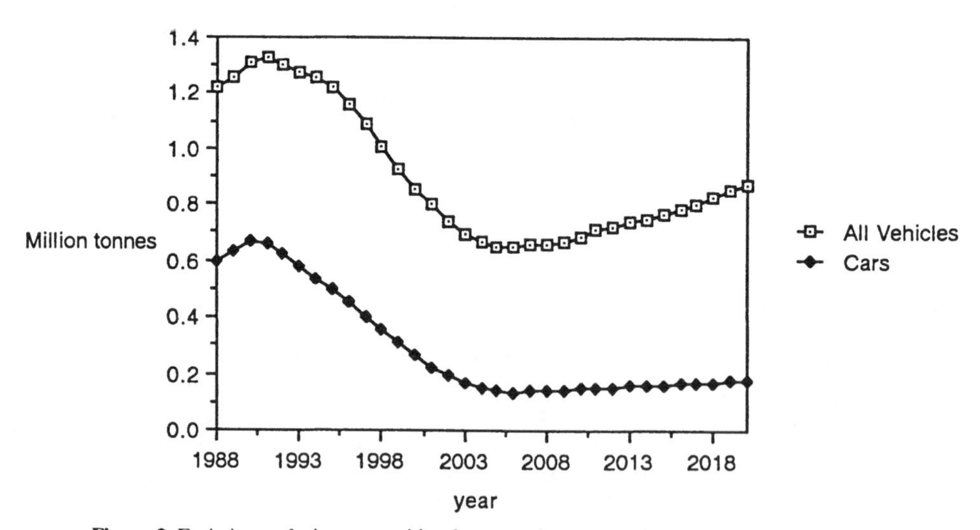

Figure 2 Emissions of nitrogen oxides from road transport in the UK, 1988 to 2020

3a: Low Forecast Traffic Growth

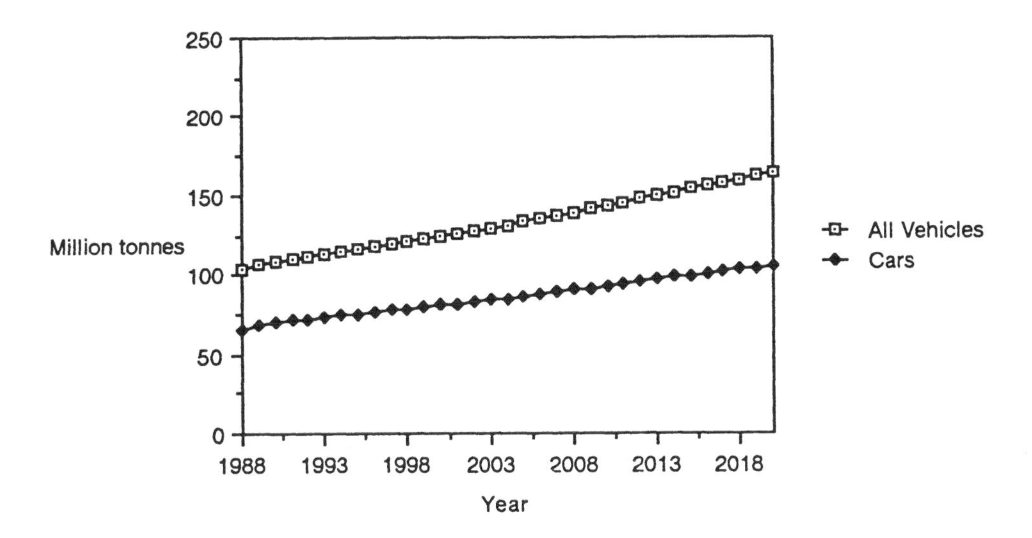

3b: High Forecast Traffic Growth

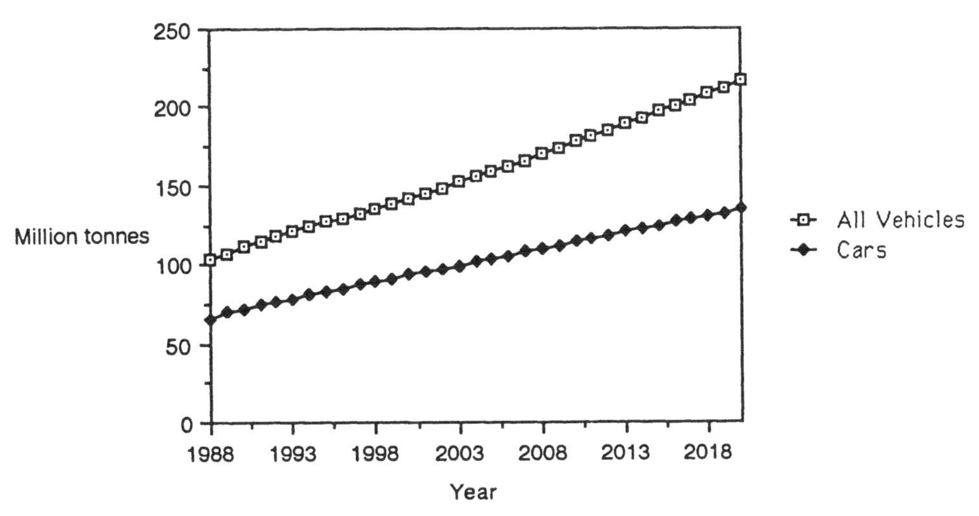

Figure 3 Emissions of carbon dioxide from road transport in the UK, 1988 to 2020

Table IV Car ownership levels in selected countries

Country	Cars per thousand population
USA	625
Western Germany	500
UK	400
EEC	385
Denmark	312
Japan	270
Eastern Germany	227
Poland	122
USSR	48
Romania	12
India	2
China	1

Source: Society of Motor Manufacturers and Traders, 1990[1]

These predictions use the UK Department of Transport's road transport forecasts,[27] covering the period to 2025. These predict far higher traffic growth rates than those previously published, particularly for car and light goods traffic and are summarized in Table V below.

Figure 2 shows that emissions of nitrogen oxides rapidly decline following the introduction of new emission limits for both cars and heavy duty vehicles, reaching a minimum in the first decade of next century. Beyond this growing traffic levels lead to an increase in nitrogen oxides emissions.

Figure 3 shows a continuous growth in carbon dioxide emissions virtually in step with the growth in vehicle use, exceeding current levels by up to 110 per cent by the year 2020. Improved fuel efficiency clearly has a part to play in reducing the level of emissions. However, current trends in vehicle purchase are towards larger and less fuel efficient models, and recent data indicate that the average fuel economy of new cars is deteriorating. It has been estimated that a radical programme of vehicle efficiency improvements would be able to stabilise emissions of carbon dioxide from cars by the year 2005, but only through government intervention, particularly in the form of fiscal incentives for the best available technology, and moderate growth in traffic.[28] Technological change alone will not be sufficient to meet future CO_2 emissions reduction targets if traffic levels continue to grow even in line with the lower traffic forecasts. The only way of reducing carbon dioxide emissions is to use some form of traffic restraint.

Table V UK traffic forecasts, 1989

	Traffic Growth 1988–2025 (%)	
	Low growth	High growth
Cars and Taxis	82	134
Light Goods Vehicles	101	215
Heavy Goods Vehicles	67	141
Buses and Coaches	0	0
All Traffic	83	142

Source: Department of Transport, 1989[27]

Globally carbon dioxide emissions from road vehicles have been estimated to have risen by 3 per cent per year (on average) from about 510 million tons of carbon in 1971 to over 830 million tons in 1987. Global emissions are currently increasing by about 19 million tons of carbon per year, of which about 10.4 million tons comes from the OECD countries and about 8.7 million tons from all the other nations combined.[25]

Carbon dioxide emissions from motor vehicles in the developing world have been estimated to be growing at about 3.5 per cent per year and account for about 45 per cent of the global growth in vehicle releases. As a result, emissions from developing countries will grow as a share of total emissions.[25]

If today's trend continues, motor vehicle emissions will increase by 50 per cent by the year 2010, to about 1.3 billion tons of carbon per annum. Controlling global climate change will prove extremely difficult if emissions from just one energy end-use sector is allowed to increase by such a large amount.

5. TRAFFIC RESTRAINT

Studies have shown that even the extremely rapid introduction of advanced technology would be insufficient to reduce carbon dioxide emissions from cars to the level needed to stabilise atmospheric concentrations.[29] The only way to reduce carbon dioxide emissions sufficiently is to initially limit the growth in traffic and then, in the longer term, to reduce traffic levels.

Vehicles perform better in free flowing traffic compared to the stop start conditions of congested roads, with lower fuel consumption and emissions. Traffic management policies may have some potential, particularly in the short term, to reduce traffic congestion by improving flow. However, improvements in traffic flow may release suppressed demand in congested areas, and recreate the very congestion they sought to combat. In the longer term, with the rapid increase in traffic demand, it is inevitable that congestion will reappear. This applies equally to the building of new roads as to traffic management schemes.

To reduce carbon dioxide emissions from road transport, traffic restraint policies would be required to tackle the problem of the number of vehicles on the roads. Unless such traffic restraint measures result in people shifting from car use to walking, cycling or public transport, and reducing both journey length and the number of journeys, the reduction in emissions will be negligible or even negative.

Recent research into land use and transport patterns[30] has shown that average distances travelled by people in the United Kingdom have increased by 30 per cent over the past decade and that only a small proportion of this increase can be attributed to changes in land use patterns or other structural changes. It seems that this increase has arisen primarily because people now have access to and can afford more travel, and have altered their domestic, business and leisure activities accordingly. In principle, there are no bounds to the distances which people might wish to travel; only the time and money which they chose to allocate to travelling. Therefore if travelling becomes cheaper and faster, greater demand for travel is very likely to follow—as has been illustrated in recent times by the dramatic increase in distances travelled in both cars and aircraft. Transport policies need to address the problem of increased accessibility without unduly increasing total mobility.

There is a wide range of techniques for managing and limiting the use of cars, and encouraging other modes of transport such as public transport, walking and cycling. These techniques range from regulatory and physical traffic controls to pricing and economic incentives, and include the use of public education campaigns and planning instruments. To reduce emissions from traffic will require a package of policies. No one policy alone will have a large enough effect on its own. Disincentives to car use need to be accompanied by incentives to use public transport, walk or cycle. For example, the introduction of road pricing in Singapore was accompanied by a 'park and ride' scheme and investment in public transport to provide suitable alternatives for the private motorist.

The flexibility of many traffic management techniques means that experiments and innovations can be introduced quickly, at low cost, and with short term impacts. As such these techniques supplement improvements in vehicle technology and infrastructure changes which typically take longer to implement and have medium to long term impacts.

In California twenty measures have recently been introduced to lower emissions by reducing vehicle use.[31] These include the banning of new drive-through facilities, improving public transport, managing the growth of new developments, promoting car pooling, introducing parking restrictions and introducing new ways of working including flexitime, alternative work weeks and telecommuting.

In the Netherlands, the Government aims to stabilise carbon dioxide emissions at 1989–90 levels by 1995, and reduce it by 10 per cent by 2010. It recognises that reducing car use will have to play a role in meeting these targets. In their Second Transport Structure Plan, the Government has proposed a number of similar traffic restraint measures as in the South Coast Air Quality Management Plan. In addition, it proposes raising fuel tax, road pricing, and ensuring that the cost of public transport does not increase faster than the variable costs of running a car.[32]

6. CONCLUSIONS

Road transport gives rise to high levels of air pollution. The main pollutants are carbon monoxide, volatile organic compounds, nitrogen oxides, particulates and carbon dioxide. Private car use is the largest single source in the sector and its predominance is projected to grow over time. Road use for trade and haulage is also a major contributor; while buses, coaches and motorcycles add a small and ever-decreasing proportion of the overall pollution burden.

Catalytic converters will reduce unit emissions of carbon monoxide, volatile organic compounds, nitrogen oxides from petrol cars, while improvements to heavy duty engines and particulate traps will reduce the pollution burden of buses and trucks. However, these measures alone will not by any means eliminate the problem of these pollutants. Indeed, soon after the turn of the century, total emissions are predicted to grow once again in line with traffic growth.

Three-way catalysts and particulate traps do nothing whatever to mitigate emissions of carbon dioxide and may actually add to the emissions through a small fuel penalty. The only way to reduce carbon dioxide emissions is to burn less fuel. Over the decade since the last oil price shock new vehicle fuel economy, as measured in the official fuel consumption tests, has improved, but due to declining load factors it still takes more energy to move a person a kilometre than it did in the 1950s.[18]

The most effective single group of measures to reduce carbon dioxide emissions is the use of fiscal incentives to encourage the adoption of the best available vehicle technology. However, it is noteworthy that this alone would do little more than restore carbon dioxide emissions from cars to current levels by the target year of 2005. That is, technological change alone will not be sufficient to meet future carbon dioxide emissions targets.

It is clear that, if current policies are pursued, road transport growth will overtake any improvements in the emission levels of pollutants. Meanwhile, the rise in carbon dioxide emissions from the transport sector will continue virtually unchecked. No foreseeable technical change will reverse this conclusion, or even mitigate it to a significant degree.

As is apparent from this brief review of fuel cycle emissions from alternative transportation fuels, none of the alternatives currently offers a viable solution to increasing carbon dioxide emissions. However, some alternative fuels may have a more limited role to play in the near future in abating high levels of emissions in particularly polluted areas, thus playing a part in a local clean air strategy.

The predicted growth in carbon dioxide emissions seems virtually certain to prove unacceptable in a climate of increasing concern over global warming, and if so other remedies will be required. Technical standards to the exclusion of a comprehensive appraisal of transport demands and their associated environmental impacts will be insufficient to improve the environment.

References

1. Society of Motor Manufacturers and Traders, *Motor Industry of Great Britain 1990: World Automotive Statistics* (Society of Motor Manufacturers and Traders, London, 1990).
2. Department of Transport, *Transport Statistics Great Britain 1979–1989* (Her Majesty's Stationery Office, London, 1990).
3. Organisation for Economic Cooperation and Development, *Transport and the Environment* (OECD, Paris, 1988).
4. J. M. Dunne, 'A Comparison of Various Emission Control Technology Cars and their Influence on Exhaust Emissions and Fuel Economy,' *Report No. LR 770 (AP)* (Warren Spring Laboratory, Stevenage, 1990).
5. B. Lübkert and W. Schöpp, *A Model to Calculate Natural VOC Emissions from Forests in Europe* (International Institute for Applied Systems Analysis, Laxenburg, Austria, 1989).
6. J. C. Bailey and M. L. Williams, "On the Road Emissions of Selected Individual Hydrocarbons from Current and Low Emissions Technology Vehicles and of Nitrous Oxide from the Latter" Paper presented at 'Automotive Power Systems: Environment and Conservation', a conference organised by the Institution of Mechanical Engineers, held in Chester, 10–12 September 1990.
7. World Health Organization, *Air Quality Guidelines for Europe* (WHO Office for Europe, Copenhagen, 1987).
8. K. D. van den Hout and R. C. Rijkeboer, *Diesel Exhaust and Air Pollution* (TNO, Research Institute for Road Vehicles, Delft, 1986).
9. International Agency for Research on Cancer, "Diesel and Gasoline Engine Exhausts and Some Nitroarenes" *IARC Monograph on the Evaluation of Carcinogenic Risks to Humans. No. 46* (World Health Organization, Lyon, 1989).
10. Department of the Environment, *Digest of Environmental Protection and Water Statistics* (Her Majesty's Stationery Office, London, 1991).
11. Photochemical Oxidants Review Group, *Ozone in the United Kingdom* (Harwell Laboratory, Oxfordshire, 1987).
12. G. McInnes, "The Effectiveness of Policy Measures to Promote Energy Efficiency in Road Transport" Paper presented at 'Tomorrow's Clean and Fuel-Efficient Automobile: Opportunities for East-West Cooperation', an international conference organised by OECD, IEA and ECMT in Berlin, March, 1991.
13. C. D. Holman, "FEAT: Remote Sensing of Vehicle Exhaust Emissions" *Clean Air* **21**, 27–31 (1991).
14. Royal Automobile Club, "RAC Study Casts New Light on Car Pollution" *Press Notice* (1990).

15. National Audit Office, *National Energy Efficiency* (Her Majesty's Stationery Office, London, 1989).
16. R. M. Heavenrich and J. D. Murrell, *Light Duty Automotive Technology and Fuel Economy Trends Through 1989* (U.S. Environmental Protection Agency, Ann Arbor, Michigan, 1989).
17. C. A. Amann, "The Automotive Engine—A Future Perspective" Paper presented to the Society of Automotive Engineers 'Future Transportation Technology' Conference and Exhibition held 8–12 August 1989 in Vancouver (Warrendale, Pa. 1989).
18. D. J. Martin and R. A. W. Shock, *Energy Use and Energy Efficiency in UK Transport up to the Year 2010* (Energy Technology Support Unit, Department of Energy, Her Majesty's Stationery Office, London, 1989).
19. D. L. Bleviss, *The New Oil Crisis and Fuel Economy Technologies* (Quorum Books, New York, 1988).
20. D. R. Blackmore and G. B. Toft, "Shell's Mileage Marathon Competition" Paper presented at 'Automotive Power Systems: Environment and Conservation', a conference organised by the Institution of Mechanical Engineers, held in Chester, 10–12 September 1990.
21. D. Sperling and M. A. DeLuchi, "Transportation Energy Futures" *Annu. Rev. Energy*, **14**: 375–424 (1989)
22. C. Holman, C. Mitchell and M. Fergusson, "Road Transport and Air Pollution," *Discussion paper 25, 'Transport and Society' Project* (Rees Jeffreys Road Fund, Oxford, 1990).
23. M. A. DeLuchi, R. A. Johnston and D. Sperling, "Methanol vs. Natural Gas Vehicles: a Comparison of Resource Supply, Performance, Emissions, Fuel Storage, Safety, Costs and Transitions" *SAE Technical paper No. 881656* (Society of Automotive Engineers, Warrendale, Pa. 1988).
24. S. P. Ho, *Global Warming Impact of Ethanol versus Petrol* (R&D Department, Amoco Oil Company, Naperville, Il. 1989).
25. J. J. MacKenzie and M. P. Walsh, *Driving Forces: Motor Vehicle Trends and their Implications for Global Warming, Energy Strategies and Transport Planning* (World Resources Institute, Washington, 1990).
26. M. Fergusson, C. Holman and M. Barrett, *Atmospheric Emissions from the Use of Transport in the United Kingdom. Volume One: The Estimation of Current and Future Emissions* (World Wide Fund for Nature, Godalming, Surrey, 1989).
27. Department of Transport, *Roads for Prosperity* (Her Majesty's Stationery Office, London, 1989).
28. M. Fergusson and C. Holman, *Atmospheric Emissions from the Use of Transport in the United Kingdom. Volume Two: The Effects of Alternative Transport Policies* (World Wide Fund for Nature, Godalming, Surrey, 1990).
29. C. Holman, *Transport and Climate Change: Cutting Carbon Dioxide Emissions from Cars* (Friends of the Earth, London, 1991).
30. J. P. Steadman and M. Barrett, *The Potential Role of Town and Country Planning in Reducing Carbon Dioxide Emissions* (Open University/Earth Resources Research, London, 1990).
31. South Coast Air Quality Management District/Southern California Association of Governments, *Air Quality Management Plan* (South Coast Air Basin, California, 1989).
32. Dutch Government, *Second Transport Structure Plan* (The Hague, 1990).

9. "BUYING THE SKY": ACID RAIN CONTROLS IN THE U.S. UNDER THE 1990 CLEAN AIR ACT

MILDRED B. ARCHER and RICHARD O. BROOKS

". . . Buy the sky and sell the sky,
And lift your arms up to the sky,
And ask the sky and ask the sky
Don't fall on me,
Don't fall on me! . . ."

Fall on Me, R.E.M.[1]

The U.S. 1990 Clean Air Act sets out to achieve acid rain control by using market-oriented regulations. Recent research pertaining to the causes and impacts of acid rain underlie the 1990 Clean Air Act. The United States has a history of failed attempts at regulating acid rain under the authority of the 1977 Clean Air Act, which was written before the consequences of acid rain were widely acknowledged. The new acid rain control provisions are the most significant effort to date to utilize market principles in regulating difficult pollution problems. This chapter tentatively evaluates the progress of this effort in promoting efficient pollution control.

1. INTRODUCTION

The U.S. 1990 Clean Air Act, in addition to being the most detailed regulatory amendment in the history of United States environmental law, ushers in the next chapter in the United States' effort to promote regulatory efficiency.[2] In addition to numerous other market-oriented provisions in the law, the new section on acid rain control is this nation's most significant effort to employ market principles in regulating difficult pollution problems.[3]

Since Earth Day in 1970, the United States government has been concerned with finding the most efficient way of achieving regulation for public interest. Part of that effort has been both the privatization of selected public institutions and the deregulation of selected industries, particularly the airline industry.[4] However, the deregulatory approach is not suited to the recognized need to achieve public purposes such as the restoration and maintenance of clean air. A second related approach was to rely upon non-regulatory incentives and information in the form of subsidies and right-to-know laws which were proposed and, in some cases, employed.[5] Economists and others, however, believe that a subsidy approach does not require regulated industries to internalize their costs. Informational approaches are also limited in their effectiveness.[6] A third alternative to outright deregulation is to adopt regulations, subject to conducting cost-benefit analyses of them.[7] Such analysis requires the centralized estimation of market costs and benefits and consequently, although such analyses are still undertaken, they remain controversial.[8]

Another approach, more closely related to the approach taken in the new acid rain provisions of the Clean Air Act, is the creation of a new regulatory market through the use of emission or effluent fees, charges or taxes.[9] In the early 1970s, it was the charges approach to water pollution which was considered, but rejected by the U.S. Senate.[10] In the reauthorization debate over the Clean Air Act in 1975–77, the U.S. House turned its attention to market trading schemes, which lie at the heart of the new acid rain provisions.

In the 1977 Clean Air Act amendments, Congress adopted four market related provisions—the noncompliance penalty, the offset policy, the permit fee provision, and authorization of studies on economic incentive alternatives.[11] The noncompliance penalties and permit fee provisions were not widely used by the states and there was little interest in the economic incentives studies.[12] It was the offset policy, which had only the weakest of economic policy rationales at its origin, but which became the central program for marketable permits.[13] The offset provision was designed primarily to permit economic growth in dirty (non-attainment) areas by permitting the proposed development to "offset" his new pollution against reductions achieved by other polluters in the area, as long as a net reduction in pollution occurs. Although administratively adopted in 1976 and ratified by Congress, and developed under the Carter administration, a full emissions trading program was not developed until 1982. The policy consisted of four components:

—Emissions offsets allow qualified new or expanding pollution sources to operate in areas that have not achieved all emissions standards, so-called non-attainment areas, provided the new or modified sources acquire sufficient emission reduction credits from other facilities they own, or from other existing sources in a region, so as to lower actual emissions in the region.

—Bubbles allow existing sources to use emission reduction credits to meet control responsibilities within plants. The policy derives its name from its treatment of multiple emission points in a plant as if they were contained under an imaginary bubble with a single opening in the top. Control requirements are applied not to the multitude of individual emission points but only to the emissions leaving the bubble through the single opening.

—Netting allows emission reduction credits earned elsewhere in a plant to offset the increases expected from expansion or modernization. By meeting the appropriate tests, a source in either attainment or nonattainment areas may "net out" of administrative and technological requirements, such as preconstruction permits, demanded by the Clean Air Act, by compensating for emission increases generated by expansion and modernization with emission reductions achieved at other points within a facility.

—Banking allows firms to store emission reduction credits for their own future use in bubbles, offsets, or netting. Banking rules also establish regional accounting ledgers or central clearinghouses for emission reduction credits. Firms in search of emission reduction credits that they themselves cannot generate can find other sources with ERCs for the appropriate pollutant through such emission credit banks and trade for or purchase the necessary credits. States operating banks can control and encourage trading, buying, and selling of emission reduction credits between pollution sources through banking rules.[14]

In the 1980s, emissions trading had an opportunity to be implemented and assessed. The need to determine the flexibility of states to implement the program, the need to adopt key legislative language originally not changed to accept the offset program, the lack of state data on actual pollution levels, and battles over determining the baseline amounts slowed down the adoption of the offset program and raised the question as to whether the transition costs to a market system would be too great.[15] Trading through the bubble concept was upheld by the courts in 1984.[16]

EPA undertook one other market-based approach to environmental regulation under Section 211 (c) of the Clean Air Act, which enabled EPA to control motor vehicle fuel additives resulting in air pollution.[17] EPA adopted a regulation which called for the phasedown of the content of lead in gasoline over a period of time.[18] In order to provide producers and importers with flexibility during the phasedown period, they would be credited for excess phasedown efforts.

It would be a mistake to view the market-oriented regulations of acid rain as emerging only out of the offset or lead marketing experiences. In 1983, EPA Administrator Ruckelshaus specifically denied the applicability of offsets to acid rain. He noted that offsets were applied on a small regional level within a state, and questioned the transfer of the technique to the acid rain problem of national scale.[19] But the concern to apply market approaches to the acid rain problem did arise in various ways before the late 1980s. In the 1975–77 amendment process of the Clean Air Act, legislators, interest groups and scholars suggested that the mandating of scrubbers for midwestern utilities was economically irrational.[20] During the mid-1980s legislators and utilities urged a subsidy approach, according to which government subsidies would be given for coal cleaning.[21]

The economic incentives for clean coal technology were not the only economic concerns posed by the proposed control of acid rain. Deep concerns were expressed about the loss of employment in high sulfur coal areas, and about the high electricity rates which might result from acid rain controls on midwestern utilities.[22] These concerns were reflected in the ultimate provisions of the 1990 Acid Rain provisions. But it was EPA and a Council of Economic Advisors study which gave the immediate impetus to the market approach which was adopted.[23] The specific economic rationale will be discussed below.

The purpose of this paper is to tentatively evaluate the progress of the new acid rain control effort to date as an example of market-oriented regulations. As part of this introduction, let us list the policies which an economic incentive program might serve. They include promotion of efficient pollution control, flexibility, i.e. leaving the choice of control options to the industry, effective enforcement of the controls through positive incentives, increased technological innovation to take advantage of possible tradable savings, better information regarding costs and benefits of control through market price of permits, reduced transaction costs of bargaining over pollution control standards, reduced administrative complexity, reduced intrusiveness of public regulation into private decisionmaking, and "market egalitarianism." Opponents contest most of these alleged advantages and also argue that the use of market tools raises fundamental moral and legitimacy issues, and problems of distributive justice.

2. THE U.S. PERCEPTION OF THE ACID RAIN PROBLEM

Acid rain is a serious environmental problem which threatens the world's industrialized nations. It is primarily a consequence of burning fossil fuels, especially coal, to produce energy. The term "acid rain" was coined by English scientist Robert Angus in his book *Air and Rain: The Beginnings of a Chemical Climatology*, published in 1872.[24]

Scientists from many countries over hundreds of years have studied and described acid rain and its effects. As early as 1661, English scientists researched the movement of pollutants between England and France and recommended locating industry outside of towns and using taller chimneys to spread "smoke" into "distant parts."[25]

Considering that the acid rain phenomenon has been acknowledged for so long and that industrialized nations have been burning fossil fuels on a large scale for over 150 years, one might ask why we are just beginning to pay attention to its consequences. The most obvious answer is that as more sources of man-made emissions of sulfur dioxide and nitrogen oxides were placed into production, the overall quantity of pollutants in the air increased, and eventually we began to notice the adverse effects. Furthermore, our emerging interest and abilities in detecting damage to the ecosystem have increased our knowledge of the types and degree of damage caused by acid rain. To understand acid rain, it is important to know where the emissions originate, where the deposition takes place, and the chemical transformation which takes place in the atmosphere and on surfaces on which acid rain falls.

Anthropogenic emissions of sulfur dioxide (SO_2) originate primarily from coal-burning power plants. Years ago, when smokestacks were only a few stories high, pollution usually stayed near the ground and settled on the land nearby causing locally unhealthy conditions. To reduce the local effects, facilities installed equipment to remove the soot, and built tall smokestacks to send the emissions into the upper atmosphere to be dispersed over a large area. This solution, however, created another more complicated problem. The particulate matter comprising the soot had acted to neutralize the sulfur emissions and helped to prevent the formation of acid rain.[26] Moreover, sending the emissions high into the atmosphere increased the time that the pollution stayed in the air; the longer it is in the air, the greater the chances that the pollutants will form acid rain.

Acid rain formation begins when sulfur dioxide is emitted and transformed in the atmosphere into sulfate. Sulfate is converted into sulfuric acid when it is washed out of the air by fog, clouds, mist or rain. Additionally, sulfate can settle out of the air in dry or gaseous form onto leaves, buildings, or other surfaces where it attracts water which converts it to sulfuric acid.[27] Hence, the common term "acid rain" has given way to the more accurate term "acid deposition," which includes acid snow, fog and dew, as well as deposits of dry acid particles and gasses.

Oxides of nitrogen (NO_x) contribute to acid deposition in a chemical conversion process very similar to that of sulfur dioxide. Nitrogen combines with oxygen when exposed to heat and pressure to form nitrogen oxides, which subsequently combine with oxygen to form nitrates. Nitrates react with water in the atmosphere to form nitric acid. The heat and pressure required to convert nitrogen into oxides of nitrogen can occur in the cylinder of a car or the furnace of a coal-fired powerplant. However, unlike sulfates, nitrates and other nitrogen-based compounds are considered beneficial to vegetation because they are plant nutrients; and for this reason some scientists and policy makers have tended to underestimate the role of NO_x in the formation of acid rain.[28] Nevertheless, there is another reason in addition to concerns over acid rain to control NO_x; it is an essential ingredient in the formation of ozone, or smog.[29]

Sulfates and nitrates are very fine particles which can be transported by the wind for tens of thousands of kilometers before they are transformed into sulfuric and nitric acids. Consequently, acid rain has become a problem in areas far from the polluting smokestacks: across local, state, and in some instances, international borders.[30] So, although the problem of acid rain is initiated by air emissions, the adverse environmental effects are a result of a water pollution problem which occurs downwind from the emission source.[31]

The best known effects of acid deposition are those on aquatic ecosystems.[32] As lakes and streams become more acidic, the numbers and types of fish and other aquatic plants and animals that live in these waters decrease. At a pH level of 5.0, most fish eggs cannot hatch, and few fish species can survive in water with a pH below 5.0.[33] Acid rain can also cause damage to aquatic ecosystems by dissolving toxic metals such as aluminum, copper, lead and mercury that were bound to soil particles and bedrock. Mass numbers of fish die when exposed to increased acidity and heavy metals mobilized during the spring snowmelt; a phenomenon referred to as "acid shock" or "acid pulse."[34]

The effects of acid rain on forests are not as well documented as the acidic effects on aquatic ecosystems. The most extreme form of damage some scientists have attributed to acid rain is the phenomenon known as "dieback," a term applied to the unexplained death of large sections of a once-thriving forest. It is difficult to make a direct correlation between acid rain and the health of a forest because of the many threats such as insects, drought, and disease which affect trees over their lifetime. Although there is little direct evidence linking acid rain per se to forest dieback, researchers suspect that the mutually aggravating stresses from ozone and acid deposition affect the availability of soil nutrients to root systems and weakens trees so they are susceptible to damage from insects and disease.[35]

The extent of damage caused by acid rain depends on the total acidity deposited in a particular area as well as the sensitivity of the area receiving it. The effects of acid rain are worse with more precipitation and at higher elevations with thinner soils. Since lakes and streams are formed by watersheds, most of the acid rain passes over and through soils and bedrock, rather than falling directly into the water. The ability for a lake and surrounding soil to resist pH change is referred to as Acid Neutralizing Capacity, or ANC. Areas with high ANC can experience years of acid deposition without problems. Soils like this are common throughout the midwestern United States. On the other hand, regions with numerous lakes and steep mountain slopes such as those in the northeastern United States and southeastern Canada have very low ANC, and are thus highly sensitive to acid deposition.[36]

Acid deposition can also accelerate the deterioration of many economically important materials. It eats away at stone, metal, paint—almost any material exposed to the weather for a long period of time. Repairing acid rain damage to houses, buildings, and monuments can cost billions of dollars annually, and some ancient monuments and buildings can never be replaced.[37] T. N. Skorlikidis, a Greek specialist on acid corrosion, has estimated that the Parthenon and Athenian monuments have deteriorated more in the past 20–25 years from pollution than in the previous 2,400 years.[38]

Emissions of SO_2 and NO_x which remain suspended in the air in the form of sulfates and nitrates can significantly impair visibility. Along with humidity, fog and dust, they reduce contrast, discolor the atmosphere, and obscure distant objects.[39]

The precursors of acid rain, sulfur dioxide and nitrogen oxides, as well as ozone are currently regulated to protect human health. However, no standards exist for airborne sulfates and nitrates which may also pose a threat to human health. When these fine particles are inhaled, the lung's own moisture can convert the sulfate or nitrate into an acid, potentially causing severe respiratory problems. Acid rain may also cause indirect health effects. Acidified waters can dissolve metals such as lead and mercury and other toxic substances such as asbestos from soils, rocks, conduits and pipes and result in human exposure through drinking water supplies.[40]

Since the 1977 amendments, considerable research on acid rain and its impact was undertaken in the United States. That recent research underlies the 1990 Clean Air Act and is summarized in the legislative history.[41] It has been documented that electric utilities contribute an increasing amount of the sulfates which are uniformly spread in an aerosol mist over the Northeast United States and Canada.[42] The eastern dispersion patterns from the midwest utilities are now much better understood.[43] The acidification of lakes,[44] the impact upon brook trout and eastern migrating ducks,[45] the dieback of red spruce,[46] the impairment of visibility,[47] and the health impacts from sulfates, as well as lead and mercury leaching have been recently documented and summarized for Congress.[48] Less certain is the impact of nitrates, although it is believed that they contribute to "over-fertilization."[49]

3. THE HISTORY OF PAST FAILURES TO DEAL WITH ACID RAIN

In passing the Clean Air Act Amendments of 1990, the 101st Congress of the United States clearly declared the need for acid rain control measures beyond what the existing Clean Air Act could offer.[50] Although the acid rain precursors, SO_2 and NO_x are regulated under the Clean Air Act because of their adverse health effects, once they are transformed into sulfates and nitrates they escape its coverage.

The Clean Air Act was written at a time when the long-range transport and transformation of air pollutants was not widely acknowledged. It was designed to control airborne pollutants, not the deposition of their transformation products.[51] In fact, the implementation of the Clean Air Act may have helped escalate the problem of acid rain; compliance with National Ambient Air Quality Standard (NAAQS)[52] was achieved by the use of tall smokestacks which, by sending pollutants high into the atmosphere, allowed regulated emissions of sulfur and nitrogen oxides the time to transform into unregulated sulfates and nitrates and travel across state and international boundaries.[53] Although the use of tall stacks to meet NAAQS is no longer legal, the stack height regulations were not issued until 1987 and are still tied up in litigation.[54]

Under existing law, states have the primary responsibility for dealing with air pollution. In order to maintain compliance, each individual state must develop a plan for meeting the NAAQS within its borders. The plan must be approved by the Environmental Protection Agency (EPA) and enforced by the state. Each state has the choice of how to meet the standards; for example, a state may allow a higher emission limit for pollutants in an area with very clean air than in an area where the level of pollution is near the limit set by the NAAQS. States may even issue more stringent standards than the EPA if they choose. Therefore, air pollution control requirements vary from state to state and from one region to another within a state.[55]

Proponents of acid rain legislation have attempted to use the existing Clean Air Act to force reductions in SO_2 and NO_x emissions below levels considered acceptable for compliance with NAAQS. In fact, there are specific sections of the 1977 Clean Air Act Amendments which authorize, but do not mandate, the development of acid rain programs. However, efforts made by environmental organizations and several states to petition EPA to enforce the law and to persuade courts to require the implementation of acid rain programs under these provisions have been unsuccessful. For example, the 1977 Amendments added provisions which prevent a state's emissions from causing violations of NAAQS in other countries.[56] However, after being tried in court, these provisions proved to be too general. Without guidelines for determining the sources or effects of transported air pollutants, the courts were unwilling to accept the petitioners' evidence as conclusive.[57]

Another mechanism in the existing law for controlling acid rain which has proven unsuccessful involves the revision of the existing NAAQS based on "any known or anticipated adverse effects associated with the presence of such air pollutant in the ambient air."[58] The 1977 Amendments to Clean Air Act require the EPA to review and make appropriate revisions to the NAAQS every five years.[59] The EPA was sued in 1985 by a group of seven states and three environmental groups after having made no move toward revising the NAAQS for SO_2 or proposing a standard for sulfates. EPA successfully opposed the claim that it was required to revise the SO_2 NAAQS, and the case was dismissed.[60]

A further provision which spawned litigation on behalf of acid rain control pertains to visibility protection. High sulfate concentrations result in "regional haze,"[61] which is especially prevalent in the eastern United States during the summer months. The Clean Air Act mandates the prevention of any future impairment of visibility and the restoration of good visibility in "Class I" areas which include national parks and federal wilderness areas.[62] EPA responded to their non-discretionary duty to promulgate regulations to protect visibility by addressing "plume blight" which is described as "an identifiable, coherent plume which is observable against a background sky or other objects." In many cases, the plume may be visually traced back to a single source.[63] Proponents of acid rain were outraged at EPA's decision to delay regulatory action on the more widespread problem of regional haze, and to resist states' efforts at regulating the sulfate pollution which was impairing the visibility of Class I areas within their borders.[64] Despite the EPA's inaction, the courts declined to intervene in EPA's policymaking authority regarding regional haze.

From the time Congress amended the Clean Air Act in 1977 until it was amended again in 1990, the message has been clear: the EPA is not willing or able to defend creative uses of existing authority to regulate acid rain, and courts are very willing to defer to the agency on seemingly discretionary public policy issues.

4. THE CLEAN AIR ACT AMENDMENTS OF 1990—ACID DEPOSITION CONTROL

As the failure of the existing Clean Air Act to adequately address acid rain became more and more apparent, Congress sought to introduce a bill which responded to the problem nationally rather than locally. After thirteen years of debate and opposition from members of congress from the midwest, electric utilities and the coal industry, the Clean Air Act Amendments were signed into law on November 15, 1990. Title IV of the 1990 Clean Air Act Amendments addresses the control of pollutants associated with acid rain through utilization of an innovative program which will allow power plants to "buy" emission allowances on the open market.

The goal of Title IV is to achieve a 10 million ton reduction of SO_2 and an approximately 2 million ton reduction in NO_x, relative to 1980 levels by the year 2000.[65] To achieve the SO_2 reductions, EPA will allocate allowances to utilities to emit a specified amount of SO_2 on an annual basis. One allowance represents the authorization to emit one ton of SO_2 during a specified year.[66] Allowances will be allocated to utilities in two phases: Phase I begins in 1995, and affects 111 high-emitting utilities specifically listed in the 1990 Amendments.[67] Phase II, which begins in the year 2000, tightens emissions imposed on Phase I plants and requires emission limits on smaller, cleaner plants as well.[68] In general, beginning January 1, 2000, the total allowances allocated to power plants must result in an emission rate of SO_2 which does not exceed 8.9 million tons annually.[69]

The 1990 Clean Air Act Amendments allow utilities to buy and sell SO_2 allowances among themselves.[70] In addition, EPA will hold allowances in reserve to be sold at annual auctions, or sold directly for $1,500 per allowance.[71] Each emitting source must have sufficient SO_2 allowances to cover its annual emissions. If not, the source becomes subject to a penalty of $2,000 per ton of excess emissions and a requirement to compensate for the excess emissions in the following year.[72] The proposed regulations for auctions and sales were to be finalized in November, 1991; and proposed regulations for an emission trading system were scheduled for finalization in May, 1992.[73]

The 1990 Amendments also include specific requirements for reducing NO_x emissions. In order to achieve a reduction of approximately 2 million tons, EPA will establish new NO_x emission rates for certain types of utility boilers by May of 1992, and for all remaining boilers by 1997. In addition, EPA was directed to propose revisions to the New Source Performance Standards (NSPS)[74] for NO_x emissions from fossil fuel fired steam generating units by January 1, 1994.[75]

Both Phase I and Phase II utilities are required to install continuous emission monitoring systems (CEMS) to track progress and assure compliance. The EPA was to issue regulations specifying the requirements for the CEMS by May 14, 1992. Phase I sources must have operational CEMS by November 14, 1993, and Phase II sources must install and operate CEMS by January 1, 1995.[76]

The acid rain provisions in the 1990 Clean Air Act Amendments give utilities a wide variety of emission reduction alternatives from which to choose. For example, some utilities may reduce future emissions by burning lower sulfur coal or natural gas, while others may install scrubbers and continue to use high sulfur coal. If utilities reduce their emissions below required levels, they can bank the excess allowances for future use, or sell them directly to other utilities or to the reserve maintained by EPA. If a utility's cost of compliance through emission reduction is greater than the market price for emission allowances, the utility can purchase allowances.[77]

The 1990 Act creates incentives for power plants using scrubbers, energy conservation, or clean coal technologies. Plants that use scrubbers to meet their Phase I reduction requirements may either postpone compliance until 1997 or receive an early-reduction bonus allowance for reductions achieved between 1995 and 1997.[78] Likewise, plants that use energy conservation or renewable energy will receive special incentive allowances for making early reductions in SO_2 reductions.[79] Clean coal technology (CCT) includes processes which achieve significant reductions in SO_2 or NO_x emissions associated with the utilization of coal in the generation of electricity and which are not currently in widespread use. The 1990 Amendments exempt CCT projects from certain requirements in the existing law, and provide for a 4-year extension to meet Phase II emission requirements for repowering a power plant with clean coal technology.[80]

The 1990 Act provides for differences in regional compliance costs and energy requirements. Additional allowances are allocated to specific plants in states which are especially dependent on the use of high sulfur coal. Plants that have experienced increases in their utilization in the past five years also receive bonus allowances.[81] In addition, extra allowances are provided to accommodate growth in "clean" states, where SO_2 emissions averages are very low.[82]

The acid rain provisions in Title IV of the Clean Air Act Amendments of 1990 represent a departure from the traditional command and control approach to environmental regulation. The Act provides a market-based approach that offers utilities flexibility to comply with SO_2 emissions reductions. Through the use of monetary incentives, the 1990 Act proposes to "encourage energy conservation, use of renewable and clean alternative technologies, and pollution prevention as a long-range strategy for reducing air pollution and other adverse impacts of energy production and use."[83]

5. THE PRECIPITATING LEGISLATIVE HISTORY

The Market Allowance system of the new Clean Air Act emerged out of a detailed report prepared by EPA in 1989, reports by the consultants of the electric generating industry and a report by the Council of Economic Advisors in 1990.[84] All of these reports argued that there would be substantial cost savings by adopting the market allowance approach. The CEA report offered a broad rationale for market approaches to environmental pollution controls by claiming that command and control regulations did not permit the flexibility or incentives for control and lacked the centralized knowledge as to what controls were appropriate.

After documenting the problem of acid rain, the official legislative history provided the following more specific rationale for the program. According to this rationale, acid rain pollution could be controlled by a variety of technologies including improving overall efficiency, changing or cleansing fuels, adopting alternative combustion technologies, installing flue gas cleansing devices or establishing end use conservation programs, as well as a combination of these techniques. At the same time, these technologies are under continuous development. Rather than seeking to have the government choose and freeze the technology to be adopted, (as in the case of the 1990 air toxics program), it was decided to let the polluter choose the appropriate technology, offer incentives for low cost over control by some utilities, and permit utilities with potentially high costs of control to avoid costly outlays by purchasing the allowances rather than controlling the pollution.[85]

The legislative history gives the following example:

"Unit A emits 25,000 tons of SO_2 annually and is allocated 10,000 allowances, requiring it to remove 15,000 tons of emissions to meet a 10,000 ton emissions limit. Unit A can remove 18,000 tons of emissions at a cost of $500.00 per ton. If it did so, it would need only 7000 annual allowances to cover its own operations, leaving it with 3,000 unused allowances. Unit B emits 15,000 tons per year and is allocated 12,000 allowances. To remove 3,000 tons to meet its 12,000 ton/allowance limit would cost it $1,000 per ton. Unit B would clearly save money by purchasing unit A's 3,000 allowances at a price somewhere between $500.00 (unit A's cost) and $1,000 (Unit B's cost) rather than incurring the $1,000 per ton cost of removing the emissions itself. If A sold its allowances to unit B in that postulated price range, then Unit A could use the proceeds to lower its net compliance costs and Unit B would be meeting its requirements for a lower cost as well."[86] The legislative history offers support for the notion that utilities would act efficiently, and that the market is determinate enough and not concentrated so that a robust market for allowances would take place. The history outlines ways in which the administrator can promote market conditions and interprets the law as permitting the speculation and banking of allowances.[87]

Although the market allowance approach embodied in the 1990 Clean Air Act would appear to permit the flexible competition among clean up technologies and presumably encourage lower cost approaches such as low sulfur coal, the Act was amended to encourage the adoption of scrubbers. Under the Act, polluting utilities can obtain a delay in compliance if they directly or indirectly adopt a "qualifying Phase I technology" which is a scrubber, as well as bonus allowances in 1997 through 1999.[88] By encouraging the use of scrubbers, the new Clean Air Act continues the policy of encouraging the use of eastern high sulfur coal, thus presumably protecting the jobs of eastern coal miners. This policy, combined with the encouragement of clean coal technology development means that the acid rain provisions are hardly pure regulatory market mechanisms. To give the reader a notion of how these complex provisions work in practice, here is an example, offered by the authors, of these scrubber amendments.[89]

To understand how these Phase I scrubber incentives might reduce the impact of the law on the high-sulfur coal industry, consider the following hypothetical. Assume that the U.S. Power Company (US Power) has the following affected Phase I units:

Table I Affected Phase I Units of US Power

	Projected annual emissions 1995/1996*	Phase I allowances	Emission reduction obligation
Big Bear			
1	160,000	67,000	93,000
Green			
1	120,000	60,000	60,000
2	80,000	40,000	40,000
3	50,000	25,000	25,000
Red Spruce			
1	36,000	20,000	16,000
2	36,000	20,000	16,000
Bass Lake			
1	40,000	25,000	15,000
2	32,000	20,000	12,000

*Uncontrolled and assumed to be equal to or less than annual emissions in 1988 and 1989.

Further assume that US Power intends to reduce its emissions at Big Bear 1 through use of a "qualifying Phase I technology," thereby reducing its emissions at the unit to a level well below that allowed in Phase I. By developing a compliance plan based upon the installation of a scrubber at Big Bear 1, US Power would be eligible to receive a two-year extension of the Phase I compliance deadline at the unit. In addition, once the scrubber was made operational, US

Power would be issued allowances, from Table I above, in excess of the number necessary to cover the unit's actual post-compliance emissions. Finally, by reducing the emissions rate at Big Bear 1 to a level less than 1.2lb/mmBtu through the use of a qualifying technology, US Power would also be eligible to receive "bonus" allowances in calendar years 1997 through 1999.

The following two scenarios show how US Power might adjust its compliance strategy and use the excess allowances generated at Big Bear 1 to offset or cover the emissions reduction obligations of other Phase I units within its system.

Scenario 1: Assume that US Power, electing to take full advantage of the two year delay, begins operation of its "qualifying Phase I technology" at Big Bear 1 on January 1, 1997.

In both 1995 and 1996, as an "eligible Phase I extension unit," Big Bear 1 would be provided with additional emission allowances sufficient to allow the unit to operate without any new emission controls. Beginning in 1997, with its scrubber operational, Big Bear 1 would emit only sixteen tons of sulfur dioxide annually, a level 51,000 tons below its allowable Phase I emissions. In addition, Big Bear 1 would also generate 16,000 "bonus" allowances each year between 1997 and 1999, as a result of its emissions rate falling to a level below 1.2 lb/mmBtu. Altogether, Big Bear 1 would generate 67,000 excess allowances a year between 1997 and 1999. It would thus generate a supply of excess allowances sufficient to cover the combined Phase I emission reduction obligations of Green Tree 2, Bass Lake 1, and Bass Lake 2. As a result, US Power would be able to transfer the emission reduction obligations of all three units to Big Bear 1, thereby also qualifying Green Tree 2, Bass Lake 1 and Bass Lake 2 for treatment as "eligible Phase I extension units."

Overall, US Power's Phase I compliance strategy would entail:

—scrubbing Big Bear 1 in 1997;
—fuel-switching Green Tree 1, Green Tree 3, Red Spruce 1, and Red Spruce 2 in 1995; and
—taking no action to reduce emissions at Green Tree 2, Bass Lake 1 or Bass Lake 2

Scenario 2: Assume US Power, electing to earn credit for making early reductions, begins operations of its "qualifying Phase I technology" at Big Bear 1 on January 1, 1995.

As under scenario 1, between 1997 and 1999, scrubbing Big Bear 1 would generate a sufficient number of excess allowances to cover the combined emissions reduction obligations of Green Tree 2, Bass Lake 1, and Bass Lake 2, thereby allowing US Power to transfer the emission reduction obligations of those units to big Bear 1. As a result, as under the previous scenario, Big Bear 1 and the three "transfer" units would qualify for treatment as "eligible Phase I extension units" and scrubbing Big Bear 1 would allow US Power to avoid having to reduce emissions at Green Tree 2, Bass Lake 1, or Bass Lake 2 during Phase I.

In addition, by scrubbing Big Bear 1 early, US Power would also be able to generate an additional 144,000 excess allowances in both 1995 and 1996. The decision to scrub Big Bear 1 early would not invalidate its eligibility for a two-year extension of the Phase I compliance deadline. While emissions at Big Bear 1, with a scrubber in operation, would be reduced to only 16,000 tons in both 1995 and 1996, the unit would be awarded 160,000 allowances in each of those years: 67,000 from the Table I allocation of allowances, and an additional 93,000 from the special Phase I allowance reserve set aside for eligible extension units. As a result, Big Bear 1 would generate a total of 288,000 additional excess allowances during the 1995–1996 period. Spread over five years, these additional allowances could be used to offset an additional 58,000 tons of emission reduction obligations during Phase I. US Power could thus cover the combined Phase I emission reduction obligations of Green Tree 3, Red Spruce 1, and Red Spruce 2 with the excess allowances generated by scrubbing Big Bear 1 early.

Overall, US Power's Phase I compliance strategy would include:

—scrubbing Big Bear 1 in 1995;
—fuel-switching Green Tree 1 in 1995; and
—taking no action to reduce emissions at Green Tree 2, Green Tree 3, Red Spruce 1, Red Spruce 2, Bass Lake 1, or Bass Lake 2.

If US Power elected not to scrub Big Bear 1, it would be required either to fuel-switch all eight of its affected Phase I units beginning in 1995, or to go into the market and purchase allowances from another utility to cover the excess emissions for any unit at which it does not fuel-switch. Clearly, the high-sulfur coal industry would suffer a significant loss of market share if utilities, such as US Power in the above hypothetical, chose not to scrub any affected Phase I units. Yet, as shown in the above hypothetical, the scrubber incentives incorporated in the Clean Air Act Amendments will enable a utility to maintain existing high-sulfur coal contracts at a larger number of affected units than those at which an FGD unit would be installed. By scrubbing Big Bear 1 in 1997, as in scenario 1, US Power would be able to cover the emission reduction obligations of three other affected units, thereby avoiding the need to fuel-switch those units. In scenario 2, scrubbing Big Bear 1 in 1995 would enable US Power to cover the emission reduction obligations of six other affected units, leaving it with only one unit at which it would be forced to fuel-switch.

6. THE LINGERING ISSUES

The market allowance approach to acid rain control, indeed the entire 1990 Clean Air Act approach to acid rain control leaves many questions unanswered. Despite the National Acid Precipitation Assessment Program which seeks an integrated research efforts on the effects of acid rain, we have no solid evidence supporting the targeted amount of sulfur dioxide controls. Second, there is evidence that sources and areas other than those targeted by the Act may be important parts of the acid rain problem. Third, the fairness of targeting only electrical utilities can and has been seriously questioned. Fourth, there remains doubt that a vigorous market will be established for market allowances. The ambivalence of the Act towards the property status of such allowances is one of many factors which contribute to the problem of the market. Fifth, those amendments to the legislation which encourage the use of scrubbers may delay reaching the goals set and may affect the workings of the market. Sixth, particular proponents for one or another approach to resolving the acid rain problem, e.g. the environmentalist favoring conservation or West Virginians favoring clean coal technology development, doubt that the present Act will offer sufficient incentives for the adoption of their specific answer to the acid rain problem. A seventh problem which remains is the broader geographical distributive consequence of the acid rain marketing scheme. Some commentators fear that the trading will encourage new pollution in areas previously relatively untouched by pollution problems.

Finally, the feasibility of implementing the program in a prompt and effective manner is also still in doubt. One ominous sign of the problems of the implementation is the failure to undertake regulatory negotiation to arrive at agreed-upon rules for control of nitrogen oxides. The delayed issuance of the market rules for sulfur dioxides is a second sign. Furthermore, the problems of harmonizing other pollutant controls within the Clean Air Act, e.g., the PSD controls for sulfur oxides, with the acid rain provisions remain to be resolved.

7. CONCLUSION

The acid rain provisions are the "great experiment" in the market-oriented control of pollution. If successful, they may lead to the extension of tradable permits to the fields of recycling and pesticides. Other incentives such as taxes for water toxics and VOC emissions are under consideration. A third area might be the expansion of deposit and return systems. The expansion of market controls into the international field is currently under consideration.[90] As a consequence, an important new economics and law literature is emerging in this field.[91] Meanwhile, the problems of the morality and legitimacy of such an effort are being shunted aside.[92] We are entering the era of buying and selling the sky.

References and Notes

1. R.E.M., *Life's Rich Pageant*. I.R.S. Inc., 1986.
2. Although this article focuses upon the market controls for acid rain, another important issue left undiscussed is the incredible legal complexity of this 1990 Act, indeed the enitre Clean Air Act. [42 USC § 7400 et seq.] The acid rain provisions were enacted as Pub. L. 101–549, § 401 [Nov. 15, 1990] and will be Title IV of the codified law.
3. Other market-related clauses include, but are not limited to § 182 [offset provisions; creditability for reductions, economic incentive programs]; § 218 [user fees]; § 219 [marketable oxygen credits]; § 205 [civil penalties]; § 249 [credits for excess compliance]; § 301 [hazardous air pollutant offsets]; § 808 [renewable energy and conservation incentives]; and § 901 [research as pollution preventions].
4. For a discussion of privatization, see E. Savas, *Dismantling the State: The Theory and Practice of Privatization* (Chatham House, Chatham, N.J., 1985); and S. Bryer, *Regulation and its Reform* (Little, Brown and Co., Boston, MA, USA, 1982).
5. J. Krier, *Environmental Law and Policy* (Bobbs-Merrill, Indianapolis, IN, pp. 418–421).
6. But see D. Roe "Barking Up the Right Tree: Recent Progress in Focusing the Toxics Issue," *Cal. J. of Envtl.* **13**, 275 (1988).
7. D. Swartzman, R. Liroff, K. Croke, *Cost-Benefit Analysis and Environmental Regulation* (Conservation Foundation, Washington, D.C. 1982).
8. For a modified approach, see cost effectiveness analyses as set forth in Paul Portney, ed., *Public Policies for Environmental Protection* (Resources for the Future, Washington, D.C., 1990).
9. Anderson, et al, *Environmental Improvement Through Economic Incentives* (Johns Hopkins University Press, Baltimore, Maryland, 1979).
10. B. Cook, *Bureaucratic Politics and Regulatory Reform: The EPA and Emission Trading* (Greenwood Press, New York, N.Y., 1988) pp. 40–42.
11. *Id.* pp. 42–48. Part D of Title I of the Clean Air Act, entitled Plan Requirements for Non-Attainment Areas authorizes offsets. Permit fees were authorized in 110(a)(2)(K) of the Clean Air Act. Economic incentive alternatives are set forth in § 405. Non-compliance penalties are contained in § 120.
12. *Id.* p. 47.
13. *Id.* pp. 48–56.

14. *Id.* p. 63.
15. *Id.* pp. 59–85.
16. *Chevron, U.S.A., Inc. vs. Natural Resources Defense Council, Inc.* (467 US 837, 1984).
17. § 211(c), Clean Air Act.
18. Once the phasedown was complete, the regulations were inoperative.
19. *Legislative History* (1991 U.S. Code Congressional and Administrative News, Clean Air Act Amendments, 2399, p. 290).
20. Ackerman and Hassler, *Clean Coal and Dirty Air* (Yale Univ. Press, New Haven, CT, 1981).
21. M. Keeler, *Acid Rain Controls or Clean Coal Technology? (DOE v. EPA);* "A Political Football," August 2, 1991 (unpublished on file, Environmental Law Center, Vermont Law School).
22. Concern about unemployment resulted in Title XI of the new Act, "Clean Air Employment Transition Assistance."
23. *Legislative History*, pp. 318–319.
24. Principal source: Ellis B. Cowling, "Acid Precipitation in Historical Perspective," *Environmental Science and Technology*, 16 (1982), reprinted in "An Acid Rain Chronology," *EPA Journal*, **12**, 18–19 (1986).
25. *Id.*
26. D. Firestone and F. Reed, *Environmental Law for Non-Lawyers* (Soro Press, South Royalton, Vt., U.S.A, 1983, p. 80).
27. S. Rep. No. 228, 101st Cong. (1990) reprinted in 1991 U.S. Code Cong. and Admin. News at 3645 [hereinafter S. Rep. No. 228] (West Publ. Co., St. Paul, MN, U.S.A., 1990/1991).
28. *Id.* at 3646.
29. Elevated levels of ozone are produced through the chemical interaction of nitrogen oxides and hydrocarbons in the presence of sunlight. Ozone poses a threat to human health, man-made materials and in combination with acid rain may result in forest decline.
30. *Id.* at 3647.
31. Firestone, supra note 26 at 82.
32. S. Rep. No. 228, supra note 27, at 3649–3656.
33. Acidity is measured on the pH scale which runs from zero to 14. A pH value of 7 is neutral; a solution with a pH greater than 7 is basic, less than 7 is acidic. The pH scale is logarithmic, so that water with a pH of 5 is ten times more acidic than water with a pH of 6, and water with a Ph of 4, is one hundred times more acidic than water with a pH of 6. All rain is naturally slightly acidic, with a pH between 5.6 and 5.7. See "The Acid Rain Phenomenon," *EPA Journal*, **12**, 16 (1986).
34. S. Rep. No. 228, supra note 27 at 3652, 3666.
35. *Id.* at 3656–3658.
36. *Id.* at 3652.
37. *Id.* at 3658–3659.
38. French, "Cleaning the Air," in *State of the World 1990* (W.W. Norton, New York, U.S.A., 1990, page 102).
39. Office of Technology Assessment, U.S. Congress, *Acid Rain and Transported Air Pollutants* (1984), in M. Squillance, *Environmental Law, Volume Three,* "Air Pollution" [hereinafter OTA] (Anderson Publishing Co., Cincinnati, Ohio, U.S.A, 1988).
40. S. Rep. No. 228, supra note 27 at 3645, 3666.
41. *Legislative History* (Clean Air Act P.L. 101-549 pp. 1–489, U.S. Code Cong. and Admin. News).
42. *Id.* p. 288, 278.
43. *Id.* p. 263.
44. *Id.* p. 269.
45. *Id.* p. 266.
46. *Id.* pp. 273, 274.
47. *Id.* p. 275.
48. *Id.* p. 282.
49. *Id.* p. 262.
50. Clean Air Act, 42 USC § 7402 et seq. (1982). The most notable amendments to the Clean Air Act prior to the 1990 amendments were those made in 1970 and 1977.
51. OTA supra note 39 at 406.
52. National Ambient Air Quality Standards (NAAQS), authorized under § 109 of the Clean Air Act, provide a basic standard for air quality across the United States. State and local standards need not be uniform across the country as long as the overall ambient standard is met. NAAQS have been issued for the following air contaminants: carbon monoxide, hydrocarbons, nitrogen oxides, total suspended particulates, ozone, sulfur oxides, and lead. The standards are codified in 40 CFR 50.

53. S. Rep. No. 228, supra note 27 at 3672.
54. *Id.* See also Clean Air Act, § 123(a), (b); 40 CFR §§ 51.1, 51.12, 51.18; *Sierra Club v. EPA*, 719 F.2d 436 (D.C. Cir. 1983), cert. denied, 468 U.S. 1204 (1984); *Natural Resources Defense Council, Inc. v. Thomas*, 838 F.2d (D.C Cir.), cert. denied, 109 S. Ct. 219 (1988).
55. Clean Air Act § 110, 42 USC § 7410 (1982).
56. Clean Air Act, § 110 (a)(2)(E) and § 115 respectively.
57. EPA declined to consider air quality effects more than fifty kilometers from the sources under review or the cumulative impacts of multiple sources to determine if emissions in one state were interfering with other states' compliance programs. The court deferred to EPA's discretion to review state plans with regard to interstate transport of pollutants. See *New York v. Administrator, USEPA*, 710 F.2d 1200; *New York v. USEPA,* 852 F.2d 574 (D.C. Cir.), cert. denied (1988). Additionally, Sec. 115 entitled "International Air Pollution" has not been used successfully to eliminate transboundary acid deposition. See *New York v. Thomas,* 613 F. Supp. 1472; *Thomas v. State of New York*, 802 F.2d 1443.
58. Clean Air Act § 109(b)(2); 42 USC § 7409(b)(2) (1983).
59. Clean Air Act; §109(d), 42 USC § 7409(d) (1983).
60. *EDF v. Thomas,* No. 85 Civ. 9507 (S.D.N.Y., 1988). The plaintiffs in the suit were the Environmental Defense Fund (EDF), the Natural Resources Defense Council (NRDC), the Sierra Club, the National Parks and Conservation Association, New York, Connecticut, New Hampshire, Vermont, Minnesota, Rhode Island and Massachusetts.
61. Regional haze is described is 45 Fed. Reg. 80,085 col. 3 (1980) as "widespread reduction in visibility resulting from a polluted air mass, and frequently occurs on a scale of hundreds of miles and lingers for long periods of time. [It] may move over long distances and cause visibility impairment in areas which have few or no man-made emission sources."
62. Clean Air Act §169A; 42 USC § 7491 (1982).
63. "Plume blight" is defined in 44 Fed. Reg. 69, 119 col. (1979); EPA promulgated visibility regulations addressing plume blight on December 2, 1980, codified in 40 CFR 51.300.
64. See *Maine v. Thomas*, 690 F. Supp. 1106 (D. Me. 1988); *Vermont v. Thomas*, 850 F.2d 99 (2d Cir. 1988).
65. Clean Air Act Amendments of 1990, Title IV, § 401(b).
66. *Id.* at § 402(3).
67. *Id.* at § 404(e)(3), Table A.
68. *Id.* at § 405.
69. *Id.* at § 403(a).
70. *Id.* at § 403(b).
71. *Id.* at § 416(c) and (d).
72. *Id.* at § 411(a) and (b).
73. *Id.* at § § 416(d)(2); 403(b) and (d) respectively.
74. NSPS established by EPA for fossil fuel fired steam generating units reflect the degree of emission reduction achievable through the application of the best technology available.
75. Clean Air Act Amendments of 1990, Title IV, § 407(b) and (c).
76. *Id.* at § 412.
77. D. Weinstein, "The New Utility Contract: Fuel and Sulfur Dioxide Allowances," *Pub. Utilities Fort.* 53 (May 1, 1991).
78. Clean Air Act Amendments of 1990, Title IV, § 404(d).
79. *Id.* § 404(f)(2).
80. *Id.* § § 415 and 409(a) and (b).
81. *Id.* § § 404 and 405.
82. *Id.* § 406.
83. *Id.* at § 401(b).
84. EPA Economic Analysis of Title V (Acid Rain Provisions) of the Administration's Proposed Clean Air Act Amendments (H.R.3030/S.1490). September 1989; Council of Economic Advisors 1990 Economic Report of the President, pp. 187–198; Economic Evaluation of H.R. 3030/S1490 "Clean Air Act amendments of 1989, Title V, The Acid Rain Control Program" prepared for the Edison Electric Institute by Temple, Barker and Sloane, Inc.
85. *Legislative History*, pp. 291–301.
86. *Id.* p. 318.
87. *Id.* pp. 319, 320.
88. Clean Air Act Amendments of 1990, Title IV, § 404(d).
89. R. Byrd "The Clean Air Act Amendments of 1990: An Innovative But Uncertain Approach to Acid Rain Control," *West Virginia Law Review* **93**, 482–86 (1991).

90. M. Breger, R. Stewart, E. Donald Elliott, D. Hankins, "Providing Economic Incentives in Environmental Regulations," *Yale Journal on Regulation* **8**, 463 (1991).
91. C. Sunstein, "Paradoxes of the Regulatory State," *U. of Chi. L. Rev.* **57**, 407 (1990); Hahn & Hester, "Marketable Permits: Lessons for Theory and Practice," *Ecology L. Q.* **16**, 361 (1989); Ackerman & Stewart, "Reforming Environmental Law: The Democratic Case for Market Incentives," *Columb. J. Envtl. L.* **13**, 171 (1988);
   R. Stewart & Weiner, "A Comprehensive Approach to Climate Change," *Am. Enterprise* **1**, No. 6 at 75 (Nov/Dec, 1990);
   U. S. Task Force on the Comprehensive Approach to Climate Change: A Comprehensive Approach to Addressing Potential Climate Change (Feb. 1991);
   Elliott, "Goal Analysis vs. Institutional Analysis of Toxic Compensation Systems," *Geo. L. Jou.* **73**, 1357 (1985).
   Elliott, "Toward Incentive-Based Procedure," *B.U.L. Rev.* **69**, 487 (1989);
   U.S.G.A.O., "A Market-Approach to Air Pollution Compliance Cost Without Jeopardizing Clean Air Goals," (1982);
   Oates, Portney & McGartland, "The Net Benefits of Incentive-Based Regulation: A Guide to Enviromental Standard Setting," *Am. Econ. Rev.* **79**, 1233 (1989);
   Stewart, "Economics, the Limits of Legal Control," *Harv. Envtl. Law* **9**, 6 (1985);
   Blake, "The Economic Impacts of Environmental Regulation," *Nat. Resources J.* **5**, 23 (1990);
   D. Roe, "Barking Up the Right Tree: Recent Progress in Focusing the Toxics Issue," *Columbia J. of Envtl. L.* **13**, 275 (1988);
   Hahn & Hester, "Where Did All The Markets Go?" *Yale J. Reg.* **6**, 109 (1989);
   Latin, "Activity Levels, Due Care and Selective Realism in Economic Analysis of Tort Law," *Rutgers L. Rev.* **39**, 487 (1987);
   Hahn & Hird, "The Costs and Benefits of Regulations," *Yale J. Reg.* **8**, 233 (1991).
92. For two excellent critiques of the use of market techniques in environmental protection, see S. Kellman, *What Price Incentives*? (Auburn Publishing House, Boston, MA, 1981).

10. ACID RAIN AND TRANSPORT: SOME LEGISLATIVE ASPECTS

JAN McHARRY

"When I see an adult on a bicycle, I have hope for the human race." (H.G. Wells, 1866–1946)

Motor vehicles are the primary source of acid-forming pollutants such as nitrogen oxides, and are responsible for growing levels of ground-level ozone, a substance which hastens the formation of acid rain.

Technologies to reduce air pollution from motor vehicles will do little more than maintain the status quo if the underlying cause—the huge increase in the number of vehicles on our roads—is not tackled. Policy measures such as road pricing, combined with efficient public transportation programmes, demand urgent attention.

1. INTRODUCTION

Acid rain is acknowledged to be a widespread problem in the industrialized nations—with damaged trees, corroded buildings and dead lakes highlighting the urgency of remedial measures. Motor vehicles are major sources of carbon monoxide, hydrocarbons and nitrogen oxides. These, with sulphur dioxide, are the cause of acid rain and oxidant smog. Eighty percent of these emissions come from Europe and North America. All forms of motorized transport have some environmental impact but road vehicles are by far the most polluting.

Concern about vehicle emissions has increased sharply in recent years. International legislation to control toxic emissions is currently being enacted, and drafted, at EEC and UN levels. Whilst concern has centred around acid rain, the combined effects of vehicle emissions and other pollutants are linked with a number of atmospheric pollution issues. The synergistic effects of vehicle emission pollutants are often more significant than their individual effects. For example, ozone, which has been linked to tree damage, is formed from a chemical reaction between other pollutants, such as nitrogen oxides, and hydrocarbons emitted from exhausts. This "cocktail" effect must not be forgotten when assessing hazards from vehicle emissions. Transport planning cannot be considered in isolation from other environmental factors—if it is, the result will continue to be cities choking in fumes, damaged vegetation and acidification of soils.

"Acid rain" is the umbrella term used to describe a number of different pollutants that cause acidification of the environment. The term "rain" is misleading since it also covers fogs and mists and there can be both "wet" and "dry" deposition. Damage to ecosystems can be direct or indirect.

Acid-forming primary pollutants (sulphur dioxide and nitrogen oxides) and secondary pollutants (corrosive ground-level ozone and trace gases like chlorofluorocarbons—CFCs) pose a growing threat to global ecosystems and human health. Acid rain has been described as the "silent killer", insidious in its effects, often causing irreversible damage. Whilst the finger of guilt may point at power stations and high chimney stacks as an obvious source of this transboundary pollution, more attention needs to focus on the impact of the motor vehicle and, in particular, of the car (Table I).

2. THE LINK BETWEEN ACID RAIN AND TRANSPORT

The growth in road traffic is now coming up against real environmental limits; pollution, congestion and conflicts in land use are all increasing. Despite policies and techniques aimed at cleaning up emissions, air pollution has reached serious levels in a number of cities throughout the industrialized world, and is a major problem in most developing countries.

The air within cities frequently does not meet the guidelines laid down by the World Health Organization for substances such as nitrogen oxides. These arise directly from vehicle emissions and as the numbers of cars on our roads increase, so does the pollution problem. Secondary pollutants, including ground-level ozone, are formed when products from the exhausts of vehicles undergo a chemical reaction in sunlight. Ground-level ozone is responsible for the photochemical smogs that hang over many of our cities. Primary and secondary pollutants have a detrimental effect on people's health, their effects spreading far beyond our cities.

Motor vehicles generate more air pollution than any other single human activity.[1] They pollute because their internal combustion engines are inefficient, leading to incomplete combustion of fuel. The products of incomplete combustion—hydrocarbons and carbon monoxide—are released through the exhaust. In OECD countries road transport produces 75% of carbon monoxide emissions, 48% of nitrogen oxides and 40% of hydrocarbons. In 1987, road transport produced 85% of the UK's total emissions of carbon monoxide; 28% of all hydrocarbons and 18% of all CO_2. The average car is estimated to produce more than a 1/4 tonne of pollution every year.[2]

"The exhaust of an average car emits enough nitrogen oxides in a year to dissolve a 20 lb steel cannonball. As there are more than 20 million cars in Britain and over 400 million worldwide—that washes out into a flood of forest-killing acid rain." (Roger Bell, *The Independent,* February, 1990.)

The position in the USA is similar. In 1985 transport was responsible for 70% of the carbon monoxide, 45% of the nitrogen oxides and 34% of the volatile organic compounds (VOCs) in the air. And, for 1991, it is estimated that motor vehicles in Europe alone will have emitted an estimated 5.5 million tonnes of hydrocarbons, 28 million tonnes of carbon monoxide and 6.5 million tonnes of oxides of nitrogen.[3]

3. VEHICLE EMISSIONS—THE COCKTAIL OF POLLUTANTS

Over 10,000 different components/chemicals are found in vehicle exhausts. Many contribute to acid deposition as well as being important greenhouse gases. For this reason acid rain and potential climate change through global warming cannot be considered in isolation when determining policies for action.

Greenhouse gases attributable to the motor vehicle include carbon dioxide,

Table I Black smoke: estimated emissions[1] by emission source and by type of fuel.

United Kingdom												Thousand tonnes
	1979	1980	1981	1982	1983	1984	1985	1986	1987	1988	1989	Percentage of total in 1989
a) By emission source												
Domestic	382	316	297	292	271	212	285	300	247	223	191	37
Commercial/public service[2]	7	6	6	6	6	6	6	5	5	4	4	1
Power stations	33	29	27	26	24	33	28	26	26	26r	25	5
Refineries	4	4	3	3	3	3	2	2	2	2	2	—
Agriculture	1	1	1	1	1	1	1	1	1	1	1	—
Other industry[3]	102	93	90	91	89	89	89	91	87r	89r	88	17
Railways	2	2	2	2	1	1	1	1	1	1	—	—
Road transport	121	118	112	116	124	135	141	155	167	184	198	39
Civil aircraft	1	1	1	1	1	1	1	1	1	1	1	—
Shipping	3r	3r	2r	3r	3r	3r	3r	2	2r	2	3	1
Total	655r	572r	541r	539r	521r	482r	555	585r	538r	533	512	100
b) By type of fuel												
Coal	413	341	322	315	294	228	306	326	274	251	218	43
Solid smokeless fuel	21	21	19	19	19	15	20	17	17	16	14	3
Petroleum:												
Motor spirit	12	12	12	12	13	13	13	14	14	15	15	3
Derv	109	105	100	103	111	122	128	142	152	169	182	36
Gas	11	10	9	9	8	9r	8	8	7	7	7	1
Fuel oil	32r	24r	20r	20r	16r	31r	19r	15	12	14r	14	3
Other petroleum	1	1	1	1	1	1	1	1	1	1	1	—
Other emissions	56	58	58	60	59	64	60	62	60r	60r	60	12
Total	655r	572	541r	539r	521r	482r	555	585r	538r	533	512	100
c) Emissions (tonnes)/GDP (£2 million)[4]	2.0	1.8	1.7	1.7	1.6	1.4	1.6	1.6	1.4	1.3	1.2	

Source: Warren Spring Laboratory, Department of Trade and Industry; Central Statistical Office

[1] Includes miscellaneous emission sources.

[2] Excludes power stations, refineries and agriculture.

[3] Power stations, refineries and a proportion (57%) of other industry.

[4] GDP measured at 1985 market prices.

nitrous oxides, methane and the precursors to tropospheric ozone—hydrocarbons and nitrogen oxides. Carbon monoxide is also released in significant quantities. This gas has been found to interact chemically with other compounds, notably hydroxyl radicals, reducing their presence in the air. As these substances react with and reduce other greenhouse gases, such as methane, their presence in the atmosphere is important. The subsequent reduction in the level of hydroxyls means that methane is allowed to rise to the upper atmosphere, where it can increase global warming.[1]

3.1. *Sulphur Dioxide*

Power stations are the main source of SO_2, although there remains a link between this chemical and transport. The most serious impact on health occurs when sulphur dioxide is present with increased levels of particulates or smoke. Here sulphur dioxide can be adsorbed onto the surface of the particulates and if these are inhaled, they will reach the deepest recesses of the lung. In London, airborne levels of SO_2 regularly exceed the one-hour World Health Organization (WHO) guideline of 350 micrograms per cubic metre. Other cities also have problems with high peak concentrations of sulphur dioxide: in 1987 Berlin suffered half-an-hour peak levels of 1,400 micrograms.[4]

3.2. *Nitrogen Oxides (NO_x) and Hydrocarbons (Fig. 1)*

Nitrogen oxides and hydrocarbons are major contributors to acid rain. The main nitrogen oxides (NO and NO_2) are formed in all combustion processes from the reaction of oxygen and nitrogen at high temperatures—the higher the temperature, the greater the amount of nitrogen oxides (NO_x) produced. The main form of NO_x from car exhausts and aircraft engines in nitric oxide, which is rapidly oxidized in the air to form nitrogen oxide. This is eventually removed from the atmosphere by dry deposition: in its original form or as gaseous nitric acid; or as wet-deposited nitric acid, dissolved in precipitation. Combustion also produces nitrous oxide (familiar as laughing gas), a "greenhouse" gas which is formed naturally in soil and water by the denitrification process. Additionally, nitrogen oxides combine with hydrocarbons in the presence of sunlight to form ozone. Hydrocarbons are responsible for soiling problems on buildings in urban centres, as well as for their impact on human health. OECD research confirms that the primary source responsible for most NO_x emissions is road transportation: "mobile sources, mainly road traffic, produce around 50% of anthropogenic VOC emissions, therefore constituting the largest man-made VOC source category in all European OECD countries."

Various nitrogen compounds have the potential to damage trees and forests—they act as stressing agents. Scientists working for the Nordic Council of Ministers have estimated that at least a 75% reduction in NO_x emissions will be needed to protect the most sensitive habitats. Nitrogen dioxide (NO_2) is generally assumed to be the most toxic form of this compound. The highest outdoor concentrations of NO_2 are found in urban areas with busy roads, and where the geography or climate enhance the slow dispersion of air pollutants.

Nitrogen dioxide levels from traffic fumes have been increasing since the 1970s. Speed plays a part in the formation of potential pollutants—nitrogen oxide emissions are six times greater at speeds of 80 miles per hour than at 30 miles per hour (Table II).

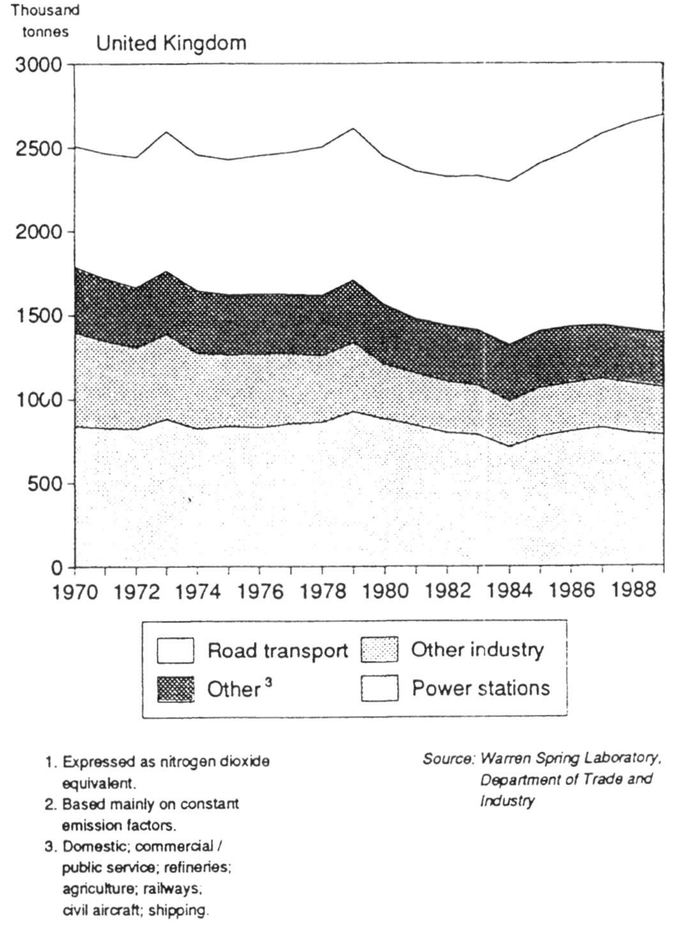

Figure 1 Nitrogen oxides (NO_x):[1] estimated emissions[2] by emission source

3.3. *Carbon Monoxide (Fig. 2)*

Carbon monoxide (CO) is produced by the incomplete combustion of fuel. It is a poisonous gas and even at moderate concentrations will cause drowsiness, impairing mental and physical alertness.

Carbon monoxide is not an acid precursor, or regarded as a "greenhouse gas." However, its presence removes the molecule that would otherwise eliminate methane, a potent "greenhouse gas" from the air. Carbon monoxide contributes to smog production.

Road traffic is the largest source of carbon monoxide—in the UK about 85% of the nation's emissions come from this one source. An average car emits five lungfuls of poisonous carbon monoxide gas per mile. In the USA, the National Ambient Air Quality Standard for carbon monoxide is exceeded in about 50 cities, exposing over 29 million Americans to increased levels of carbon monoxide. Emissions are rising over 1% annually, most of it due to motor vehicles.[5]

Table II Volatile organic compounds (VOCs):[1] estimated emissions[2] by emission source and by type of fuel

United Kingdom												**Thousand tonnes**
	1979	1980	1981	1982	1983	1984	1985	1986	1987	1988	1989	Percentage of total in 1989
a) By emission source												
Domestic	97	80	76	75	69	55	73	77	64	58	50	2
service[3]	1	1	1	1	1	1	1	1	1	1	1	—
Power stations	14	14	13	12	12	10	12	13	13	13	12	1
Other industry[4]	43	42	42	42	42	41	42	42	42	42	42	2
Processes and solvents[5]	1,031	1,033	1,031	1,034	1,035	1,039	1,040	1,046	1,050	1,056	1,059	51
Gas leakage[6]	31	31	31	30	31	32	32	34	36	34	34	2
Forests[7]	80	80	80	80	80	80	80	80	80	80	80	4
Railways	11	10	10	9	10	9	9	9	9	9	8	—
Road transport[8]	490r	580r	595r	614r	607r	623r	621r	640r	674r	705r	762	37
Civil aircraft	3r	3r	3r	3r	3r	3r	3r	3r	3r	4r	4	—
Shipping	14r	13r	11r	13r	12r	13r	13r	12r	11r	11r	14	1
Total	1,813r	1,887r	1,893r	1,912r	1,903r	1,907r	1,926r	1,957r	1,984r	2,013r	2,066	100

b) By type of fuel												
Coal	105	88	83	81	76	58	78	84	72	65	57	3
Solid smokeless fuel	5	5	4	4	4	3	4	4	4	3	3	—
Petroleum:												
Motor spirit	356r	448r	470r	491r	480r	492r	488r	503r	524r	544r	590	29
Derv	134r	131r	125r	123r	127r	131r	133r	137r	150r	161r	173	8
Gas	22r	21r	19r	20r	20r	20r	21r	20r	19r	19r	20	1
Fuel oil	5r	4r	3r	4r	3r	5r	3r	3r	3r	3r	3	—
Other petroleum	3r	3r	3r	3r	3r	3r	3r	4r	4r	4r	4	—
Other gas	4	4	4	4	4	4	5	5	5	5	5	—
Other emissions	1,180	1,182	1,179	1,182	1,184	1,189	1,190	1,198	1,204	1,208	1,211	59
Total	1,813r	1,887r	1,893r	1,912r	1,903r	1,907r	1,926r	1,957r	1,984r	2,013r	2,066	100
c) Emissions (tonnes)/GDP (£2 million)[9]												
	5.4	5.8	5.9	5.9	5.7	5.5	5.4	5.3	5.1	5.0	5.0	

Source: Warren Spring Laboratory, Department of Trade and Industry; Central Statistical Office
[1]Excluding methane.
[2]Most of the figures in this table are based on constant emission factors.
[3]Includes miscellaneous emission sources.
[4]Excludes power stations.
[5]Including evaporation of motor spirit during production, storage and distribution.
[6]Gas leakage is an estimate of losses during transmission along the distribution system.
[7]An order of magnitude estimate of natural emissions from forests.
[8]Includes evaporative emissions from the petrol tank and carburettor of petrol-engined vehicles.
[9]GDP measured at 1985 market prices.

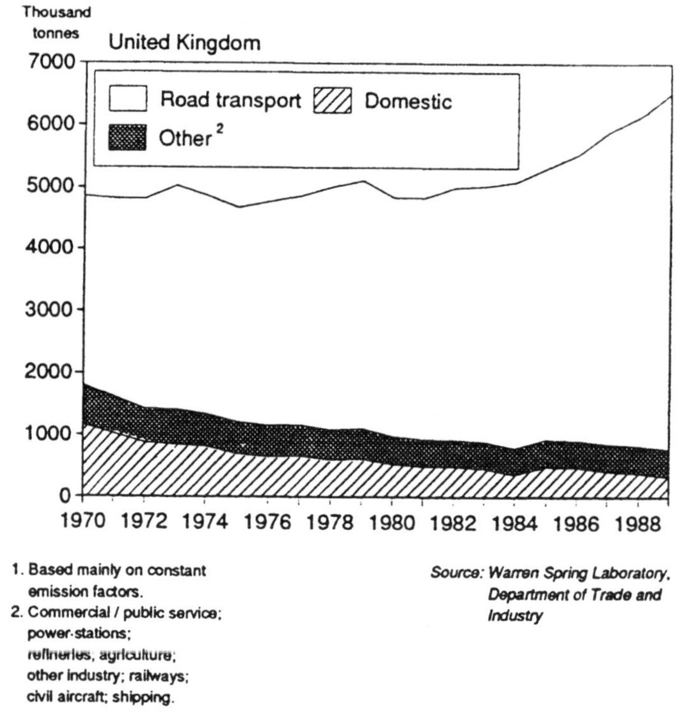

Figure 2 Carbon monoxide: estimated emissions[1] by emission source

3.4. *Ozone (Fig. 3)*

Ozone hastens the formation of acid rain. Vehicle exhausts are directly responsible for rising levels. Ozone is produced naturally in the stratosphere yet its breakdown process is hindered by hydrocarbons from vehicle exhausts. In sunlight, and with the presence of nitrogen oxides, a photochemical reaction occurs, causing the level of ozone in the lower atmosphere to increase. Acid deposition which occurs in the presence of tropospheric ozone is much more likely to cause forest death, die-back, and crop losses than deposition occurring in the absence of ozone. It can also reduce the productive capacity of land through increased acidification of the soil, and affect fish in rivers and streams.

It has been estimated that the amount of background ozone close to the ground (tropospheric) doubled in Europe between 1957 and 1987. Saving Europe's forests means tackling the rising number of motor vehicles and their associated pollution.

3.5. *Volatile Organic Compounds (Fig. 4)*

These consist of a large number of compounds including hydrocarbons which arise from a variety of industrial processes and motor vehicle activity—they evaporate from the fuel tank and fuel system. VOCs are directly implicated in the formation of ground-level (tropospheric) ozone and photochemical smog and, for

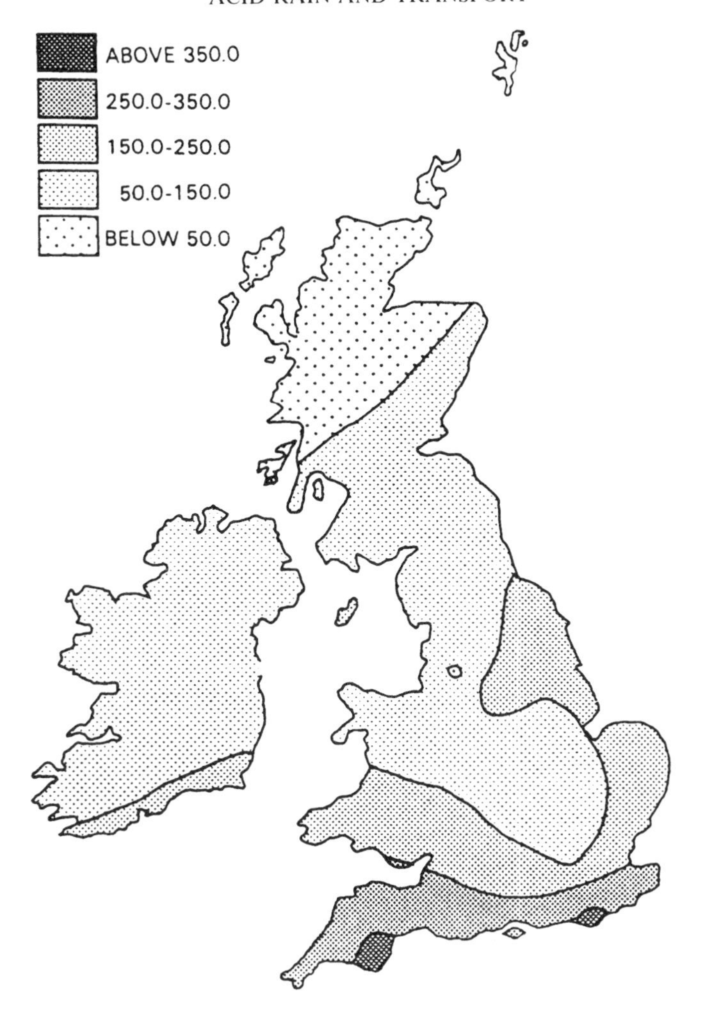

Source: Warren Spring Laboratory
Department of Trade and Industry

Figure 3 Number of exceedences of 60 ppb hourly average tropospheric ozone concentrations during summer (April–September) 1989

this reason, need to be controlled for public health considerations, as well as to reduce damage to ecosystems.

The two major sources of VOCs are car emissions and solvents. Since 1970 the total emissions of VOCs has been climbing steadily, to reach 2.1 million tonnes in 1989. Figures from the Department of the Environment (UK) show the increased volume of VOCs in the air is a direct result of emissions from road transport—overall emissions from road transport have increased by over 50% since 1979.

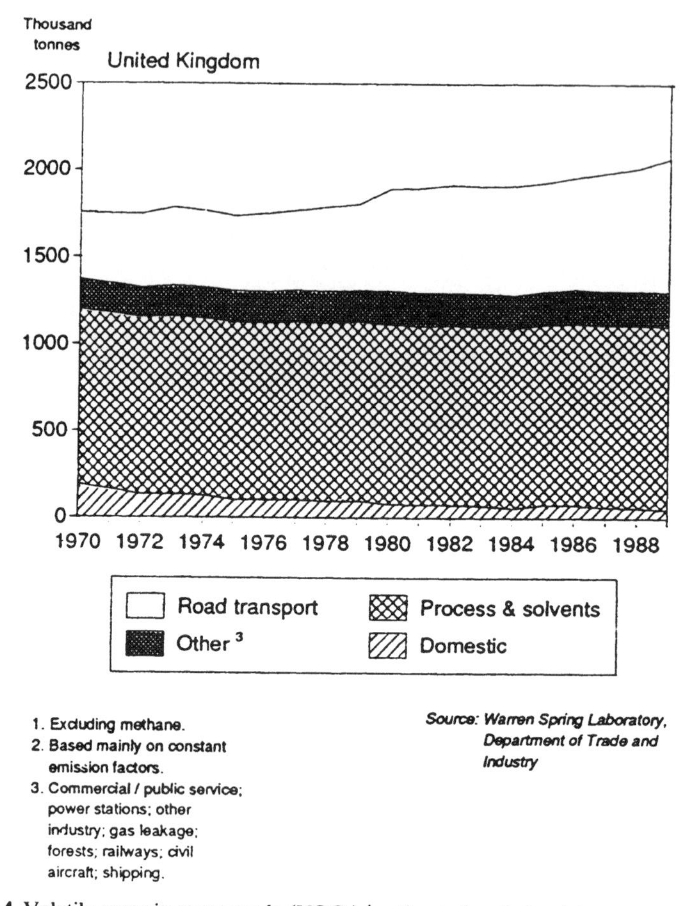

Figure 4 Volatile organic compounds (VOCs):[1] estimated emissions[2] by emission source

3.6. *Carbon Dioxide (CO_2)*

For more than a decade scientists have warned that increases in the level of carbon dioxide in the atmosphere is leading to enhanced climate change through global warming. Carbon dioxide arises from many modern industrial developments, as well as from the destruction of the world's forests which act like a "sink" for carbon dioxide. CO_2, along with other gases such as CFCs and methane, acts to trap the heat of the sun's rays like glass in a greenhouse.

Carbon dioxide from vehicular sources is rising. Between 1971 and 1987 the increase in the USA was 70%, in the OECD countries it was 70% and in the non-OECD countries is was 120%.[7] As each motor vehicle produces about 19 pounds of carbon dioxide per gallon of petrol consumed and with the increasing number of cars and other vehicles in existence, a larger proportion of carbon dioxide emissions will in future come from this source of fossil-fuel burning.[5]

In December, 1990, the US-based World Resources Institute, made the following recommendations concerning the need to stabilize, and indeed substantially reduce, carbon dioxide emissions:

- The efficiency of new vehicles must be improved—the goal being to attain an efficiency of 100 miles per gallon of petrol.
- The need for an accelerated phase-out of older vehicles.
- Advance pollution control on all vehicles including controls over CFCs used in motor air conditioning systems.
- The need for enhanced public transport systems.
- The ultimate aim must be the production of non-polluting vehicles—substantial research and development work is already underway on electric vehicles and hydrogen-powered cars.

3.7. *Pollution From Diesel*

Diesel-powered vehicles are the most visible polluters on the UK's roads. They emit nearly one third of all black smoke—in towns and cities, this figure can rise to 90%. Particulate emissions, or smoke, is a universal problem in many Third World countries especially where badly maintained or old engines are in use.

Whilst diesel engines release lower concentrations of some gaseous pollutants, they emit on average about 100 times the weight of particulates produced by gasoline engines of comparable performance.[8] Evidence from experimental studies has revealed that these particulate emissions, including polyaromatic hydrocarbons, can be mutagenic and carcinogenic.[9] Accordingly, the British Lung Foundation has now recommended the use of masks for occupational groups exposed to high levels of particulates such as bicycle couriers.[10,11] The effect of diesel particulates will vary depending upon the concentrations present and whether susceptible groups within the population are affected. The Swedish Environmental Protection Board has estimated that emissions from diesel-fuelled cars and lorries are some 100 times more carcinogenic than emissions from a catalyst-equipped petrol car, or 10 times more carcinogenic than emissions from a non-catalyst equipped car.[12] Switching to diesel in order to reduce pollution is not necessarily beneficial to the environment.

Back in 1984, the Royal Commission on Environmental Pollution stated that "We consider that smoke emissions from diesel vehicles in the United Kingdom are in many circumstances at an unacceptable level. We believe there is scope for improvements in both vehicle construction standards and enforcement in use."[7]

Meeting the challenge at the international level, companies like Mercedes-Benz and Volkswagen have developed emission-control devices that permit diesel cars to meet the strict 1986 California particulate limit of 0.2 grams per mile.[10]

4. DAMAGE TO ECOSYSTEMS

Acid deposition is recognized to cause severe stress to many ecosystems, although the precise interrelationships have yet to be fully understood. There is little scientific disagreement on the impact of acid rain as it increases the acidity of many lakes in Scandinavia, north-east USA and south-east Canada. Studies have linked acid rain with direct damage to forests through tree die-back, whilst other research has highlighted the indirect effects of acid precipitation. Some researchers believe that acid rain increases the acidity of the forest soil, causing essential nutrients to be washed away whilst leaving roots of trees exposed to toxic aluminium.

One theory still regarded as the most plausible is the "stress hypothesis," put forward by Professor Schutt and colleagues at the Forestry Department of Munich University. This hypothesis suggests that the cocktail of chemical pollutants in the air disturbs the essential process of photosynthesis. The pollutants acting together, or synergistically, weaken the trees, making them less resistant to disease. The leaves or needles are affected, becoming unable to help the tree build up reserves of nutrients. Down at root level the beneficial fungi, which supply nutrients, die. This constant stress affecting the vital functions of the tree means that it is less able to withstand various climatic changes such as frost or drought. Widespread tree damage and death is inevitable.

Such effects have been most pronounced in Scandinavia and Eastern European countries where hillsides of spindly, diseased, and dead trees may be seen. However, acid rain is not just a problem for these areas—it is now affecting substantial areas of the industrialized nations, as well as the developing nations. The statistics make stark reading. The third annual inventory of forest health published by the European Commission reveals that 10% of EEC forests in 1989 showed some signs of damage—that is they have lost more than 25% of their leaves or needles. Discoloration was also observed in 16% of the trees surveyed. Whilst atmospheric pollution is a likely attributable cause, the study noted that more information was needed on pollution levels at the specific sites monitored.[6]

The first comprehensive study attempting to quantify the effects of acid air pollution on all European forests, and to estimate the economic costs of damage, was published in December 1990 by the International Institute for Applied Systems Analysis (IIASA) in Austria. Overall, they conclude that the cost of damage will be at least $30 billion per year, over the next 100 years, unless drastic measures are taken to curb pollution. The estimates are regarded as conservative as they are based on the effects of sulphur emissions alone.[13]

Although tree damage had been apparent for some years, the true scale only became apparent in the early 1980s from surveys initiated by the government of the Federal Republic of Germany. By 1985, 52% of forests showed signs of damage, a huge increase from the 1982 estimates when damage to the trees was put at just 8%.

Tree die-back and other effects were observed in pine and beech trees—these findings were attributed to air pollution (acid rain and ozone) which attacked the trees through the air and through the soil. In the area formerly known as West Germany, the term "Waldsterben" or "the death of forests" is used to describe the impact of acid rain.

In Czechoslovakia, acidification had affected 500,000 to 1 million trees by 1985—27% of all forest areas were damaged. In Sweden, about 2000 lakes are nearly lifeless and a further 15,000 are said to be too acid to support aquatic life. In southern Norway, more than 13,000 square kilometres of lakes are devoid of fish and stocks are reduced in a further 20,000 square kilometres.

A similar picture is seen in North America. Acid rain has been a major issue in Canada for over a decade. Damage in eastern Canada is estimated at about $1 billion annually (*Canada Green Plan*, 1990). Over 150,000 lakes are already suffering from acid damage and more than 14,000 are considered "dead." More than 15 million hectares of forests receive high levels of acid rain. 84% of Canada's most productive farmland suffers "unacceptable" acid deposition. Over 80% of Canadians live in areas of high acid rain-related pollution—there is evidence that this is contributing to respiratory illnesses in children and other susceptible groups. More than half of the acid deposition in eastern Canada originates in the US. Although the USA's cars have emission-control equipment, the sheer rise in the number of vehicles on the road, results in a net increase of air pollutants.[14]

In the UK, widespread acid deposition exists in lowland Scotland (Galloway), Cumbria and Wales. The Warren Spring Laboratory has recorded markedly different acid rainfall patterns in parts of Britain. There is a clear graduation from the west to the east of Britain in terms of acidity (pH), sulphate and ammonium ion concentrations. Acid rain deposition tends to be higher in the high rainfall areas. The harshness of the effects depends to some extent upon the underlying geology and climate. Where soils are thin and poorly developed, for example, in granite areas, there are often higher rates of acidification.

4.1. *Exceeding the Tolerance Level*

Nitrogen compounds are seriously stressing Europe's forests. The estimated total nitrogen inputs to forests in central Europe and southern areas of the Nordic countries usually exceed the calculated tolerance limits of nature several times over. Whilst the maximum tolerable level is about 5–10 kg/ha, total deposition to forests is in the range of 20–50 kg/ha. In areas of older forests this level can be even higher.[15]

Too little is known about how damage to trees actually happens. But it is thought that several stresses work together. There can be direct damage through gases or acids attacking the leaf surfaces or penetrating the leaf thereby making it less resistant to parasites, and/or indirect damage through the disturbance of nutrient and water uptake. In addition, many experiments have shown that damage is caused to leaf surfaces by ozone, especially in combination with other pollutants. One suggestion is that there is a "melting" of the fibrillous wax which in many species surrounds the stomata pores in the leaf—this makes it more difficult for plants to stop water losses through their leaves.[16] Higher concentrations of gases may cause direct cell damage, affecting the parts where essential photosynthesis takes place. Leaf yellowing, discoloration and damage makes leaves or needles more susceptible to frosts and drought stress.

It has been suggested that conifers are more susceptible to damage as they retain their leaves throughout the year. Older trees, and those at high altitudes open to prevailing winds, will be more sensitive to damage.

In the Black Forest of Germany a direct link between transport and acid rain damage was made by the forestry authorities in the mid 1980s. Information boards placed in the forest were part of a local authority campaign to "clean the air." The information strategy focused on five main points:

a) The need to use unleaded petrol
b) The need to keep to certain speed limits
c) Natural gas should be used in preference to coal or oil
d) It is better to transfer freight and move people by rail
e) The need for public information on energy conservation and related issues.

Beyond information campaigns, direct action has also been used in Germany to publicize the issue. The environmental lobby has become increasingly vociferous, joining forces with the foresters to make the general public aware of the issue. Notices pinned to trees bore the message "I'm dying—help me" and "When trees die, so do people." Another innovative idea involving the general public came from the Netherlands in the mid-1980s. Dutch environmentalists decided that people should be taken to the forests and shown damage firsthand.

Consequently on the first "Forest Alert" 10,000 people, including the Dutch Environment Minister, took part in the exercise to raise awareness over the issue. In the UK, Friends of the Earth held a similar nationwide exercise several years later, attracting over 4,000 people.

But it is not just trees that suffer. Lichens and mosses are often used as an indicator of pollution. In a 1988 report, *Acid rain and nature conservation in Britain*, the Nature Conservatory Council warned that "lichens and mosses may be suffering damage in the areas most affected by acid rain." The report also noted that "there are fewer species of water plants, invertebrates and fish in acidic lakes and rivers. This may be due to acidification in some cases. Birds such as the dipper are less successful on acidic streams."

While the effects of air pollution on vegetation may be clear, the policies for action are not. In some countries, including the UK, it has taken years for some scientists and governments to acknowledge the problem. In 1985 Friends of the Earth (FoE) UK launched its Tree Die-back Survey, revealing that over half the UK's beech and yew trees showed signs of damage—the survey suggested that air pollution was one of the contributory factors. The FoE survey was heavily criticized by the Forestry Commission, yet continued pressure from FoE resulted in a shift of attitudes several years later. By December 1987, the Forestry Commissions publicly stated that "the British Government should *immediately* introduce far more rigorous restrictions on power station emissions and vehicle exhausts." Opinions however waivered and, whilst giving evidence to a House of Commons Environment Committee in 1988, the Forest Commission advised caution, stressing that there were big differences in the way countries understood monitoring programmes. However, in its report, published in June 1988, the House of Commons Environment Committee criticized the Forestry Commission—"it stands alone in its refusal to accept a nexus between air pollution and tree damage." By 1989 the Forestry Commission confirmed that their own monitoring research had shown the levels of air pollution in Britain to be high enough to damage trees and that the situation was worsening rapidly.

Evidence that too little is still known about the mechanisms that cause acid rain came with an agreement, in February 1991, between the Forestry Commission and the Department of the Environment. They publicly agreed that tree planting in certain areas can actually lead to acidification—the tree canopies "scavenge" pollutants from the air. In areas where rainfall is heavy and where there are high concentrations of sulphur dioxide, nitrogen oxides and ammonia, soils often have reduced capacities to neutralize acids. The "scavenging effect" can increase the loading of sulphur dioxide and nitrogen oxides by 50–100%. As a result of these findings the report recommended that sensitive water catchments around the country should now be assessed to see how much pollution their ecosystems will tolerate.[17]

5. BUILDING DAMAGE

Transport-related air pollution causes continuous and costly damage to buildings and structures. A range of materials, including stone, metals and synthetic materials can be damaged—directly, indirectly or through corrosion. Many of the mechanisms that trigger building damage are not yet fully understood.

Direct damage is a result of atmospheric corrosion by gaseous nitrogen pollutants. "Acid rain" has the potential to eat deep into the surface of materials such as limestone, sandstone and brick. Acid deposition also reacts with calcium carbonate to form gypsum—the black crust seen on many urban buildings. Measurements reveal that 2.5 centimetres of stonework on St. Paul's Cathedral in the UK has been eaten away over the past 100 years. In Greece, the Acropolis is suffering the ravages of air pollution—accelerating the effects of natural decay. A Greek specialist on acid corrosion has estimated that the Athenian monuments have deteriorated more in the past 20–25 years from pollution than in the previous 2,400 years.[18] Corrosion from acidification is also thought to be responsible for damage to railway tracks—in the Katowice region of southern Poland, trains have to slow down in certain places due to damaged tracks.[18]

Throughout Europe, stone structures are being affected and often irreparably damaged. Stone statues are losing recognizable features. The original stonework at Westminster Abbey, one of Europe's most visited "medieval" buildings, is having to be replaced by modern Portland stone. Renovation work in the late 1980s cost around £5 million.

In 1984 a House of Commons Environment Committee report noted: "It is beyond doubt that acid rain due to sulphur dioxide emissions is damaging British buildings, and slowly but surely is dissolving away our historical heritage."

For a long time sulphur dioxide was regarded as the prime suspect, but it is now recognized that nitrogen oxides have a significant part to play—particularly in urban areas. It is extremely difficult to put an economic cost on acid rain damage, but the link between vehicle emissions and subsequent damage by acid deposition is not disputed.

6. DAMAGE TO HEALTH

The direct effects of acid rain upon human health have not yet been fully identified. However, clear evidence does exist on the health effects and risks from many of the individual components that form the acid cocktail. One in five British people are at risk from polluted air. The elderly, infants, pregnant women and those suffering from lung and heart illnesses are particularly vulnerable. In the UK, guidelines set by the World Health Organization for carbon monoxide, sulphur dioxide, nitrogen dioxide and ozone are often exceeded.

Smog or pollution alerts are already common in many other countries; the UK has belatedly adopted similar warning schemes. Pollution alerts are issued through the media in the USA, Japan, Germany, Netherlands, France, Sweden and Denmark. Those at risk are advised to avoid strenuous exercise or to stay indoors; when pollution is particularly bad, local authorities may temporarily close down polluting factories or halt car use. But overall there needs to be far tougher enforcement of national air quality standards in line with World Health Organization guidelines, backed by an expanded monitoring network and public data registers.[4]

7. REACHING THE LIMITS

Despite the increasing burden of pollution, the number of vehicles on our roads continues to escalate. There are now 408 million cars in the world—this figure is expected to rise to half a billion by the end of this century. The UK Government's official White Paper *Roads to Prosperity* (18 May 1989) forecast an increase in traffic of 142% by the year 2025, if the economy grows at a rate of 3% until that year. This would lead to an extra 27.5 million vehicles on our roads. One estimate calculates that these vehicles would form a queue 104,000 miles long. Such a jam could be accommodated, if stationary, on a new motorway stretching from London to Edinburgh—but only if it were 257 lanes wide.[8]

Clearly something has to be done. Besides the sheer physical impact of such a number of vehicles, there is the whole infrastructure accompanying it—the large scale road building and widening programmes that create their own significant impacts upon the environment.

7.1. *Transport and Sustainable Development*

The concept of sustainable development is vital for future environment protection. Transport cannot be exempt. Sustainable development formed the basis of the 1987 Brundtland report *Our Common Future.* Defined as "development that meets the needs of the present without compromising the ability of future generations to meet their own needs", sustainable development does not simply mean tinkering with the existing systems of economic growth, but demands a complete rethink and the emergence of new patterns of behaviour. Uncontrolled growth, however "green", is not tenable.

Enshrined within the concept of sustainable development is a stark fact. If we want a cleaner environment we are going to have to pay for it. Proper economic values must be placed on the services provided by the environment. For if these services are to be perceived as free, there is little incentive to protect them. This encourages their use in a manner which is detrimental to the development prospects of future generations.[19]

Unfortunately, it has taken too long for decision makers and policy shapers to recognize this. In future, environmental issues need to be fully incorporated into our economic decision making and the full costs of production, use and disposal of a resource will need to be assessed, to ensure that actual or potential damage to the environment is costed out. There will be increasing emphasis on the "Polluter must Pay" concept.

An example of the urgent need for this integration already exists. Throughout the European Community the annual cost of acid rain damage to crops is put at over one billion dollars. Yet one of its primary sources, the motor vehicle, is reflected as only generating net addition to the Gross National Product (GNP).

8. CONTROLLING EMISSIONS

Initiatives to control sulphur emissions and nitrogen oxides are part of a long-term plan—the implementation of the ECE Convention on Long-range Transboundary Air Pollution. This covers Europe and North America. Governments need to commit themselves to at least a 90% reduction of sulphur dioxide (based on 1980 levels) and a 75% reduction in emissions of nitrogen oxides (based on 1985 levels) by the year 2000. Eastern European countries will need help to meet these targets.

In 1988, after much stalling, the UK finally agreed to the European Community Large Combustion Plants Directive. This aims to control emissions of the main acid rain gases produced in fossil-fuel fired power stations and industrial complexes. Emissions of sulphur dioxide have to be reduced from such facilities by 20% by 1993, and 40% by 1998 and 60% by 2003. Nitrogen oxides must be reduced by 15% by 1993 and 30% by 1998. Much controversy surrounded this Directive and, although welcomed by the environmental lobby, influential groups such as the Worldwatch Institute (USA) have criticized the agreement as being "too little, too late." They suggest that cuts of up to 90% in the two main acid rain forming gases are needed and 75% reductions in ozone levels.

Focusing specifically on nitrogen oxides (NO_x), the Sofia Protocol concerning the Control of Emissions of Nitrogen Oxides or their Transboundary Fluxes was signed in November 1988 and became internationally binding in February 1991. Signatories to the protocol agreed a package of measures designed to freeze nitrogen oxide emissions from 1994. The protocol is based on the "critical load" concept which provides a scientific rationale for emission reductions.

An EC Directive for controlling emissions from small cars has already been adopted. In addition, legislation has also been targeted at medium to large size cars—this required that emissions of NO_x from passenger cars, by the end of 1992, be reduced by 75%. The standards require all new petrol-engined cars be fitted with catalytic converters to maintain this reduction.[6] The amendments to the EC Directive on emissions from large diesel vehicles will reduce oxides of nitrogen as well as smoke. Diesel vehicles are now responsible for half of the emissions of oxides of nitrogen from road transport in Britain. The features of diesel engines which make them fuel efficient rule out using catalysts to remove NO_x. Instead high standards need to be set for heavy diesels, and engineering solutions for reducing pollution need to be found.

In June 1991 the European Commission published a draft Directive as a first step towards setting up air quality standards for ozone. Existing international guidelines designed to protect human health and vegetation from damage by exposure to ozone are routinely exceeded throughout Europe. The primary pollutants which form ozone in a photochemical reaction—nitrogen oxides (NO_x) and volatile organic compounds (VOCs)—will not be tackled effectively until the end of this century. These substances are known to contribute to widespread forest decline. Originally the European Commission wanted the Directive to propose fixed air quality standards for ozone—this was changed after disagreement from Member States that such a move was too premature. Instead the draft Directive proposes to establish common procedures for ozone monitoring and information exchange. Four ozone thresholds are proposed with notification to Brussels and to the public, if certain levels are reached. Mandatory air quality standards have been introduced in a number of countries to improve air quality in polluted areas. They are also used in areas where the air is relatively clean, in an attempt to maintain air quality.

The prevention of air pollution needs to be based on four principles:

- The polluter pays.
- The precautionary principle—prevention is better than cure.
- Best available technology—for pollution control, ensuring that diversion from one source does not result in damage to another.
- Freedom of information—available data on pollution levels and limits, as well as monitoring results, should be kept in a register open to public inspection.

9. POLLUTION REDUCTION TECHNOLOGIES

Many of the technologies already exist for reducing air pollution but these will do little more than maintain the status quo if the underlying cause—the huge increase in vehicles—is not tackled.

Pressure is needed to ensure these technologies are implemented for, as one commentator has noted, "left to its own devices, the motor vehicle industry shows no inclination to adopt state-of-the-art pollution controls or advanced efficiency technology" . . . "industrialized countries have actually begun to slip backward in terms of new vehicle efficiency."[1] The same pressure needs to be applied to prevent the other main stumbling block—the lack of political will.

Catalytic converters are devices which substantially reduce some emissions from vehicles running on lead-free petrol. They do not cut air pollution completely. Their real role is to minimize the release of oxides of nitrogen from the exhaust, by passing the waste gases through a bed of rhodium and platinum precious metals. This regulates the oxidation process, controlling nitrogen oxides. A report from the Intergovernmental Panel on Climate Change, concerned with global warming and its associated changes on climate, suggests that, to make a real impact, catalytic converters to reduce emissions of nitrogen oxides and carbon monoxide would have to be fitted to all cars in the industrialized world by 2000, and to all vehicles in the world by 2025.

Catalytic converters have been standard fittings on all new cars in the USA and Japan for a number of years. The more environmentally-conscious European countries have been adopting them; the UK has been slow to accept them. New European Community emission limits for large cars require all new cars to be fitted with a controlled three-way catalyst.

However worthy, catalytic converters do have their limitations. They do nothing to reduce carbon dioxide emissions—the major "greenhouse gas." In fact, they have the potential to increase it since catalytic converters change toxic gases like carbon monoxide into less harmful substances including carbon dioxide. Overall, transport is currently responsible for almost 30% of the total CO_2 emissions in the world.

Catalytic converters are responsible for fostering the impression that cars can be "green." Few people are aware that catalytic converters do not start to work until the engine reaches a working temperature of 300 degrees celsius and optimal performance is not achieved until 1000 degrees celsius. As the majority of car journeys are five miles or less (Department of Transport National Travel Survey 1985–6), engines often do not reach the higher temperature.

The principle advantage of another device, the lean burn engine, is that it uses less fuel—between 5%–10% less than an equivalent car fitted with a controlled three-way catalyst. Lean burn engines are low NO_x emitters (the reduction is achieved in the engine) and they emit less CO_2 than catalyst technology.[6] However, adoption of lean burn technology is not widespread.

Any benefits achieved by catalyst technology and lean burn engines could be cancelled out by the current rate of growth in traffic.

9.1. *Energy Efficiency*

Over the next 40 years emissions of carbon dioxide are expected to escalate. The only solution is to reduce fuel consumption through energy conservation measures—less miles per vehicle, more efficient vehicles and, even better, less vehicles.

All fossil fuels release carbon dioxide when burnt; energy efficiency in vehicles is vital to save non-renewable petroleum reserves and to cut air pollution. But increasing energy efficiency is only a technical fix; on present forecasts the reduction is not sufficient to make up for increasing traffic. Official predictions from the Department of Energy forecast that improvements in car efficiency between now and the year 2010 will reach 28%. Yet over the same time span the Department of Transport expects the increase in car traffic to be at least 41%.

The ideal high efficiency car is one where the engine controls combustion electronically, its tyres have reduced rolling resistance and the car body, built in strong but lightweight materials, is designed to minimize air resistance. According to Bateele, the international research organization, these measures would double the fuel economy of an American car for only about US$500 per car; even at the low US petrol prices the extra investment would pay off at the petrol pump within a year or two.[20]

However, in the USA falling oil prices mean less interest by car owners in fuel efficiency. Such a situation is causing the average fuel economy of new cars to decrease again.

9.2. *Alternative Vehicle Fuels*

Concern about air pollution and energy security has prompted considerable interest in alternative fuels and in vehicles that can use different combinations of fuel. However, switching from gasoline to alternatives such as methanol, diesel, and gas is a short-term solution to the growing air pollution problem from transport. Gains could be wiped out by the huge projected increases in vehicles upon the roads.

Alternative fuels are not without their own environmental impacts. Methanol and ethanol, for instance, burn without VOCs and some of the other toxic emissions, but they do result in higher emissions of carbon dioxide and other greenhouse gases. Hydrogen burns cleanly; the only appreciable emission is water vapour. However, in areas prone to fog several thousand cars could reduce visibility to near zero. Solar-powered vehicles could be promising but the low levels of research mean these remain merely a future development.

But even short-term measures that reduce air pollution are to be welcomed. In the USA air quality programmes are already operating in areas like Los Angeles. Alternatives to "conventional" hydrocarbon reduction strategies are being investigated because of the concern that the conventional strategies might not allow emission targets to be met where there is continued growth in traffic.

One study claims the answer could lie in methanol substitution.[21] A new photochemical modelling study of the Los Angeles Air Basin shows that methanol substitution in motor vehicles would provide important reductions in peak-hour ozone levels, and even larger benefits in basin-wide ozone measures, such as total hours above episode levels. The results of the study demonstrate that photochemical reactions are reduced by 25% to 50% compared to gasoline vehicles. When these results are combined with available methanol cost projections, they indicate that methanol could be an effective ozone-control strategy.

Technical fixes, such as three-way catalysts are only short-term solutions. True long-term solutions to acid rain must be policies of planned restraint and improved public transport systems. National and international regulations and agreements, as well as market mechanisms such as taxes, grants and incentives, are crucial to reducing the impact from transport.

10. POLICY MEASURES

Too many problems experienced in urban centres relate to the sheer volume of traffic upon the roads. More road building is not the answer, but improved public transportation systems are.

Belatedly cities like Los Angeles are now exploring measures to curb the motor vehicle in an attempt to solve their air pollution problems, as well as traffic congestion.

10.1. *Tackling the Problem—The LA Experience*

Los Angeles is renowned for the photochemical smogs that hang over the city. But in 1989 stringent new regulations were introduced, in an attempt to cut air pollution fivefold over the next twenty years. The Air Quality Management Plan introduces nearly 150 regulations that will cost $3 billion a year to implement. However, air pollution alone currently causes an estimated $13 billion in damage every year.[16]

By the year 2000 all new cars and all buses will have to use alternative fuels—within 17 years there will be a complete ban on petrol and diesel engines. America's two largest car makers, Ford and General Motors, anticipated the clean fuel era with their plans to have 10,000 non-petrol cars on the road by 1993. Two leading oil companies are setting up methanol stations.

The Air Quality Plan means that companies will have to look to drastically reduce the number of car journeys their employees make. Car pooling is becoming mandatory for many businesses, with $25,000 a day fines for non-compliance. Other suggestions involve 4-day weeks and more working at home by employees. All fleet vehicles will shortly have to switch to methanol or other clean-burning fuels.

Whilst emission reductions in the past have led to significant improvement in air quality, continuous traffic growth means that the Californian South Coast District now violates the air quality standards for ozone, carbon monoxide, nitrogen dioxide and particulate matter. Reductions of around 80% in NO_x and volatile organic compounds are needed to bring the district into compliance with all air quality standards in the next 20 years.[22]

10.2. *The Netherlands*

In the Netherlands 80% of all nitrogen oxides, the key component of acid rain and photochemical smog, comes from Dutch cars. Their National Environmental Policy Plan, due for implementation by 1994, makes stringent 90% cuts in the emissions responsible for acidification of the environment. This action is targeted against cars that are producing more and more frequent smog alerts in the Netherlands.

A press release from the Dutch Government in May 1989 states: "It is clear technical measures alone will be insufficient—the use of motor cars will have to be reduced." Elements of their National Environment Plan include the use of vehicles that are as clean, quiet, efficient and safe as possible. They should be built in such a manner that facilitates full recycling of materials and components. Full encouragement needs to be given to the use of public transport and bicycles; municipal planning needs to be sensitive to a more efficient alignment between residential, work, shopping and recreational areas to limit the need for transport.[23]

The Dutch hope to achieve these objectives by increased use of catalytic converters, reducing dangerous substances in vehicles, enforcing land use measures to reduce the need for transport, by expansion of the public transport network, including cycle routes, and by influencing the price mechanism.

10.3. *Sweden*

In Sweden, a total of 390,000 tonnes of nitrogen oxides is emitted annually by transportation, energy production and industrial processes.[15] Light vehicles emit 120,000 tonnes of NO_x a year: proposals from the Swedish Environmental Protection Agency aim to impose stricter limits on nitrogen oxide emissions from cars and light lorries/buses from the 1994 and 1995 models. Heavy vehicles are responsible for around 75,000 tonnes of NO_x: proposals aim to tighten these by 1996. A system of levies and grants for environmental performance is under consideration as a means of introducing an incentive for developing and fitting advanced emission control equipment.

Studies in Sweden have even found that aircraft emit about 7,000 tonnes of nitrogen oxides every year. Domestic airlines already pay an environmental tax which has resulted in favourable developments to engines.

10.4. *Germany*

In Germany, the extensive monitoring programmes of the North Rhine-Westphalian State Agency for Air Pollution Control and Noise Abatement have confirmed that ambient air concentrations for pollutants like NO_x, CO and hydrocarbons are considerably higher at traffic-related sites than at other urban measuring points. At one busy junction in Düsseldorf the average concentration for CO was three times the average, and for NO_x four times the average, for the overcrowded Rhine-Ruhr area.

Germany has followed the example of the Netherlands by introducing traffic calming measures in certain urban areas to improve local environments. Traffic is permitted in areas displaying "woonerf" signs, but vehicles must travel only at a walking pace. Pedestrians may use the whole width of the street. There are clear environmental benefits—air pollution is reduced if drivers drive more calmly. The types of driving behaviour commonly found in urban areas—the acceleration, constant slowing down and braking—are all practices which increase the amount of air pollution emitted from a vehicle.

10.5. *Elsewhere in the World*

In Italy, traffic is banned from the centre of Florence in daylight hours and in Rome traffic is banned from the centre during seven hours of the day. Budapest bans motor traffic from all but two of its downtown streets. In Santiago, a fifth of all cars are kept off the road, by rotation, each day. In Zurich, Switzerland, penalties are imposed on car travel, making it faster to go by bicycle. Elsewhere, engine switch-off at traffic lights is mandatory to reduce emissions.

In Canada a market-based approach is being considered in addition to tougher regulations to control smog. The *Canada Green Plan* (1990) notes that emission trading incentives are already part of the acid rain abatement programme approved by the U.S. Congress. These programmes operate by issuing companies with permits to emit nitrogen oxides and VOCs up to specified emission levels. Companies can then trade these permits—if an individual company can reduce its emissions more cost-effectively than another company, it has an incentive to do so and can sell the unused permits. This method achieves overall emission target at the least cost. The number of permits in overall circulation can be reduced to achieve environmental quality targets. Canada is looking seriously at introducing such a scheme.

10.6. *Road Pricing*

Another method of reducing congestion and traffic jams is to charge motorists extra when they drive in congested streets—a policy known as road pricing. Revenue can then be redirected into improving public transportation systems and local environments.

According to Phil Goodwin, director of the transport studies unit at Oxford University, road pricing is the only way of improving city environments without vast public expenditure and very tight restrictions on traffic, such as banning cars from whole areas.[24]

Road pricing already works in other cities. In Singapore the practice was introduced in 1975. Motorists wanting to drive into the centre of the city at rush-hour have to buy and display an additional license. After the introduction of road pricing one in ten motorists immediately gave up driving to work in the city centre, taking the bus instead. There was a drop in reported accidents and some improvement in air quality. By 1983 fewer than one in four commuters was driving to work. To make road pricing even more efficient, modern electronic schemes are in operation.

Elsewhere the introduction of road pricing schemes has been subject to delay and political controversy. The Netherlands have abandoned the proposed plans to introduce it on motorways from 1992; instead they are investigating a small scheme involving Amsterdam, Rotterdam, and Utrecht. Meanwhile in the UK the Department of Transport remains opposed to its introduction in London, saying that implementation would create far too many difficulties.[24]

10.7. *Introducing Policies at the Corporate Level*

A key component in many traffic jams is the large number of cars containing just one person—this is particularly the case at peak rush-hours. Reducing the number of cars and the accompanying pollution is an objective that needs tackling at the corporate level.[25]

Car users play an important role in air pollution. Their choice of vehicle and subsequent behaviour, including driving practice, can substantially increase the amount of fuel used—and pollution emitted. Cruising at 70 mph can use up to 25% more fuel than driving at 55 mph.[25] So better driving techniques can improve fuel consumption and cut pollution by up to 20%. Therefore all educational programmes should be considered useful in creating increased driver awareness.

All company cars should be converted to run on unleaded fuel and, where possible, catalytic converters should be fitted to existing models. Catalytic converters cut out up to 90% of the nitrogen oxides, hydrocarbons and carbon monoxide from exhausts. From 1993 these are now required by law on all new cars. Measures to cut unnecessary car use are vital. In Los Angeles the implementation of the 1989 Air Quality Management Plan makes car pooling a viable option—new car pool lanes are being provided on highways, allowing cars with two passengers or more to speed past the traffic jams. Car pools operate most efficiently where people, who live in the same area and work the same hours, share a car.

A recent publication from the UK Trades Union Congress entitled *Greening the Workplace* provides a useful checklist for companies trying to reduce the environmental impact of transport. Key questions include: the need for advice on "economical driving"; improvement of the environmental performance of a vehicle by careful assessment of type of vehicle, its fuel efficiency, load factors, maintenance and route planning; and questions concerning the organization's policy on company cars. Public transport is not forgotten: companies need to consider what incentives they offer to encourage staff to use public transport.[26]

11. FORECASTS FOR THE FUTURE

Whilst this chapter has considered the impact of the motor vehicle in industrialized countries upon acid deposition, the impact of growing numbers of vehicles in the developing nations should not be underestimated. In Mexico City—rapidly becoming the most polluted urban area in the world—vehicles are responsible for as much as 80% of total emissions by weight. In future, industrial countries must be prepared to take a lead in implementing emission-control strategies to help ease acid rain and ozone problems. Higher-income countries should help with the transfer of the technology that will be needed.

References

1. Walsh M., *Global Warming*, Greenpeace, London, 1990.
2. World Wide Fund for Nature. *The Route Ahead—Vehicle Pollution, Causes, Effects, Answers?* WWF, London, 1990.
3. Read R. and Read C. "Breathing can be Hazardous to Your Health." *New Scientist*, 34–37, 23 Feb. 1991.
4. Holman, C. *Air Pollution and Health*, Friends of the Earth, London, 1989.
5. McKenzie, J., ed. *Driving Forces: motor vehicle trends and their implications*. WRI, Washington DC, Dec. 1990.
6. Department of the Environment. *Digest of Environmental Protection and Water Statistics*, HMSO, London, 1991.
7. Royal Commission on Environmental Pollution. *Tenth Report—Tackling Pollution: experience and prospects*, HMSO, London, 1990.
8. Adams, J. "Car ownership forecasting. Pull the ladder up or climb back down?" *Traffic Engineering and Control*, March, 1990.
9. World Resources Institute. *Energy for a Sustainable World*, WRI, Washington DC, 1987.
10. Blevis, D. *Preparing for the 1990s: the world automobile industry and prospects for future fuel economy innovation in light vehicles*, Federation of American Scientists, Washington DC, Jan. 1987.
11. Green, M., and Read, R. *Brit. Med. J.*, **300**:761–762, 24 March 1990.
12. Sheeland, K. "Lung cancer and diesel exhaust: a review." *Amer. J. Ind. Med.*, **10**:177–189, 1986.
13. Rose, J. "Comprehensive study of American forests." *Env. Management & Health*, **2**, No. 1, 35, 1991.
14. Canadian Government. *Canada Green Plan*, Ottawa, 1990.
15. Hanneby, P. *Enviro (Acid) Magazine*, Sweden, 7–9, June 1990.
16. Myers, N. *Future Worlds*, Gaia Books, London, 1990.
17. *New Scientist*, 9 Feb. 1991.
18. French, H. F. "Clearing the air: a global agenda." *The State of the World*, Worldwatch Institute, Washington DC, 1990.
19. Pearce, D., *et al. Blueprint for a Green Economy*, Earthscan, London, 1989.

20. Patterson, W. *The Energy Alternative,* Boxtree, London, 1990.
21. Meyer, C., Unnasch, S. and Jackson, M. "Air quality programmes as driving forces for a transition to methanol use. (Special Issue 'Alternative Fuels')" *Transp. Res.*, **23A**:209–216, No. 3, May 1989.
22. Anon. "Highway pollution symposium." *Science Total Environment,* 93, March 1989.
23. Ministrie van Volkshuisteshing (Netherlands) Press Release. *Drastic Decisions Inevitable to Improve the Environment,* 25 May 1989.
24. Hamer, M. "Pricing cars off city streets." *New Scientist,* 40–42, 2 March 1991.
25. Ralston, K. and Church, C. *Working Greener,* Greenprint, London, 1991.
26. Trades Union Congress. *Greening the Workplace,* T.U.C., London, 1991.

INDEX